Carboxylic Acid Production

Special Issue Editor
Gunnar Lidén

MDPI • Basel • Beijing • Wuhan • Barcelona • Belgrade

MDPI

Special Issue Editor
Gunnar Lidén
Lund University
Sweden

Editorial Office
MDPI AG
St. Alban-Anlage 66
Basel, Switzerland

This edition is a reprint of the Special Issue published online in the open access journal *Fermentation* (ISSN 2311-5637) from in 2017 (available at: http://www.mdpi.com/journal/fermentation/special_issues/carboxylic-acid).

For citation purposes, cite each article independently as indicated on the article page online and as indicated below:

Author 1; Author 2. Article title. *Journal Name* **Year**, *Article number*, page range.

First Edition 2017

ISBN 978-3-03842-552-6 (Pbk)
ISBN 978-3-03842-553-3 (PDF)

Cover photo courtesy of: Aikaterini Papadaki, Nikolaos Androutsopoulos, Maria Patsalou, Michalis Koutinas, Nikolaos Kopsahelis, Aline Machado de Castro, Seraphim Papanikolaou and Apostolis A. Koutinas.

Table of Contents

About the Special Issue Editor

Gunnar Lidén, PhD, is Professor in Chemical Engineering at Lund University, Sweden. He studied Chemical Engineering at Chalmers University, Göteborg, where he obtained both his MSc (1985) and his PhD (1993). In 1999, he moved to Lund University to take up his current position. His research interests are fermentation and enzyme technology for lignocellulose conversion into fuels and chemicals. Most of his work has concerned processes using yeasts, but bacterial hosts have also been studied. He has co-authored more than 100 papers, and also a textbook in Biochemical engineering, a topic which he regularly teaches.

fermentation

MDPI

Editorial

Carboxylic Acid Production

Gunnar Lidén

Department of Chemical Engineering, Lund University, P.O. Box 124, 221 00 Lund, Sweden;
gunnar.liden@chemeng.lth.se

Received: 21 August 2017; Accepted: 12 September 2017; Published: 14 September 2017

Keywords: biorefinery; natural carboxylic acid producers; fungi; bacteria; fermentation technology;
downstream processing

Carboxylic acids are central compounds in cellular metabolism, and in the carbon cycle in nature. Carbon dioxide is captured from the atmosphere and enters living cells through the formation of carboxylic groups, and it is released from living cells by the decarboxylation of carboxylic acids. The aerobic extraction of free energy from sugars in cellular respiration hinges on the ingeniously designed tricarboxylic acid cycle involving a range of carboxylic acids, and the reactivity of the carboxylic group with amino- or hydroxyl-groups enables the formation of peptide and ester bonds in macromolecules. The functionality of the carboxylic group is, not surprisingly, also of huge importance in the industrial world for a wide range of applications. The loosely bound hydrogen provides weak acid functionality, much desired for food industry applications in preservatives and flavor compounds, and citric acid is one of our oldest and quantitatively largest industrial fermentation products. The presence of two carboxylic groups, or a combination of one carboxylic group and another functional group, make the compounds interesting building blocks for polymer production. Several carboxylic acids, including e.g., lactic, succinic, 3-hydroxypropionic and itaconic acids, have been recognized as suitable platform chemicals for a foreseen growing non-fossil based industry. Economic margins are, however, narrow when competing with petroleum-based products. Microbial production strains, fermentation technology and—not least—downstream processing, all need to be improved to enable a viable commercial production and speed up the transition towards non-fossil-based production of carboxylic acids. This special issue is devoted to the microbial production of carboxylic acids.

Several reviews give updates on the production of some of the most interesting acids. Succinic acid is one of the products for which bio-based production has increased most rapidly in recent years. Nghiem et al. [1] give an overview of both the microbial production and the state of commercialization of this acid. Malic acid, a close neighbor to succinic acid in the tricarboxylic acid cycle, is treated by West [2], with a special focus on use of biofuel related co-products as substrates (e.g., corn stover, straw, and glycerol, but also syngas). Interestingly, one way to produce malic acid is via poly β-L-malic acid, which accumulates intracellularly in *Aspergillus pullulans*. 2-oxopropanoic acid is the proper (but rarely used) name for pyruvic acid. Many metabolites are at metabolic crossroads, but few are more centrally localized in terms of major pathways than pyruvate. Maleki and Eiteman [3] give a detailed account of the involved pathways, and review engineering and process strategies for pyruvate production. Propionic acid is a fermentation product which has received less attention than the previously mentioned acids. Gonzalez-Garcia et al. [4] cover this acid in their review, which includes an in-depth analysis of the energetics of various possible pathways towards propionic acid. A broader review by Murali et al. [5], covering the production of several acids, such as acetic, butyric, lactic, and propionic acid using microbial consortia, is also included in this issue.

The choice of substrate is clearly crucial in bio-based production, primarily for economic reasons but also for environmental reasons. Couvreur et al. [6] report work on response surface optimization of growth media for *Lactobacillus reuteri* based on agro-industrial by-product streams, and the subsequent use of *L. reuteri* for the bioconversion of glycerol to 3-hydroxypropionic acid. A strong impact on the product distribution was found in comparison to a standard medium. A two-stage process was also used by Alcantara et al. [7] in their work on succinic acid production by a mixed fungal culture of *Aspergillus niger*, *Trichoderma reesei*, and *Phanerochaete chrysosporium*. Here, the objective was to use the enzyme production ability of the fungi in solid state fermentation, with *A. niger* and *T. reesei* grown on soy bean hulls and *P. chrysosporium* grown on birch wood chips, as well as their metabolic capacity, to form succinate in a second stage slurry fermentation. Papadaki et al. [8] also examined fungal carboxylic acid production in their study on fumaric acid production by *Rhizopus arrhizus*. The focal point of the investigation was the effect of the fungal morphology on productivity and yield. The authors found that higher titers and yields were obtained with dispersed mycelia rather than pellets. Corn stover is another widely available agro-residue, which after hydrolysis can be used as a substrate. Nelson et al. [9] report on the production of a mixture of butyric and hexanoic acid by the anaerobic bacterium *Megasphaera elsdenii* isolated from sheep rumen. A common problem in carboxylic acid production is the end-product inhibition, which limits final titers. Through in situ product removal by a reactive extraction system, end product inhibition could be minimized, allowing total titers of more than 55 g/L to be reached in the extract in a glucose fed-batch process.

Anaerobic digestion is an established waste treatment method, which primarily gives methane as a valuable product. Anaerobic digestion is a complex multistage process involving an adapted and selected consortium of microbes, which is normally not fully characterized. The overall process is an initial formation of short chain carboxylic acids, also known as "volatile fatty acids", which are in turn converted to methane. Queiros et al. [10] selected an inoculum and process conditions such that the methanogens would be inhibited, thereby stopping the process at carboxylic acids. A mixture of carboxylic acids—acetate, propionate, butyrate, valerate, and lactate—was obtained in long-term trials using spent sulfite liquor from hardwood.

Purification is a crucial problem for process economics, in particular when using hydrolyzates or waste streams as substrates. Figueira et al. [11] present a method for the purification of fumaric acid from Eucalyptus spent sulfite liquor reaching sufficient purity for polymer production. With a two-step precipitation method, based on the low solubility of fumaric acid, followed by activated carbon treatment, sufficient purity could be obtained with recovery yields of about 80% from broths holding 50 g/L of fumarate.

To conclude, there is clearly progress towards a significant increase in commercial bio-based carboxylic acid production, although many challenges remain. Titers, yields, and productivities must continue to increase, and to this end a combination of targeted engineering and evolutionary engineering will likely be used. Screening for new natural producers will be important—primarily to supply novel enzymes, but new host organisms or strains may also be found. A shift in carbon source from glucose to biomass-derived sugars will give additional requirements on host organisms in terms of robustness to impurities and utilization of multiple sugars. Downstream processing will also be strongly affected by such a shift, where it is important to keep in mind that additional separation costs must not exceed the price difference between the substrates.

Acknowledgments: The author is grateful for research support in this field through the EU contracts BRIGIT (grant number FP7-311935) and Biorefine-2G (grant number FP7-613771).

Conflicts of Interest: The author declares no conflict of interest.

References

1. Nghiem, N.P.; Kleff, S.; Schwegmann, S. Succinic Acid: Technology Development and Commercialization. *Fermentation* **2017**, *3*, 26. [CrossRef]
2. West, T.P. Microbial Production of Malic Acid from Biofuel-Related Coproducts and Biomass. *Fermentation* **2017**, *3*, 14. [CrossRef]
3. Maleki, N.; Eiteman, M.A. Recent Progress in the Microbial Production of Pyruvic Acid. *Fermentation* **2017**, *3*, 8. [CrossRef]

4. Gonzalez-Garcia, R.A.; Tim McCubbin, T.; Navone, L.; Stowers, C.; Nielsen, L.K.; Marcellin, E. Microbial Propionic Acid Production. *Fermentation* **2017**, *3*, 21. [CrossRef]

5. Murali, N.; Srinivas, K.; Ahring, B.K. Biochemical Production and Separation of Carboxylic Acids for Biorefinery Applications. *Fermentation* **2017**, *3*, 22. [CrossRef]

6. Couvreur, J.; Teixeira, A.R.S.; Allais, F.; Henry-Eric Spinnler, H.-E.; Claire Saulou-Bérion, C.; Clément, T. Wheat and Sugar Beet Coproducts for the Bioproduction of 3-Hydroxypropionic Acid by Lactobacillus reuteri DSM17938. *Fermentation* **2017**, *3*, 32. [CrossRef]

7. Alcantara, J.; Mondala, A.; Hughey, L.; Shields, S. Direct Succinic Acid Production from Minimally Pretreated Biomass Using Sequential Solid-State andSlurry Fermentation with Mixed Fungal Cultures. *Fermentation* **2017**, *3*, 30. [CrossRef]

8. Papadaki, A.; Androutsopoulos, N.; Patsalou, M.; Koutinas, M.; Kopsahelis, N.; de Castro, A.M.; Papanikolaou, S.; Koutinas, A.A. Biotechnological Production of Fumaric Acid: The Effect of Morphology of Rhizopus arrhizus NRRL 2582. *Fermentation* **2017**, *3*, 33. [CrossRef]

9. Nelson, R.S.; Peterson, D.J.; Karp, E.M.; Beckham, G.T.; Salvachúa, D. Mixed Carboxylic Acid Production by Megasphaera elsdenii from Glucose and Lignocellulosic Hydrolysate. *Fermentation* **2017**, *3*, 10. [CrossRef]

10. Queirós, D.; Sousa, R.; Pereira, S.; Serafim, L.S. Valorization of a Pulp Industry By-Product through the Production of Short-Chain Organic Acids. *Fermentation* **2017**, *3*, 20. [CrossRef]

11. Figueira, D.; Cavalheiro, J.; Sommer Ferreira, B. Purification of Polymer-Grade Fumaric Acid from Fermented Spent Sulfite Liquor. *Fermentation* **2017**, *3*, 13. [CrossRef]

fermentation

MDPI

Review

Recent Progress in the Microbial Production of Pyruvic Acid

Neda Maleki [1] and Mark A. Eiteman [2,*]

[1]   Department of Food Science, Engineering and Technology, University of Tehran, Karaj 31587-77871, Iran; nmaleki@uga.edu

[2]   School of Chemical, Materials and Biomedical Engineering, University of Georgia, Athens, GA 30602, USA

*   Correspondence: eiteman@engr.uga.edu; Tel.: +1-706-542-0833

Academic Editor: Gunnar Lidén

Received: 10 January 2017; Accepted: 6 February 2017; Published: 13 February 2017

Abstract: Pyruvic acid (pyruvate) is a cellular metabolite found at the biochemical junction of glycolysis and the tricarboxylic acid cycle. Pyruvate is used in food, cosmetics, pharmaceutical and agricultural applications. Microbial production of pyruvate from either yeast or bacteria relies on restricting the natural catabolism of pyruvate, while also limiting the accumulation of the numerous potential by-products. In this review we describe research to improve pyruvate formation which has targeted both strain development and process development. Strain development requires an understanding of carbohydrate metabolism and the many competing enzymes which use pyruvate as a substrate, and it often combines classical mutation/isolation approaches with modern metabolic engineering strategies. Process development requires an understanding of operational modes and their differing effects on microbial growth and product formation.

Keywords: auxotrophy; *Candida glabrata*; *Escherichia coli*; fed-batch; metabolic engineering; pyruvate; pyruvate dehydrogenase

1. Introduction

Pyruvic acid (pyruvate at neutral pH) is a three carbon oxo-monocarboxylic acid, also known as 2-oxopropanoic acid, 2-ketopropionic acid or acetylformic acid. Pyruvate is biochemically located at the end of glycolysis and entry into the tricarboxylic acid (TCA) cycle (Figure 1). Having both keto and carboxylic groups in its structure, pyruvate is a potential precursor for many chemicals, pharmaceuticals, food additives, and polymers. For example, pyruvate has been used in the biochemical synthesis of L-DOPA [1], *N*-acetyl-D-neuraminic acid [2], (R)-phenylacetylcarbinol [3], butanol using a three-enzyme cascade [4], and has been proposed as a starting material for the enzymatic synthesis of propionate [5]. A recent assessment of *Escherichia coli* as a cell factory concluded that pyruvate was one of the most useful metabolic precursors to a wide range of non-native commercial products [6]. Microbial pyruvate production has been the subject of previous reviews [7,8].

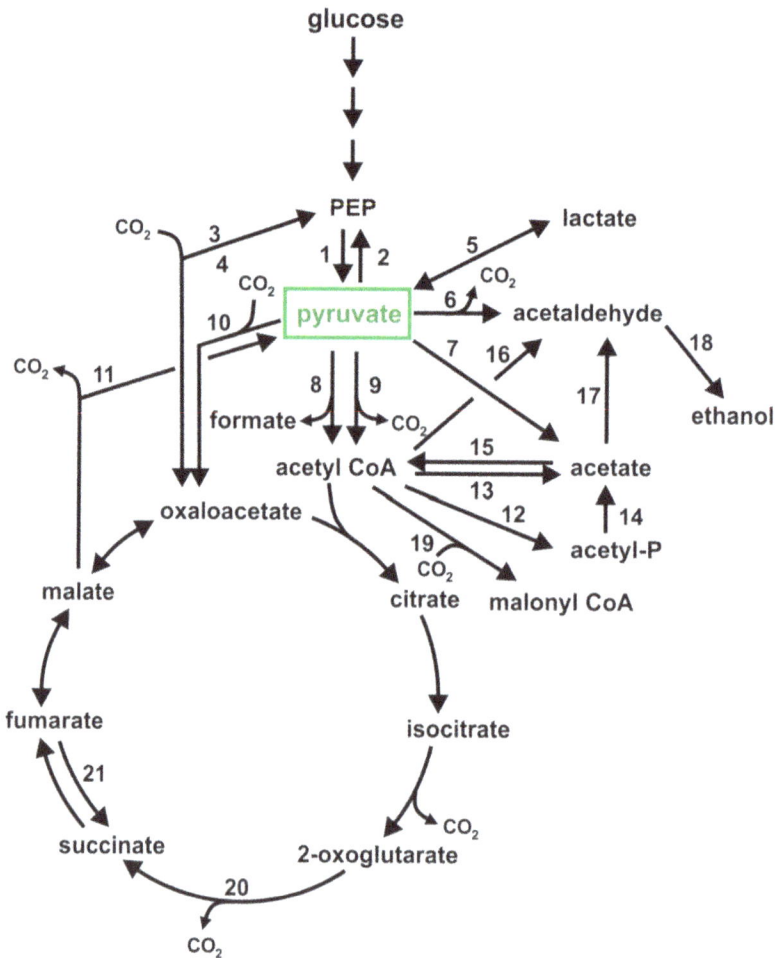

Figure 1. The key metabolic pathways of microorganisms involved in the formation and consumption of pyruvate. Enzyme cofactors (e.g., NAD and NADH) and compounds involved in energy transfer (e.g., ATP) are not shown. Not all organisms express each enzyme shown. The enzymes indicated by numbers are detailed in Table 1.

Pyruvate itself has also long been studied for a wide range of health benefits. For example, pyruvate protects against oxidative stress in human neuroblastoma cells [9,10] and rat cortical neurons [11], protects retinal cells against zinc toxicity [12], improves cerebral metabolism during hemorrhagic shock [13], and protects the brain from ischemia-reperfusion injury [14]. Pyruvate improves myocardial function and increases ejection fraction without increasing heart rate [15,16]. In one double blind study, supplementation with 6 g pyruvate per day for six weeks in conjunction with mild physical activity resulted in a significant decrease in body weight and fat mass [17], and a similar study involving 10 g calcium pyruvate daily for one month with supervised exercise increased very low-density lipoprotein (VLDL) cholesterol and triacylglycerol, and decreased HDL cholesterol [18]. Calcium or magnesium pyruvate is now accepted as a food supplement [19]. Interestingly, pyruvate appears to detoxify hydrogen peroxide in the environment and stimulate the growth of ammonia-oxidizing archaea [20].

Table 1. The key enzymes associated with the metabolism of pyruvate, including gene designations for *Escherichia coli*, *Corynebacterium glutamicum*, *Saccharomyces cerevisiae*, and *Candida glabrata*.

Figure 1 (ref)	Enzyme Accepted Name	EC [1] Number	Reaction	E. coli [2]	C. glutamicum [3]	S. cerevisiae	C. glabrata
1	pyruvate kinase	2.7.1.40	PEP + ADP → pyruvate + ATP	pykA, pykF	pyk	PYK1	CAGL0E05610g
						PYK2	CAGL0M12034g
2	PEP synthase	2.7.9.2	pyruvate + ATP + H$_2$O → PEP + AMP + Pi	ppsA	cg0642, cg0644	–	–
3	PEP carboxylase	4.1.1.31	PEP + HCO$_3^-$ → oxaloacetate + Pi	ppc	ppc	–	–
4	PEP carboxykinase	4.1.1.49	PEP + CO$_2$ + ADP → oxaloacetate + ATP	pck	–	PCK1	CAGL0H06633g
		4.1.1.32	PEP + CO$_2$ + GDP → oxaloacetate + GTP	–	pck	–	–
5	D-lactate dehydrogenase	1.1.1.28	pyruvate + NADH + H$^+$ → D-lactate + NAD$^+$	ldhA	dld	–[4]	–[4]
	L-lactate dehydrogenase	1.1.1.27	pyruvate + NADH + H$^+$ → L-lactate + NAD$^+$	–	ldh	–[4]	–[4]
6	pyruvate decarboxylase	4.1.1.1	pyruvate → acetaldehyde + CO$_2$	–	–	THI3, PDC1, PDC5, PDC6	CAGL0G02937g, CAGL0L06842g, CAGL0M07920g
7	pyruvate oxidase	1.2.5.1	pyruvate + ubiquinone → acetate + CO$_2$ + ubiquinol	poxB	poxB	–	–
8	pyruvate formate lyase	2.3.1.54	pyruvate + CoA → acetyl CoA + formate	pflB	–	–	–
	pyruvate dehydrogenase complex:						
9	pyruvate dehydrogenase (E1)	1.2.4.1	pyruvate + CoA + NAD$^+$ → acetyl CoA + NADH + H$^+$ + CO$_2$	aceE	aceE	PDB1	CAGL0K06631g
						PDA1	CAGL0L12078g
	dihydrolipoamide acetyltransferase (E2)	2.3.1.12		aceF	–	PDA2	CAGL0J10186g
	dihydrolipoamide dehydrogenase (E3)	1.8.1.4		lpd	lpd	LPD1, IRC15	CAGL0R01947g
10	pyruvate carboxylase	6.4.1.1	pyruvate + HCO$_3^-$ + ATP → oxaloacetate + ADP	–	pyc	PYC1, PYC2	CAGL0R06941g
11	malate dehydrogenase (NAD$^+$, decarboxylating)	1.1.1.38	L-malate + NAD$^+$ → pyruvate + CO$_2$ + NADH	maeA	–	MAE1	CAGL0L02035g
	malate dehydrogenase (NADP$^+$, decarboxylating)	1.1.1.40	L-malate + NADP$^+$ → pyruvate + CO$_2$ + NADPH	maeB	–	–	–
12	phosphate acetyltransferase	2.3.1.8	acetyl CoA + Pi → acetyl-P + CoA	pta	pta	–	–
13	acetyl CoA hydrolase	3.1.2.1	acetyl CoA + H$_2$O → acetate + CoA	–	–	ACH1	CAGL0J04268g
14	acetate kinase	2.7.2.1	acetyl-P + ADP → acetate + ATP	ackA	ackA	–	–

Fermentation **2017**, *3*, 8

Table 1. *Cont.*

Figure 1 (ref)	Enzyme Accepted Name	EC [1] Number	Reaction	E. coli [2]	C. glutamicum [3]	S. cerevisiae	C. glabrata
15	acetyl CoA synthetase	6.2.1.1	acetate + ATP + CoA → acetyl CoA + AMP + PPi	acs	-	ACS1	CAGL0B02717g
						ACS2	CAGL0L00649g
16	acetaldehyde dehydrogenase (acetylating)	1.2.1.10	acetyl CoA + NADH + H$^+$ → acetaldehyde + NAD$^+$ + CoA	adhE	-		-
17	acetaldehyde dehydrogenase (NAD)	1.2.1.3	acetate + NADH + H$^+$ → acetaldehyde + NAD$^+$ + H$_2$O	-	xylC	ALD4-6	CAGL0X06688g
						HFD1	CAGL0H05137g
							CAGL0J03212g
							CAGL0K03509g
	acetaldehyde dehydrogenase (NADP)	1.2.1.5	acetate + NADPH + H$^+$ → acetaldehyde + NADP$^+$ + H2O	-	-	ALD2 ALD3	CAGL0R07777g
18	alcohol dehydrogenase (NAD)	1.1.1.1	acetaldehyde + NADH + H$^+$ → ethanol + NAD$^+$	adhE	adhA	ADH1-5	CAGL0I07843g
	alcohol dehydrogenase (NADP)	1.1.1.2	acetaldehyde + NADPH + H$^+$ → ethanol + NADP$^+$	-	-	ADH6-7	CAGL0H06853g
19	acetyl CoA carboxylase	6.4.1.2	acetyl CoA + HCO$_3^-$ + ATP → malonyl CoA + ADP + Pi	accABCD	accABCD	HFA1, ACC1	CAGL0L10780g
	2-oxoglutarate dehydrogenase complex:						
20	2-oxoglutarate dehydrogenase (E1)	1.2.4.2	2-oxoglutarate + CoA + NAD$^+$ → succinyl CoA + NADH + H$^+$ + CO$_2$	sucA	odhA	KGD1	CAGL0G08712g
	dihydrolipoamide succinyltransferase (E2)	2.3.1.61		sucB	sucB	KGD2	CAGL0E01287g
	dihydrolipoamide dehydrogenase (E3)	1.8.1.4		lpd	lpd	LPD1, IRC15	CAGL0R01947g
21	fumarate reductase	1.3.5.4	fumarate + quinone → succinate + quinol	frdABCD	-	-	-

(1) Enzyme Commission Number; (2). *E. coli* MG1655; (3). *C. glutamicum* ATCC 13032; (4). *S. cerevisiae* and *C. glabrata* have D-lactate dehydrogenase and L-lactate dehydrogenase (cytochromes) (EC 1.1.2.4 and EC 1.1.2.3).

2. Microbial Formation of Pyruvate

Because pyruvate is a central metabolite, small amounts of pyruvate have historically been reported in microorganisms under a variety of circumstances. The biochemical formation of pyruvate from glucose via glycolysis generally follows the stoichiometric equation:

$$\text{glucose} + 2\text{NAD} + 2\text{Pi} + 2\text{ADP} \rightarrow 2\text{pyruvate} + 2\text{NADH} + 2\text{ATP} \qquad (1)$$

Equation (1) indicates that the maximum theoretical yield of pyruvate (as the ion) is 0.966 g/g glucose, and the equation becomes balanced if the microbial process is able to regenerate NAD and ADP needed to sustain the reaction. Of course, some of the carbon/energy source glucose must also be used to form cellular materials. Nevertheless, Equation (1) indicates that pyruvate theoretically could accumulate without other carbon by-products at a high yield. Also, Equation (1) suggests that the rate of pyruvate formation is affected by the rate of NAD and ADP formation. Thus, recurring themes in research have been reducing by-product formation (including cells themselves) and increasing the availability of NAD and ADP. The key enzymes involved in pyruvate formation and catabolism are listed in Table 1.

Significant progress was made when researchers linked pyruvate generation from glucose in certain fungi to the availability of thiamine, with the observation of about 3 g/L pyruvate in the absence of thiamine but no pyruvate in the presence of excess thiamine [21–25]. Similarly, pyruvate can be observed in lipoic acid auxotrophs [26]. Coupled with increased knowledge of the mechanisms for enzyme kinetics, researchers thus began to appreciate that pyruvate could accumulate in microbes having an impaired ability to decarboxylate pyruvate oxidatively (i.e., low pyruvate dehydrogenase activity), or which were auxotrophic for thiamine or lipoic acid. The observations made with these auxotrophs result from thiamine and lipoic acid each being essential cofactors for the activity of the pyruvate dehydrogenase multienzyme complex: thiamine binds to the E1 decarboxylase domain (coded by the *aceE* gene in *E. coli*) while lipoic acid facilitates acetyl transfer by attaching via an amide linkage to a single lysyl residue of the E2 transacetylase subunit (code by the *aceF* gene in *E. coli*).

Vitamin auxotrophy has therefore often been used to isolate pyruvate-accumulating microorganisms. For example, a thiamine-requiring *Acinetobacter* isolate was able to convert 20 g/L 1,2-propanediol into about 12 g/L pyruvate [27], *Schizophyllum commune* converted glucose into 19 g/L pyruvate in 5 days at a yield of 0.38 g/g [28], and *Debaryomyces coudertii* generated 9.7 g/L pyruvate in 48 h from pectin-containing citrus peel extract [29]. After screening 18 yeasts for pyruvate formation from glucose or glycerol, one thiamine-auxotrophic *Yarrowia lipolytica* generated over 61 g/L pyruvate from glycerol in 78 h at a yield of 0.71 g/g [30]. Thiamine or lipoic acid *Enterobacter* auxotrophs isolated after exposure to the mutagen *N*-nitrosoguanidine generated 4.7 g/L pyruvate from 20 g/L glucose after 72 h [26], and an *E. coli* lipoic acid auxotroph generated over 25 g/L pyruvate from 50 g/L glucose in 40 h in a controlled fermenter [31]. A study of 132 strains isolated *Trichosporon cutaneum* which generated nearly 35 g/L pyruvate from glucose at a yield of 0.43 g/g [32]. Another investigation of several genera of yeasts used oxythiamine, an analogue of thiamine, to select strains for pyruvate productivity [33]. These researchers isolated a strain of *Candida glabrata* (formerly *Torulopsis glabrata*) auxotrophic for thiamine, nicotinate, pyridoxine and biotin which generated 57 g/L pyruvate from glucose and 40 g/L peptone in 59 h at a yield of 0.57 g/g. Mutagenesis of a pyruvate-producing *C. glabrata* generated arginine and isoleucine/valine auxotrophs which accumulated about 60 g/L pyruvate from glucose in 43 h at a yield of 0.60 g/g [34]. Because of this strain's natural predisposition at accumulating pyruvate, *C. glabrata* remains the principal microbe used for pyruvate production [35]. More recently, a *Blastobotrys adeninivorans* isolate generated 43 g/L pyruvate from glucose in 192 h at a yield of 0.77 g/g [36].

3. Medium Optimization

Because auxotrophy specifically and enzyme activity more generally are important to pyruvate accumulation in isolated strains, improvements in pyruvate production can be achieved by media optimization. For example, careful optimization of nitrogen sources and the key cofactors nicotinate and thiamine allowed the development of a fed-batch process using *C. glabrata* leading to nearly 68 g/L pyruvate from glucose in 63 h at a yield of 0.49 g/g [37]. A similar focused comparison of nitrogen nutrients for *C. glabrata* led to 57 g/L pyruvate from glucose in 55 h at a yield of 0.50 g/g [38], while a subsequent statistical optimization of vitamin concentration generated 69 g/L pyruvate from glucose in a 56 h batch process at a yield of 0.62 g/g [39]. Yeast extract is generally not suitable as a medium component because it is rich in thiamine [32]. These studies made clear the importance of vitamins, including those found in complex medium components, and the importance of aeration to pyruvate formation, necessary to ensure NAD availability. Recent genome-scale network analysis of *C. glabrata* confirms this yeast's propensity for glucose transport, its multivitamin auxotrophy and ability to transport organic acids, all attributes which contribute to pyruvate accumulation [40,41].

These early studies represent "classical" (e.g., pre-genomic) approaches to pyruvate formation, where production strains have been *isolated* for a particular target phenotype such as lipoic acid auxotrophy, and then the medium and environmental conditions optimized. Isolation/optimization approaches have continued to advance pyruvate formation. For example, researchers noted that urea is superior to ammonium chloride as the nitrogen source for *C. glabrata*, increasing final pyruvate concentration to about 86 g/L at a yield of 0.70 g/g [42]. The resulting increase in glucose consumption rate and pyruvate productivity was attributed to reduced futile cycling of ammonia ions and to elevated activities in enzymes which generate NADPH. Similarly, since the rate of pyruvate generation slows with an increased concentration of the base-neutralized product, researchers proposed increasing the NaCl-tolerance of the *C. glabrata* production strain [43]. Using continuous culture, a more salt-tolerant mutant was isolated which led to a 41% increase in the final pyruvate concentration compared to the parent strain to 94 g/L in 82 h. Supplementing a culture of *C. glabrata* with proline during growth also protects cells against a high osmotic pressure, and increased pyruvate production from 60 g/L to 74 g/L [44].

Being naturally tolerant to salt and osmotic pressure, halophilic microbes have recently gained attention for microbial production of organic acids [45–47]. One alkaliphilic, halophilic *Halomonas* generated 63 g/L pyruvate in 48 h under aerobic conditions using an unsterilized defined medium having high ionic strength [48]. Isolation/optimization over the recent three decades has proven quite successful in improving pyruvate titer, yield and productivity. In parallel with these classical approaches, the development of powerful genetic techniques over the last twenty years, as well as more sophisticated use of operational methodology, has facilitated the approach of "engineering" microbial metabolism toward the improvement of pyruvate production.

4. Metabolic Engineering of Pathways

Most broadly, metabolic engineering involves the use of genetic tools for the intentional optimization of pathways and regulatory circuitry in cells to affect the formation of an end product. Thus, the use of predictive algorithms and genetic tools to construct a strain *by design* is what distinguishes building a strain with the targeted characteristics to facilitate pyruvate formation from the simple isolation of pyruvate-accumulating strains. Like any product, improved pyruvate production primarily means increasing final concentration, yield from substrate and productivity. Since pyruvate is biochemically located at the end of glycolysis as a direct product from glucose or glycerol (Figure 1), efforts to accumulate pyruvate from these substrates ultimately involve restricting or eliminating the further metabolism of pyruvate, preventing by-product formation, and increasing the rate of glycolysis. Furthermore, any metabolic engineering approach for any product must account for other system constraints, such as the need to provide cells with sufficient NADPH and biochemical precursor molecules to satisfy biosynthetic demand.

An early metabolic engineering approach for pyruvate production focused on increasing the rate of glycolysis. Researchers have long understood that uncoupling oxidative phosphorylation by adding 2,4-dinitrophenol to the medium reduces the energy charge of a cell and increases the glucose consumption rate in *E. coli* [49]. Genetic tools have subsequently allowed the more direct uncoupling of respiration for *E. coli* by mutations in the *atp* operon, resulting in the doubling of glycolytic flux [50]. Essentially, cells compensate for reduced ATP generation from proton motive force by increasing the rate of ATP formation via glycolysis. Since pyruvate formation is so directly linked to glycolysis, researchers armed with modest gene manipulation tools proposed that introducing *atp* operon mutations would affect pyruvate formation. Thus, the introduction of an *atpA* mutation into a previously-isolated *E. coli* lipoic acid auxotroph resulted in a strain generating over 31 g/L pyruvate from glucose in 32 h at a yield of 0.64 g/g in batch culture [51,52]. Notably, as a result of the *atpA* mutation, the volumetric rate of pyruvate formation increased from about 0.8 g/L·h to over 1.2 g/L·h, and the biomass yield decreased from 0.26 g/g to 0.14 g/g. More sophisticated methods to control the intracellular ATP content continue to be developed. One example is constructing a copper-inducible F_0F_1-ATPase inhibitor, the *INH1* gene from *S. cerevisiae* [53]. When used with a pyruvate-overproducing strain of *C. glabrata*, this approach increased the volumetric productivity of pyruvate by 23% to 1.69 g/L·h.

Since increasing glycolytic flux usually has a limited impact on pyruvate yield, researchers have focused on other targets to direct more substrate to the product. Not surprisingly, an early target for knockout was the pyruvate dehydrogenase complex, which is the principal catabolic route for pyruvate (Figure 1). In contrast to earlier approaches which controlled pyruvate dehydrogenase by auxotrophy or by a naturally low enzyme activity, growth of a pyruvate dehydrogenase mutant usually necessitates introducing a secondary carbon source such as acetate or ethanol to provide cells with a source of acetyl CoA. A wide variety of *E. coli* mutants deficient in components of the pyruvate dehydrogenase complex (*aceE*, *aceF*, *lpd* genes) growing with an acetate supplement led to a high pyruvate yield from glucose [54]. The best performing strain *E. coli aceF ppc* generated 35 g/L pyruvate from glucose and acetate in 35 h at a yield of 0.78 g/g glucose [54]. The formation of acetate and lactate under certain conditions in *aceE* or *aceF* strains also implies that pyruvate oxidase (*poxB*), and, despite aerobic conditions, lactate dehydrogenase (*ldhA*) are important conduits for pyruvate metabolism and thus potential future targets for gene knockouts. The importance of pyruvate oxidase in *E. coli* lacking pyruvate dehydrogenase was shown through ^{13}C-flux analysis, which demonstrated that pyruvate oxidase as well as the Entner-Doudoroff and anaplerotic pathways are upregulated in the absence of a functional pyruvate dehydrogenase [55]. Another study using *E. coli aceEF pflB poxB pps ldhA* with a defined medium highlighted the relationship between acetate consumption, measurable CO_2 consumption and cell growth [56]. By careful control of both the acetate and glucose feeds (using online measurement, respectively, of CO_2 evolution rate and glucose concentration), these researchers achieved 62 g/L pyruvate in 30 h at a yield of 0.55 g/g. Integrating electrodialysis to separate pyruvate with a repetitive (fed)-batch fermentation process using this same *E. coli* strain reduced product inhibition and allowed an average yield of 0.82 g/g and productivity of 3.9 g/L·h over four cycles before a reduction in productivity was observed at 40 h [57]. An unstructured model incorporating pyruvate inhibition of growth and product formation was able to represent growth and pyruvate formation adequately, but did not account for the higher glucose consumption rate for pyruvate and maintenance than expected [58]. A neural network approach on the other hand was superior to predict the dynamics of substrate and product concentration changes during acetate feeding [59].

Although many researchers have exploited reduced activity in pyruvate dehydrogenase or have altogether eliminated one of the components of pyruvate dehydrogenase, metabolic engineering approaches can also use strategies which do not directly target this enzyme. In one study, *E. coli pflB poxB ackA ldhA adhE frdBC sucA atpFH* generated 52 g/L pyruvate from glucose as the sole carbon source in 43 h at a yield of 0.76 g/g [60]. This strategy combined a variety of features including (1) preventing lactate (*ldhA* gene deletion), ethanol (*adhE*) and acetate (*poxB*, *ackA*) formation; (2) curtailing both oxidative (*sucA*) and reductive (*frdABC*) tricarboxylic acid cycle; and (3) increasing the rate of glycolysis by uncoupling respiration (*atpFH*). The process was operated under reduced oxygenation (5% saturation) which would tend to increase the pool of NADH, an inhibitor of the dihydrolipamide dehydrogenase component of pyruvate dehydrogenase [61] and citrate synthase [62]. Thus, these operating conditions were critical for attaining a high yield of pyruvate (discussed in greater detail below). The study also confirmed the importance of pyruvate oxidase (*poxB*) in pyruvate accumulation. Obviously, the primary distinction between this approach and those involving a gene deletion in the pyruvate dehydrogenase complex is that maintaining some pyruvate dehydrogenase activity obviates an acetate requirement.

New and promising metabolic engineering approaches propose gene silencing rather than gene deletions. For example, silencing the *aceE* gene, particularly when combined with the silencing or deletions of other genes, resulted in 26 g/L pyruvate from glucose in 72 h [63]. A similar approach modified the promoters for the *accBC* genes coding acetyl CoA carboxylase as well as the *aceE* gene [64]. These promoters allowed the doxycycline-controlled expression of these two enzymes involved in pyruvate catabolism, leading to 26 g/L pyruvate in 73 h with a yield of 0.54 g/g [64]. Gene silencing seems particularly suited to targeting pyruvate dehydrogenase, as maintaining some residual activity in this enzyme allows glucose to be the sole carbon source. Gene silencing techniques are destined to become more widespread as methodologies are developed to tune finely the activity of targeted enzymes.

Metabolic engineering approaches have also been applied to microbes other than *E. coli*. In yeast a key enzyme in pyruvate metabolism is typically pyruvate decarboxylase (*PDC* gene), which decarboxylates and reduces pyruvate to acetaldehyde, which is itself reduced to ethanol via an alcohol dehydrogenase. Disruption of *PDC* in *C. glabrata* resulted in significantly greater pyruvate accumulation and less ethanol formation: 82 g/L pyruvate and less than 5 g/L ethanol accumulated from glucose in 52 h at a yield of 0.55 g/g using a medium with 30 g/L peptone [65]. Similarly, disruption of the genes coding the two structural enzymes (*PDC1* and *PDC5*) in *Saccharomyces cerevisiae* increased pyruvate to 25 g/L with a yield of 0.30 g/g in a 96 h shake flask culture in a medium containing 20 g/L peptone, 10 g/L yeast extract and 2.5 g/L sodium acetate [66]. *PDC*-negative *S. cerevisiae* unfortunately require acetate or ethanol for growth, and also cannot tolerate a high concentration of glucose. In one study, a *PDC*-negative strain able to grow on glucose without acetate or ethanol was isolated in a chemostat using progressively lower acetate concentration in the feed, resulting in a strain accumulating 135 g/L pyruvate in a repeated batch process from glucose in 100 h at a yield of 0.54 g/g [67]. In a similar study, a *PDC*-negative strain able to grow on glucose without acetate or ethanol was isolated after 1000 generations of sequential batch cultures incrementally exposed to reduced ethanol and increased glucose concentrations [68]. These approaches thus combine targeted metabolic engineering with more classical mutation isolation methodology.

Increasing interest in the use of lignocellulosic hydrolysates as an inexpensive source of carbohydrates has created an interest in cells which are able to metabolize glucose and xylose simultaneously. Disrupting *PDC1* and *GPD1* (coding for glycerol-3P dehydrogenase) and introducing a xylose transporter and several metabolic genes in *Kluyveromyces marxianus* led to 29 g/L pyruvate in 36 h from a sugar mixture, although glucose was still the favored substrate [69]. Recently, a study generated pyruvate from alginate using the bacterium *Sphingomonas* having an *ldh* knockout [70]. The strain generated nearly 3 g/L pyruvate in about 3 days, and the study demonstrated the importance of oxygen availability.

Corynebacterium glutamicum is widely used industrial bacterium for the production of several amino acids such as L-lysine and L-glutamate. In one study focused on the production of L-valine, formed in four enzymatic steps from pyruvate, strains having an *aceE* gene deletion were observed also to accumulate over 2 g/L pyruvate [71]. A subsequent study examined the additional deletion of pyruvate oxidase (*pqo*, or *poxB* gene) which improved L-valine production and also introduced an acetate growth requirement, suggesting that in *C. glutamicum* the combination of pyruvate oxidase, acetate kinase and phosphotransacetylase can bypass pyruvate dehydrogenase to allow acetyl CoA generation [72]. Further knockouts in the genes for lactate dehydrogenase (*ldhA*), two transaminases (*alaT* and *avtA*), and the replacement of native acetohydroxyacid synthase with an attenuated variant (C-T *ilvN*) led to 46 g/L pyruvate from glucose and acetate in 105 h at a yield of 0.48 g/g [73]. It should be noted that *C. glutamicum* and various yeasts including *S. cerevisiae* and *C. glabrata* have an active pyruvate carboxylase whereas *E. coli* do not.

5. Cofactor Engineering

Cofactor engineering focuses on altering the availability of NADH/NAD, and it is another means to improve pyruvic acid production. Since the conversion of glucose into pyruvate generates 2 mol NADH per mol glucose (Equation (1)), and NADH is an inhibitor of the dihydrolipamide dehydrogenase component of pyruvate dehydrogenase [61], the redox state has a direct impact on the pyruvate production rate and yield. Elevated NADH is one means of suppressing pyruvate catabolism in strains having pyruvate dehydrogenase activity. On the other hand, enhancing the oxidation of NADH to elevate NAD should increase the rate of glycolysis. Researchers have thus examined contrasting cofactor/metabolic engineering approaches to optimize pyruvate formation, depending on the genetic background of the strain. In one case, overexpression of water-forming NADH oxidase from *Streptococcus pneumoniae* (*nox* gene) in *E. coli* lead to a 70% increase in glucose uptake rate compared to a wild-type background [74]. However, in *E. coli aceEF pflB poxB pps ldhA* under acetate-limited conditions, no improvement in the rate of glycolysis was observed as a consequence of overexpressing *nox* [75]. Only additional knockouts in *atpFH* and *arcA* caused NADH oxidase to have a 12% increase in glucose uptake, and a modest 8% increase in the specific rate of pyruvate formation [75]. Similarly, overexpression of water-forming NADH oxidase (*nox*) or the endogenous mitochondrial alternative oxidase (*AOX1*) in *C. glabrata* increased the pyruvate yield and productivity by 15%–30% to 0.79 g/g and 1.63 g/L·h, respectively [76]. Expression of water-forming NADH oxidase from *Lactococcus lactis* (*noxE* gene) and *E. coli* transhydrogenase (*udhA*) in an evolved *PDC*-knockout *S. cerevisiae* improved pyruvate production to 75 g/L in 120 h with a yield of 0.63 g/g [68]. Another approach enhanced pyruvate formation by introducing acetaldehyde into the medium, which was then reduced to ethanol in a *C. glabrata* strain with high alcohol dehydrogenase activity. The resulting increases in pyruvate productivity, concentration and yield were attributed to elevated NAD concentration in the acetaldehyde-reducing strain [43]. The goal of each of these studies was to decrease NADH concentration. In contrast, for a strain having pyruvate dehydrogenase activity, the generation of more NADH was accomplished by expressing formate dehydrogenase from *Mycobacterium vaccae* (*fdh*) in *E. coli pta* [77]. In a medium with glucose, formate and peptone, using formate dehydrogenase increased pyruvate production from 6.8 to 9.0 g/L in 24 h with a yield of 0.48 g/g glucose.

6. Process Engineering

The formation of any microbial product relies not only on the best strain characteristics (genotype), but also on an optimal operating process. Generally, the optimal conditions for growth are different from those for product formation, and one strategy to accomplish both high growth/productivity and high yield often can be best performed by two distinct process stages. For example, low oxygenation surprisingly favors glucose uptake in *C. glabrata*, while high oxygenation favors pyruvate yield, so that a two-stage high-to-low oxygenation process improves overall performance over any single oxygenation process [78,79]. This compromise comes at the cost of the highest yield, which occurs at the highest oxygenation, demonstrating the difficulty in maximizing both yield and productivity. Because the conversion of glucose to pyruvate generates NADH as noted (Equation (1)), high oxygenation is often required in pyruvate formation. In addition to contributing to a high NAD concentration to drive glycolysis, oxygenation can prevent the formation of reduced by-products in strains lacking knockouts. For example, lactate formation by *E. coli* and pyruvate formate lyase activity can both be discouraged by high oxygenation during pyruvate generation [77]. In stark contrast, for strains containing a functional pyruvate dehydrogenase, pyruvate generation is enhanced by low oxygenation which encourages intracellular NADH accumulation [80]. pH and other environmental conditions also impact pyruvate formation. Pyruvate production by an *E. coli* lipoic acid auxotroph was maximal at a pH of 6 [31], while *E. coli* with a pyruvate dehydrogenase deletion showed greatest pyruvate generation at 32 °C and a pH of 7 [54]. Not surprisingly, the optimal conditions depend on the genetic background of the strain and the associated strategy used to accumulate pyruvate.

Processes often use nutrient-limited conditions to control growth rate and direct carbon to the desired product. In other words, a limiting nutrient is supplied at a rate below what the microbes can maximally metabolize, resulting in growth being controlled by that nutrient. Glycolytic rate and pyruvate productivity by *E. coli poxB aceEF pps pflB ldhA* were compared at steady-state using glucose, acetate, nitrogen, or phosphorus limitation [75]. The highest yield (0.70 g/g) and specific productivity (1.11 g/g·h) were attained under acetate-limited conditions, allowing the design of an exponential acetate-limited fed-batch process to maintain a constant growth rate of 0.15 h^{-1}. Using an *E. coli* strain with additional *atpFH* and *arcA* knockouts, the process achieved 90 g/L pyruvate in 44 h at a yield of 0.68 g/g glucose [75]. A similar acetate-limited fed-batch process using glycerol as the primary carbon source attained about 40 g/L in less than 40 h with a yield of 0.62 g/g glycerol [81]. The yield from glycerol was greatly improved to 0.95 g/g by the addition of a small amount of glucose into the feed, presumably because the glucose could be catabolized for the generation of NADPH and biomass precursors "above" the entry point of glycerol into glycolysis [81].

7. Other Biological Methods

Pyruvate can be formed through biotransformation of a biochemical using non-growing cells. Advantages of biotransformation over fermentation include using a medium which is typically free of salts and other impurities, resulting in a more easily purified product, and the prospect of using the biocatalysts for several cycles. Of course, the cell "biocatalysts" must be generated and harvested previously in a separate process, so the advantage of improved productivity only exists if several cycles are accomplished. The obvious choices for starting material are compounds which are more reduced than pyruvate, thus allowing the generation of NADH in the cells, which provides the cells with some maintenance energy to sustain the process. For example, *Acinetobacter* sp. at a concentration of 6 g/L (dry cells) converted 0.5 M L-lactate into pyruvate in 12 h [82]. Permeabilized yeast cells of either *Hansenula polymorpha* or *Pichia pastoris* expressing glycolate oxidase and catalase were able to oxidize 0.5 M L-lactate to pyruvate for 12 cycles with 98% conversion of the lactate [83]. An L-lactate concentration above 0.5 M showed substrate inhibition of a whole-cell biocatalyst based on recombinant *P. pastoris*, and oxygen, a substrate for glycolate oxidase, was important for the conversion [84]. These conversions based on catalase/glycolate oxidase were operated at 5 °C to prolong enzyme activities. *Pseudomonas stutzeri* can serve as a biocatalyst for the conversion of D/L-lactate into pyruvate, generating about 23 g/L pyruvate at a yield of 0.89 g/g [85]. Another study with *P. stutzeri* generated 48 g/L pyruvate at a yield of 0.98 g/g after 29 h at a pH of 8 and a temperature of 30 °C [86]. Similarly, 50 g/L D/L-alanine was converted to 14.6 g/L pyruvate in 30 h by previously-grown *E. coli* having deletions in native pyruvate and alanine transporters and overexpressing membrane-bound L-amino acid deaminase [87].

Most often glucose or glycerol is used as the carbon source for pyruvate production. However, because lignin is a large and untapped feedstock, there is growing interest in using the components of lignin for the production of commodity chemicals. Recently, researchers have demonstrated that manipulation of the aromatic catabolic pathway, replacing the *ortho*-cleavage pathway with a *meta*-cleavage pathway, in *Pseudomonas putida* significantly increases pyruvate production from *p*-coumarate and benzoate in an *aceEF* knockout strain [88].

8. Downstream Processes

Several general separations methods may be considered for the recovery of pyruvate from fermentation broth. A traditional method for many organic acids is to precipitate a poorly soluble calcium salt followed by acidification with H_2SO_4 to generate gypsum and free acid, a method disfavored because of the use of large volumes of acid, the generation of a $CaSO_4$ waste stream and the poor selectivity of the process. One alternative to precipitation is liquid-liquid extraction. To be practical, extraction should selectively transfer the acid product from a fermentation broth to an immiscible solvent, and then regenerate the product-free solvent through back extraction. A useful modification of extraction is reactive extraction, whereby an additional chemical (an "extractant") is introduced into the solvent-aqueous extraction system [89]. The extractant is selected for its ability to undergo a reversible complexation with the acid product. Typical extractants for organic acids include phosphoryl-containing extractants such as tributyl phosphate (TBP) or trioctyl phosphine oxide (TOPO) and aliphatic amines such as tri-*n*-octylamine (TOA). The chemical properties of the functional groups present on both the extractant and the diluent (i.e., the solvent) affect the extraction equilibria, and therefore impact the selectivity of the specific desired extraction. Extractant and diluent are regenerated by introducing a pH shift, a temperature shift or by changing the composition of the diluent. Aliphatic amines combined with a suitable diluent are very commonly chosen for carboxylic acids because this combination provides high distribution coefficients at pH values near or below the pK_A of the target carboxylic acid [90].

Numerous studies have compiled extraction equilibria for pyruvate in extractant/diluent mixtures. For the tertiary amine Alamine 336 as extractant, cyclic alcohols provide a high extraction efficiency as diluents [91] and among aliphatic alcohols 1-dodecanol/Alamine yielded the largest separation factors [92]. TOA diluted in decanol/kerosene resulted in 98% pyruvate recovery, with TOA/decanol being a superior system. [93]. An examination of the kinetics and mass transfer of the tributylamine/butyl acetate system demonstrated that the reaction was second order in the diffusion film and independent of hydrodynamic conditions [94]. Equilibrium and kinetics studies for the TOA/1-octanol system revealed that the reaction is first order [95]. Recent studies have also considered mixing phosphate extractants with amine extractants in decanol, resulting in >98% pyruvate recovery [96]. Extraction equilibria for TBP with several diluents suggest that *n*-heptane and toluene provide the greatest extraction [97]. Most systems show that the pyruvate and extractant react with a 1:1 molar ratio. Care must be taken for selecting a system which limits co-extraction of other organic acids and also water. Comparatively little research has been completed in the selective recovery of pyruvate from a solution containing the multiple acids or directly from a fermentation broth. One study examined TOA with various diluents focused on the separation of lactate and pyruvate, taking advantage of the pK_A differences between lactate (3.86) and pyruvate (2.49) to accomplish a pH swing extraction. Using pH swing extraction with trimethylamine as a back extractant, these researchers were able to obtain 97% pyruvic acid purity after removal of the amine with simple distillation [98]. One recent comprehensive study using the TOA/1-octanol system showed significant competition for the amine between lactic acid, acetic acid and pyruvic acid, but pyruvic acid could be recovered selectively in a process which removes acetic acid first, followed by separation of lactic acid and pyruvic acid [99]. The process is most readily accomplished when the lactate concentration is no greater than 20% of the pyruvate concentration and the initial TOA concentration is at least equivalent to the pyruvic acid concentration [99].

A similar recovery method is repulsive extraction, also known as salt-assisted solvent extraction, and this method is applicable to a wide range of fermentation-derived biochemicals [100,101]. The approach involves selectively reducing the solubility of the desired product. Acetone is the most effective extractant at reducing the solubility of pyruvate from the fermentation broth, and integrating acetone extraction with microfiltration, evaporation (to concentrate the mixture) and crystallization led to 97% pyruvate purity and 75% recovery [102].

Another general type of recovery process possible for organic acids is electrodialysis, wherein current is applied to alternating anion exchange and cation-exchange membranes. This method has the advantage of requiring no organic solvent. Electrodialysis has also been successfully applied to pyruvate recovery [103]. Ideally cells are removed prior to the electrodialysis step using ultrafiltration. Electrodialysis can be operated either in a mono-polar mode to generate sodium pyruvate or in a bipolar mode to generate pyruvic acid and a hydroxide, the latter which can potentially be recycled back to the fermenter for pH control. Thus, the fermentation process can be fully integrated with the electrodialysis [57]. Some hurdles remain to make electrodialysis fully compatible with fermentation, including membrane fouling and co-separation of other medium components.

One additional method to recover pyruvic acid is using adsorption, and a primary amine, weakly basic sorbent can be used to uptake pyruvic acid near the pK_A of pyruvic acid (2.49) [104]. Generally, ion exchange has a low capacity compared to extraction or electrodialysis [57].

9. Conclusions

Pyruvate lies at the center of metabolism, at a key junction where carbon partitions between the energy-generating tricarboxylic acid cycle, anaplerotic reactions to replenish this cycle and several biochemical branches leading to building block components such as amino acids and fatty acids. There has been a long history of producing pyruvate by both bacteria and yeast. Original production strains were based on vitamin auxotrophy, as several of the steps from pyruvate involve enzymes which use vitamins as cofactors, while more recent approaches have developed engineered strains, often starting from the high-producing auxotrophs. Approaches which combine evolution or selection of desirable characteristics with targeted engineering of strains will likely dominate future improvements.

Although a great detail of research has used glucose as the carbon source, surprisingly limited research has involved the conversion of glycerol or pentoses derived from lignocellulosic biomass (xylose, arabinose). Of particular interest would be processes which are able to generate pyruvate from mixed substrates. There also continue to be opportunities to isolate or develop strains which are tolerant not only to high osmolarity, but also to low pH, higher temperature and to lignocellulosic hydrolysates or multiple stresses, factors which could reduce production costs. We foresee increasing interest in the integration of continuous or semi-continuous fermentation processes, with their inherent higher productivity, with downstream pyruvate recovery operations. Improving existing microbial processes to produce pyruvate will necessitate an understanding of how cells respond to imposed operational conditions.

Acknowledgments: The authors express their appreciation to Sarah Lee for her continued assistance.

Author Contributions: Both authors wrote and edited this review article.

Conflicts of Interest: The authors declare no conflict of interest.

References

1. Park, H.-S.; Lee, J.-Y.; Kim, H.-S. Production of L-DOPA (3,4,-dihydroxyphenyl-L-alanine) from benzene by using a hybrid pathway. *Biotechnol. Bioeng.* **1998**, *58*, 339–343. [CrossRef]
2. Zhang, Y.; Tao, F.; Du, M.; Ma, C.; Qiu, J.; Gu, L.; He, X.; Xu, P. An efficient method for N-acetyl-D-neuraminic acid production using coupled bacterial cells with a safe temperature-induced system. *Appl. Microbiol. Biotechnol.* **2010**, *86*, 481–489. [CrossRef] [PubMed]
3. Rosche, B.; Sandford, V.; Breuer, M.; Hauer, B.; Rogers, P. Biotransformation of benzaldehyde into (R)-phenylacetylcarbinol by filamentous fungi or their extracts. *Appl. Microbiol. Biotechnol.* **2001**, *57*, 309–315. [PubMed]
4. Reisse, S.; Haack, M.; Garbe, D.; Sommer, B.; Steffler, F.; Carsten, J.; Bohnen, F.; Sieber, V.; Brück, T. In vitro bioconversion of pyruvate to *n*-butanol with minimized cofactor utilization. *Front. Bioeng. Biotechnol.* **2016**, *4*, 74. [CrossRef] [PubMed]

5.   Stine, A.; Zhang, M.; Ro, S.; Clendennen, S.; Shelton, M.C.; Tyo, K.E.J.; Broadbelt, L.J. Exploring *De Novo* metabolic pathways from pyruvate to propionic acid. *Biotechnol. Prog.* **2016**, *32*, 303–311. [CrossRef] [PubMed]

6.   Zhang, X.; Tervo, C.J.; Reed, J.L. Metabolic assessment of *E. coli* as a biofactory for commercial products. *Metab. Eng.* **2016**, *35*, 64–75. [CrossRef] [PubMed]

7.   Li, Y.; Chen, J.; Lun, S.-Y. Biotechnological production of pyruvic acid. *Appl. Microbiol. Biotechnol.* **2001**, *57*, 451–459. [PubMed]

8.   Xu, P.; Qiu, J.; Gao, C.; Ma, C. Biotechnological routes to pyruvate production. *J. Biosci. Bioeng.* **2008**, *105*, 169–175. [CrossRef] [PubMed]

9.   Mazzio, E.; Soliman, K.F. Cytoprotection of pyruvic acid and reduced beta-nicotinamide adenine dinucleotide against hydrogen peroxide toxicity in neuroblastoma cells. *Neurochem. Res.* **2003**, *28*, 733–741. [CrossRef] [PubMed]

10.   Wang, X.; Perez, E.; Liu, R.; Yan, L.-J.; Mallet, R.T.; Yang, S.-H. Pyruvate protects mitochondria from oxidative stress in human neuroblastoma SK-N-SH cells. *Brain Res.* **2007**, *1132*, 1–9. [CrossRef] [PubMed]

11.   Nakamichi, N.; Kambe, Y.; Oikawa, H.; Ogura, M.; Takano, K.; Tamaki, K.; Inoue, M.; Hinoi, E.; Yoneda, Y. Protection by exogenous pyruvate through a mechanism related to monocarboxylate transporters against cell death induced by hydrogen peroxide in cultured rat cortical neurons. *J. Neurochem.* **2005**, *93*, 84–93. [CrossRef] [PubMed]

12.   Yoo, M.H.; Lee, J.Y.; Lee, S.E.; Hok, J.Y.; Yoon, Y.H. Protection by pyruvate of rat retinal cells against zinc toxicity in vitro, and pressure-induced ischemia in vivo. *Invest. Ophthalmol. Visual Sci.* **2004**, *45*, 1523–1530. [CrossRef]

13.   Mongan, P.D.; Capacchione, J.; Fontana, J.L.; West, S.; Bünger, R. Pyruvate improves cerebral metabolism during hemorrhagic shock. *Am. J. Physiol. Heart Circ. Physiol.* **2001**, *281*, H854–H864. [PubMed]

14.   Ryou, M.-G.; Liu, R.; Ren, M.; Sun, J.; Mallet, R.T.; Yang, S.-H. Pyruvate protects the brain against ischemia-reperfusion injury by activating the erythropoietin signaling pathway. *Stroke* **2012**, *43*, 1101–1107. [CrossRef] [PubMed]

15.   Hasenfuss, G.; Maier, L.S.; Hermann, H.-P.; Lüers, C.; Hünlich, M.; Zeitz, O.; Janssen, P.M.L.; Pieske, B. Influence of pyruvate on contractile performance and Ca^{2+} cycling in isolated failing human myocardium. *Circulation* **2002**, *105*, 194–199. [CrossRef] [PubMed]

16.   Hermann, H.-P.; Arp, J.; Pieske, B.; Kögler, H.; Baron, S.; Hanssen, P.M. L.; Hasenfuss, F. Improved systolic and diastolic myocardial function with intracoronary pyruvate in patients with congestive heart failure. *Eur. J. Heart Fail.* **2004**, *6*, 213–218. [CrossRef] [PubMed]

17.   Kalman, D.; Colker, C.M.; Wilets, I.; Roufs, J.B.; Antonio, J. The effects of pyruvate supplementation on body composition in overweight individuals. *Nutrition* **1999**, *15*, 337–340. [CrossRef]

18.   Koh-Banerjee, P.K.; Ferreira, M.P.; Greenwood, M.; Bowden, R.G.; Cowan, P.N.; Almada, A.; Kreider, R.B. Effects of calcium pyruvate supplementation during training on body composition, exercise capacity, and metabolic responses to exercise. *Nutrition* **2005**, *21*, 312–319. [CrossRef] [PubMed]

19.   EFSA. Calcium acetate, calcium pyruvate, calcium succinate, magnesium pyruvate magnesium succinate and potassium malate added for nutritional purposes to food supplements. *EFSA J.* **2009**, *1088*, 1–25.

20.   Kim, J.-G.; Park, S.-J.; Damsté, J.S.S.; Schouten, S.; Rijpstra, W.I.C.; Jung, M.-Y.; Kim, S.-J.; Gwak, J.-H.; Hong, H.; Si, S.-J.; et al. Hydrogen peroxide detoxification is a key mechanism for growth of ammonia-oxidizing bacteria. *Proc. Natl. Acad. Sci. USA* **2016**, *113*, 7888–7893. [CrossRef] [PubMed]

21.   Kitahara, K.; Fukui, S. On the pyruvic acid fermentation (I): Selection of species from genus *Mucor*, and examination of production conditions. *J. Ferment. Technol.* **1951**, *29*, 227–233.

22.   Kitahara, K.; Fukui, S. On the pyruvic acid fermentation (II): Possibility of biochemical determination of vitamin B-1 by *Mucor mandshuricus* Saito. *J. Ferment. Technol.* **1951**, *29*, 287–290.

23.   Kitahara, K.; Fukui, S. On the pyruvic acid fermentation (III): Substances which inhibit the accumulation of pyruvic acid. *J. Ferment. Technol.* **1951**, *29*, 326–331.

24.   Kitahara, K.; Fukui, S. On the pyruvic acid fermentation (IV): The mechanism of pyruvate acid accumulation by *Mucor mandshuricus*. *J. Ferment. Technol.* **1951**, *29*, 378–384.

25.   Kitahara, K.; Fukui, S. Beriberi-like symptoms in microorganisms. *J. Gen. Appl. Microbiol.* **1955**, *1*, 61–76. [CrossRef]

26. Yokota, A.; Takao, S. Pyruvic acid production by lipoic acid auxotrophs of *Enterobacter aerogenes*. *Agric. Biol. Chem.* **1989**, *53*, 705–711. [CrossRef]
27. Izumi, Y.; Matsumura, Y.; Tani, Y.; Yamada, H. Pyruvic acid production from 1,2-propanediol by thiamin-requiring *Acinetobacter* sp. 80-M. *Agri. Biol. Chem.* **1982**, *46*, 2673–2679. [CrossRef]
28. Takao, S.; Tanida, M. Pyruvic acid production by *Schizophyllum commune*: Pyruvic acid fermentation by *Basidiomycetes* (I). *J. Ferm. Technol.* **1982**, *60*, 277–280.
29. Moriguchi, M. Fermentative production of pyruvic acid from citrus peel extract by *Debaryomyces coudertii*. *Agric. Biol. Chem.* **1982**, *64*, 955–961. [CrossRef]
30. Morgunov, I.G.; Kamzolova, S.V.; Perevoznikova, O.A.; Shishkanova, N.V.; Finogenova, T.V. Pyruvic acid production by a thiamine auxotroph of *Yarrowia lipolytica*. *Process Biochem.* **2004**, *39*, 1469–1474. [CrossRef]
31. Yokota, A.; Shimizu, H.; Terasawa, Y.; Takaoka, N.; Tomita, F. Pyruvic acid production by a lipoic acid auxotroph of *Escherichia coli* W1485. *Appl. Microbiol. Biotechnol.* **1994**, *41*, 638–643. [CrossRef]
32. Wang, Q.; He, P.; Lu, D.; Shen, A.; Jiang, N. Screening of pyruvate-producing yeast and effect of nutritional conditions on pyruvate production. *Lett. Appl. Microbiol.* **2002**, *35*, 338–342. [CrossRef] [PubMed]
33. Yonehara, T.; Miyata, R. Fermentative production of pyruvate from glucose by *Torulopsis glabrata*. *J. Ferment. Bioeng.* **1994**, *78*, 155–159. [CrossRef]
34. Miyata, R.; Yonehara, T. Breeding of high-pyruvate-producing *Torulopsis glabrata* and amino acid auxotrophic mutants. *J. Biosci. Bioeng.* **2000**, *90*, 137–141. [CrossRef]
35. Li, S.; Chen, X.; Liu, L.; Chen, J. Pyruvate production in *Candida glabrata*: Manipulation and optimization of physiological function. *Crit. Rev. Biotechnol.* **2016**, *36*, 1–10. [CrossRef] [PubMed]
36. Kamzolova, S.V.; Morgunov, I.G. Biosynthesis of pyruvic acid from glucose by *Blastobotrys adeninivorans*. *Appl. Microbiol. Biotechnol.* **2016**, *100*, 7689–7697. [CrossRef] [PubMed]
37. Miyata, R.; Yonehara, T. Improvement of fermentative production of pyruvate from glucose by *Torulopsis glabrata* IFO 0005. *J. Ferment. Bioeng.* **1996**, *82*, 475–479. [CrossRef]
38. Li, Y.; Chen, J.; Liang, D.-F.; Lun, S.-Y. Effect of nitrogen source and nitrogen concentration on the production of pyruvate by *Torulopsis glabrata*. *J. Biotechnol.* **2000**, *81*, 27–34. [CrossRef]
39. Li, Y.; Chen, J.; Lun, S.-Y.; Rui, X.-S. Efficient pyruvate production by a multi-vitamin auxotroph of *Torulopsis glabrata*: Key role and optimization of vitamin levels. *Appl. Microbiol. Biotechnol.* **2001**, *55*, 680–685. [CrossRef] [PubMed]
40. Xu, N.; Liu, L.; Zou, W.; Liu, J.; Hua, Q.; Chen, J. Reconstruction and analysis of the genome-scale metabolic network of *Candida glabrata*. *Mol. Biosyst.* **2013**, *9*, 205–216. [CrossRef] [PubMed]
41. Xu, N.; Ye, C.; Chen, X.; Liu, J.; Liu, L.; Chen, J. Genome sequencing of the pyruvate-producing strain *Candida glabrata* CCTCC M202019 and genomic comparison with strain CBS138. *Sci. Rep.* **2016**, *6*, 34893. [CrossRef] [PubMed]
42. Yang, S.; Chen, X.; Xu, N.; Liu, L.; Chen, J. Urea enhances cell growth and pyruvate production in *Torulopsis glabrata*. *Biotech. Prog.* **2014**, *30*, 19–27. [CrossRef] [PubMed]
43. Liu, L.; Xu, Q.; Li, Y.; Shi, Z.; Zhu, Y.; Du, G.; Chen, J. Enhancement of pyruvate production by osmotic-tolerant mutant of *Torulopsis glabrata*. *Biotechnol. Bioeng.* **2007**, *97*, 825–832. [CrossRef] [PubMed]
44. Xu, S.; Zhou, J.; Liu, L.; Chen, J. Proline enhances *Torulopsis glabrata* growth during hyperosmotic stress. *Biotechnol. Bioproc. Eng.* **2010**, *15*, 285–292. [CrossRef]
45. Calabia, B.P.; Tokiwa, Y.; Aiba, S. Fermentative production of L-(+)-lactic acid by an alkaliphilic marine microorganism. *Biotechnol. Lett.* **2011**, *33*, 1429–1433. [CrossRef] [PubMed]
46. Yokaryo, H.; Tokiwa, Y. Isolation of alkaliphilic bacteria for production of high optically pure L-(+)-lactic acid. *J. Gen. Appl. Microbiol.* **2014**, *60*, 270–275. [CrossRef] [PubMed]
47. Yin, J.; Chen, J.-C.; Wu, Q.; Chen, G.-Q. Halophiles, coming stars for industrial biotechnology. *Biotechnol. Adv.* **2015**, *33*, 1433–1442. [CrossRef] [PubMed]
48. Kawata, Y.; Nishimura, T.; Matsushita, I.; Tsubota, J. Efficient production and secretion of pyruvate from *Halomonas* sp. KM-1 under aerobic conditions. *AMB Express* **2016**, *6*, 22. [CrossRef] [PubMed]
49. Dietzler, D.N.; Leckie, M.P.; Magnani, J.L.; Sughrue, M.J.; Bergstein, P.E. Evidence for the coordinate control of glycogen synthesis, glucose utilization, and glycolysis in *Escherichia coli*. II. Quantitative correlation of the inhibition of glycogen synthesis and the stimulation of glucose utilization by 2,4-dinitrophenol with the effects on the cellular levels of glucose 6-phosphate, fructose, 1,6-diphosphate, and total adenylates. *J. Biol. Chem.* **1975**, *250*, 7195–7203. [PubMed]

50. Jensen, P.R.; Michelsen, O. Carbon and energy metabolism of *atp* mutants of *Escherichia coli*. *J. Bacteriol.* **1992**, *174*, 7635–7641. [CrossRef] [PubMed]

51. Yokota, A.; Terasawa, Y.; Takaoka, N.; Shimizu, H.; Fusao, T. Pyruvic acid production by an F_1-ATPase-defective mutant of *Escherichia coli* W1485lip2. *Biosci. Biotechnol. Biochem.* **1994**, *58*, 2164–2167. [CrossRef] [PubMed]

52. Yokota, A.; Henmi, M.; Takaoka, N.; Hayashi, C.; Takezawa, Y.; Fukumori, Y.; Tomita, F. Enhancement of glucose metabolism in a pyruvic acid-hyperproducing *Escherichia coli* mutant defective in F_1-ATPase activity. *J. Ferment. Bioeng.* **1997**, *83*, 132–138. [CrossRef]

53. Zhou, J.; Huang, L.; Liu, L.; Chen, J. Enhancement of pyruvate productivity by inducible expression of a F_0F_1-ATPase inhibitor INH1 in *Torulopsis glabrata* CCTCC M202019. *J. Biotechnol.* **2009**, *144*, 120–126. [CrossRef] [PubMed]

54. Tomar, A.; Eiteman, M.A.; Altman, E. The effect of acetate pathway mutations on the production of pyruvate in *Escherichia coli*. *Appl. Microbiol. Biotech.* **2003**, *62*, 76–82. [CrossRef] [PubMed]

55. Li, M.; Ho, P.Y.; Yao, S.; Shimizu, K. Effect of *lpdA* knockout on the metabolism in *Escherichia coli* based on enzyme activities, intracellular metabolite concentrations and metabolic flux analysis by ^{13}C-labeling experiments. *J. Biotechnol.* **2006**, *122*, 254–266. [CrossRef] [PubMed]

56. Zelić, B.; Gerharz, T.; Bott, M.; Vasić-Rački, Đ.; Wandrey, C.; Takors, R. Fed-batch process for pyruvate production by recombinant *Escherichia coli* YYC202 Strain. *Eng. Life Sci.* **2003**, *3*, 299–305. [CrossRef]

57. Zelić, B.; Gostović, S.; Vuorilehto, K.; Vasić-Rački, Đ.; Takors, R. Process strategies to enhance pyruvate production with recombinant *Escherichia coli*: From repetitive fed-batch to in situ product recovery with fully integrated electrodialysis. *Biotech. Bioeng.* **2004**, *85*, 638–646. [CrossRef] [PubMed]

58. Zelić, B.; Vasić-Rački, Đ.; Wandrey, C.; Takors, R. Modeling of the pyruvate production with *Escherichia coli* in a fed-batch bioreactor. *Bioprocess Biosyst. Eng.* **2004**, *26*, 249–258. [CrossRef] [PubMed]

59. Zelić, B.; Bolf, N.; Vasić-Rački, Đ. Modeling of the pyruvate production with *Escherichia coli*: Comparison of mechanistic and neural networks-based models. *Bioprocess Biosyst. Eng.* **2006**, *29*, 39–47. [CrossRef] [PubMed]

60. Causey, T.B.; Zhou, S.; Shanmugam, K.T.; Ingram, L.O. Engineering the metabolism of *Escherichia coli* W3110 for the conversion of sugar to redox-neutral and oxidized products: Homoacetate production. *Proc. Natl. Acad. Sci. USA* **2003**, *100*, 825–832. [CrossRef] [PubMed]

61. Hansen, H.G.; Henning, U. Regulation of pyruvate dehydrogenase activity in *Escherichia coli* K12. *Biochim. Biophys. Acta* **1966**, *122*, 355–358. [CrossRef]

62. Weitzman, P.D.J. Regulation of citrate synthase activity in *Escherichia coli*. *Biochim. Biophys. Acta* **1966**, *128*, 213–215. [CrossRef]

63. Nakashima, N.; Ohno, S.; Yoshikawa, K.; Shimizu, H.; Tamura, T. A vector library for silencing central carbon metabolism genes with antisense RNAs in *Escherichia coli*. *Appl. Environ. Microbiol.* **2014**, *80*, 564–573. [CrossRef] [PubMed]

64. Akita, H.; Nakashima, N.; Hoshino, T. Pyruvate production using engineered *Escherichia coli*. *AMB Express* **2016**, *6*, 94. [CrossRef] [PubMed]

65. Wang, Q.; He, P.; Lu, D.; Shen, A.; Jiang, N. Metabolic engineering of *Torulopsis glabrata* for improved pyruvate production. *Enzyme Microb. Technol.* **2005**, *36*, 832–839. [CrossRef]

66. Wang, D.; Wang, L.; Hou, L.; Deng, X.; Gao, Q.; Gao, N. Metabolic engineering of *Saccharomyces cerevisiae* for accumulating pyruvic acid. *Ann. Microbiol.* **2015**, *65*, 2323–2331. [CrossRef]

67. Van Maris, A.J.A.; Geertman, J.-M.A.; Vermeulen, A.; Groothiuzen, M.K.; Winkler, A.A.; Piper, M.D.W.; van Dijken, J.P.; Pronk, J.T. Directed evolution of pyruvate decarboxylase-negative *Saccharomyces cerevisiae*, yielding a C2-independent, glucose-tolerant, and pyruvate-hyperproducing yeast. *Appl. Environ. Microbiol.* **2004**, *70*, 159–166. [CrossRef] [PubMed]

68. Wang, Z.; Gao, C.; Wang, Q.; Liang, Q.; Qi, Q. Production of pyruvate in *Saccharomyces cerevisiae* through adaptive evolution and rational cofactor metabolic engineering. *Biochem. Eng. J.* **2012**, *67*, 126–131. [CrossRef]

69. Zhang, B.; Zhu, Y.; Zhang, J.; Wang, D.; Sun, L.; Hong, J. Engineered *Kluyveromyces marxianus* for pyruvate production at elevated temperature with simultaneous consumption of xylose and glucose. *Bioresour. Technol.* **2017**, *224*, 553–562. [CrossRef] [PubMed]

70. Kawai, S.; Ohashi, K.; Yoshida, S.; Fujii, M.; Mikami, S.; Sato, N.; Murata, K. Bacterial pyruvate production from alginate, a promising carbon source from marine brown macroalgae. *J. Biosci. Bioeng.* **2014**, *3*, 269–274. [CrossRef] [PubMed]

71. Blombach, B.; Schreiner, M.E.; Holátko, J.; Bartek, T.; Oldiges, M.; Eikmanns, B.J. L-Valine production with pyruvate dehydrogenase complex-deficient *Corynebacterium glutamicum*. *Appl. Environ. Microbiol.* **2007**, *73*, 2079–2084. [CrossRef] [PubMed]

72. Blombach, B.; Schreiner, M.E.; Bartek, T.; Oldiges, M.; Eikmanns, B.J. *Corynebacterium glutamicum* tailored for high-yield L-valine production. *Appl. Microbiol. Biotechnol.* **2008**, *79*, 471–479. [CrossRef] [PubMed]

73. Wieschalka, S.; Blombach, B.; Eikmanns, B.J. Engineering *Corynebacterium glutamicum* for the production of pyruvate. *Appl. Microbiol. Biotechnol.* **2012**, *94*, 449–459. [CrossRef] [PubMed]

74. Vemuri, G.N.; Altman, E.; Sangurdekar, D.P.; Khodursky, A.B.; Eiteman, M.A. Overflow metabolism in *Escherichia coli* during steady-state growth: Transcriptional regulation and effect of the redox ratio. *Appl. Environ. Microbiol.* **2006**, *72*, 3653–3661. [CrossRef] [PubMed]

75. Zhu, Y.; Eiteman, M.A.; Altman, R.; Altman, E. High glycolytic flux improves pyruvate production by a metabolically engineered *Escherichia coli* strain. *Appl. Environ. Microbiol.* **2008**, *74*, 6649–6655. [CrossRef] [PubMed]

76. Qin, Y.; Johnson, C.H.; Liu, L.; Chen, J. Introduction of heterogeneous NADH reoxidation pathways into *Torulopsis glabrata* significantly increases pyruvate production efficiency. *Kor. J. Chem. Eng.* **2011**, *28*, 1078–1084. [CrossRef]

77. Ojima, Y.; Suryadarma, P.; Tsuchida, K.; Taya, M. Accumulation of pyruvate by changing the redox status in *Escherichia coli*. *Biotech. Lett.* **2012**, *34*, 889–893. [CrossRef] [PubMed]

78. Hua, Q.; Araki, M.; Koide, Y.; Shimizu, K. Effects of glucose, vitamins, and DO concentrations on pyruvate fermentation using *Torulopsis glabrata* IFO 0005 with metabolic flux analysis. *Biotechnol. Prog.* **2001**, *17*, 62–68. [CrossRef] [PubMed]

79. Li, Y.; Hugenholtz, J.; Chen, J.; Lun, S.-Y. Enhancement of pyruvate production by *Torulopsis glabrata* using a two-stage oxygen supply control strategy. *Appl. Microbiol. Biotechnol.* **2002**, *60*, 101–106. [PubMed]

80. Causey, T.; Shanmugam, K.; Yomano, L.; Ingram, L. Engineering *Escherichia coli* for efficient conversion of glucose to pyruvate. *Proc. Natl. Acad. Sci. USA* **2004**, *101*, 2235–2240. [CrossRef] [PubMed]

81. Zhu, Y.; Eiteman, M.A.; Lee, S.A. Conversion of glycerol to pyruvate by *Escherichia coli* using acetate- and acetate/glucose-limited fed-batch processes. *J. Ind. Microbiol. Biotechnol.* **2010**, *37*, 307–312. [CrossRef] [PubMed]

82. Ma, C.Q.; Xu, P.; Dou, Y.M.; Qu, Y.B. Highly efficient conversion of lactate to pyruvate using whole cells of *Acinetobacter* sp. *Biotechnol. Prog.* **2003**, *19*, 1672–1676. [CrossRef] [PubMed]

83. Eisenberg, A.; Seip, J.E.; Gavagan, J.E.; Payne, M.S.; Anton, D.L.; di Cosimo, R. Pyruvic acid production using methylotrophic yeast transformants as catalysts. *J. Mol. Catal. B: Enzym.* **1997**, *2*, 223–232. [CrossRef]

84. Gough, S.; Dostal, L.; Howe, A.; Deshpande, M.; Scher, M.; Rosazza, J.N.P. Production of pyruvate from lactate using recombinant *Pichia pastoris* cells as catalyst. *Proc. Biochem.* **2005**, *40*, 2597–2601. [CrossRef]

85. Hao, J.; Ma, C.; Gao, C.; Qiu, J.; Wang, M.; Zhang, Y.; Cui, X.; Xu, P. *Pseudomonas stutzeri* as a novel biocatalyst for pyruvate production from DL-lactate. *Biotechnol. Lett.* **2007**, *29*, 105–110. [CrossRef] [PubMed]

86. Gao, C.; Qiu, J.; Ma, C.; Xu, P. Efficient production of pyruvate from DL-lactate by the lactate-utilizing strain *Pseudomonas stutzeri* SDM. *PLoS ONE* **2012**, *7*, e40755. [CrossRef] [PubMed]

87. Hossain, G.S.; Shin, H.-D.; Li, J.; Du, G.; Chen, J.; Liu, L. Transporter engineering and enzyme evolution for pyruvate production from D/L-alanine with a whole-cell biocatalyst expressing L-amino acid deaminase from *Proteus mirabilis*. *RSC Adv.* **2016**, *6*, 82676–82684. [CrossRef]

88. Johnson, C.W.; Beckham, G.T. Aromatic catabolic pathway selection for optimal production of pyruvate and lactate from lignin. *Metab. Eng.* **2015**, *28*, 240–247. [CrossRef] [PubMed]

89. Hong, Y.K.; Hong, W.H.; Han, D.H. Application of reactive extraction to recovery of carboxylic acids. *Biotechnol. Bioproc. Eng.* **2001**, *6*, 386–394. [CrossRef]

90. Wasewar, K.L.; Neesink, A.B.M.; Versteeg, G.F.; Pangarkar, V.G. Reactive extraction of lactic acid using Alamine 336 in MIBK: Equilibria and kinetics. *J. Biotechnol.* **2002**, *97*, 59–68. [CrossRef]

91. Senol, A. Influence of diluent on amine extraction of pyruvic acid using Alamine system. *Chem. Eng. Process.* **2006**, *45*, 755–763. [CrossRef]

92. Senol, A. Liquid-liquid equilibria for mixtures of (water + pyruvic acid + alcohol/Alamine). Modeling and optimization of extraction. *J. Chem. Eng. Data* **2013**, *58*, 528–536. [CrossRef]

93. Pal, D.; Tripathi, A.; Shukla, A.; Gupta, K.R.; Keshav, A. Reactive extraction of pyruvic acid using tri-octylamine diluted in decanol/kerosene: Equilibrium and effect of temperature. *J. Chem. Eng. Data* **2015**, *60*, 860–869. [CrossRef]

94. Pal, D.; Keshav, A. Kinetics of reactive extraction of pyruvic acid using tributylamine dissolved in *n*-butyl acetate. *Int. J. Chem. React. Eng.* **2015**, *12*, 63–69. [CrossRef]

95. Marti, M.E.; Gurkan, T. Selective recovery of pyruvic acid from two and three acid aqueous solutions by reactive extraction. *Separ. Purif. Technol.* **2015**, *156*, 148–157. [CrossRef]

96. Pal, D.; Thakre, N.; Kumar, A.; Keshav, A. Reactive extraction of pyruvic acid using mixed extractants. *Separ. Sci. Technol.* **2016**, *51*, 1141–1150. [CrossRef]

97. Pal, D.; Keshav, A. Extraction equilibria of pyruvic acid using tri-*n*-butyl phosphate: Influence of diluents. *J. Chem. Eng. Data.* **2014**, *59*, 2709–2716. [CrossRef]

98. Ma, C.Q.; Li, J.C.; Qiu, J.H.; Wang, M.; Xu, P. Recovery of pyruvic acid from biotransformation solutions. *Appl. Microbiol. Biotechnol.* **2006**, *70*, 308–314. [CrossRef] [PubMed]

99. Marti, M.E.; Gurkan, T.; Doraiswamy, L.K. Equilibrium and kinetic studies on reactive extraction of pyruvic acid with trioctamine in 1-octanol. *Ind. Eng. Chem. Res.* **2011**, *50*, 13518–13525. [CrossRef]

100. Zhigang, T.; Rongqi, Z.; Zhanting, D. Separation of isopropanol from aqueous solution by salting-out extraction. *J. Chem. Technol. Biotechnol.* **2001**, *76*, 757–763. [CrossRef]

101. Birajdar, S.D.; Padmanabhan, S.; Rajagopalan, S. Repulsive effect of salt on solvent extraction of 2,3-butanediol from aqueous fermentation solution. *J. Chem. Technol. Biotechnol.* **2015**, *90*, 1455–1462. [CrossRef]

102. Ingle, U.; Rajagopalan, S.; Padmanabhan, S. Isolation and scale-up of sodium pyruvate from fermentation broth using a repulsive extraction and crystallization method. *J. Chem. Technol. Biotechnol.* **2015**, *90*, 2240–2248. [CrossRef]

103. Zelić, B.; Vasić-Rački, Đ. Process development and modeling of pyruvate recovery from a model solution and fermentation broth. *Desalination* **2005**, *174*, 267–276. [CrossRef]

104. Huang, S.; Qin, W.; Dai, Y. Recovery of pyruvic acid with weakly basic polymeric sorbents. *J. Chem. Technol. Biotechnol.* **2008**, *83*, 683–687. [CrossRef]

fermentation

MDPI

Article

Mixed Carboxylic Acid Production by *Megasphaera elsdenii* from Glucose and Lignocellulosic Hydrolysate

Robert S. Nelson, Darren J. Peterson, Eric M. Karp, Gregg T. Beckham * and Davinia Salvachúa *

National Bioenergy Center, National Renewable Energy Laboratory, 15013 Denver West Parkway, Golden, CO 80401, USA; robert.nelson@nrel.gov (R.S.N.); Darren.Peterson@nrel.gov (D.J.P.); Eric.Karp@nrel.gov (E.M.K.)
* Correspondence: gregg.beckham@nrel.gov (G.T.B.); davinia.salvachua@nrel.gov (D.S.); Tel.: +1-303-384-7699 (G.T.B. & D.S.)

Academic Editor: Gunnar Lidén
Received: 7 December 2016; Accepted: 23 February 2017; Published: 1 March 2017

Abstract: Volatile fatty acids (VFAs) can be readily produced from many anaerobic microbes and subsequently utilized as precursors to renewable biofuels and biochemicals. *Megasphaera elsdenii* represents a promising host for production of VFAs, butyric acid (BA) and hexanoic acid (HA). However, due to the toxicity of these acids, product removal via an extractive fermentation system is required to achieve high titers and productivities. Here, we examine multiple aspects of extractive separations to produce BA and HA from glucose and lignocellulosic hydrolysate with *M. elsdenii*. A mixture of oleyl alcohol and 10% (v/v) trioctylamine was selected as an extraction solvent due to its insignificant inhibitory effect on the bacteria. Batch extractive fermentations were conducted in the pH range of 5.0 to 6.5 to select the best cell growth rate and extraction efficiency combination. Subsequently, fed-batch pertractive fermentations were run over 230 h, demonstrating high BA and HA concentrations in the extracted fraction (57.2 g/L from ~190 g/L glucose) and productivity (0.26 g/L/h). To our knowledge, these are the highest combined acid titers and productivity values reported for *M. elsdenii* and bacterial mono-cultures from sugars. Lastly, the production of BA and HA (up to 17 g/L) from lignocellulosic sugars was demonstrated.

Keywords: butyric acid; hexanoic acid; caproic acid; volatile fatty acids; short-chain carboxylates; extractive fermentation; oleyl alcohol; corn stover; biochemical; biofuel

1. Introduction

The conversion of lignocellulosic sugars to bio-based products and biofuels presents an opportunity to replace non-sustainable, petroleum-based products while avoiding direct competition with food sources. Microbial conversion approaches, often enabled by the use of strains that naturally accumulate products of interest, have garnered significant interest in the emerging bioeconomy. One such strategy, the "Carboxylate Platform"—namely the production of volatile fatty acids (VFAs, specifically C2–C8 VFAs) from biomass sources via anaerobic fermentation either through mixed cultures or mono-culture—has emerged as an attractive approach to generate high yields of promising biofuel and biochemical precursors [1,2]. Most well studied VFA-producing microbes employ anaerobic chain elongation [3,4]. Both butyric acid (BA) and hexanoic acid (HA) are particularly promising VFAs that can readily be upgraded to biofuels and solvents through known chemistries [5–8]. Currently, much of the industrial production of BA and HA is from fossil-derived sources; however, the biological production of BA from non-fossil-based sources is gaining attention [9,10]. In general, homo-BA or homo-HA fermentations are pursued.

However, for downstream catalytic transformations aimed at producing biofuels, mixtures are often desirable, creating an opportunity for heterofermentative production of mixed VFAs.

Biological production of BA (e.g., by *Clostridium tyrobutyricum* or *Clostridium butyricum*) has been extensively studied relative to HA production. In the case of BA production, the maximum titers reported reached up to 300 g/L in an extractive fermentation mode from glucose by *C. tyrobutyricum* [11]. Additionally, BA fermentations typically produce considerable acetic acid concentrations [12]. To date, reports of simultaneous HA and BA production are scarce. In addition, most of the substrates utilized in these studies are not lignocellulose-derived sugars. For instance, the production of HA and BA by *Clostridium kluyveri* is from ethanol and acetate (or succinate) fermentation [13], or in the case of *Clostridium* sp. BS-1 from galactitol [14]. The latter study reported one of the highest HA titers to date, 32 g/L in 16 days of fermentation. In other studies, VFAs were produced in bacterial co-cultures [15–21], and important productivity improvements have been recently shown [3,22].

Megasphaera elsdenii is a strictly anaerobic bacterium which was isolated from sheep rumen in 1956 [23]. This bacterium is able to ferment different carbon sources (e.g., glucose, fructose, sucrose, and lactic acid) to produce VFAs (C2 to C6), CO_2 and H_2 [23]. Specifically, BA and HA are the major metabolic products from glucose [20,21,23]. Recently, it has been also demonstrated that *Megasphera* sp. MH can generate C7- and C8-carboxylic acids when adding different carbon-chain length electron acceptors to the culture media [24]. *M. elsdenii* metabolism and, in particular, the steps involved in HA production are not still well understood, but it is hypothesized that HA may be formed from chain elongation of butyrate via reverse β-oxidation [25]. Furthermore, it has been shown that *M. elsdenii* utilizes acetate and butyrate for the production of HA [20]. Despite carboxylic acids being major metabolites of *M. elsdenii*, they are also often potent fermentation inhibitors [26]. The sensitivity of *M. elsdenii* to initial HA concentrations has been investigated as well as the toxicity of different VFAs, showing that HA is more toxic than shorter carboxylic acids [27]. Some of the highest reported concentrations for HA production by *M. elsdenii* (with or without simultaneous production of BA) were from sucrose (28.4 g/L HA) [20] and glucose (up to 19 g/L of HA and 3.4 g/L BA) [21] fermentations. In both cases, fermentations were coupled to extraction systems to overcome product inhibition.

Several methods to mitigate carboxylate inhibition have been previously explored, such as initial pH adjustment [20], pH control by addition of base [21] or a buffer [14], cell immobilization [21], product removal by extraction [20], gas stripping [28], or ion exchange resin in the fermentation vessel [21]. In the case of in situ product removal via organic solvent extraction, there is potential to avoid product feedback inhibition, decrease processing costs associated with neutralization, and facilitate product purification. This strategy has been utilized and demonstrated to be effective for a variety of carboxylic acids including propionic acid, BA, and HA [11,14,20,29,30]. For these highly water soluble compounds, a reactive extraction method is preferred. In a reactive extraction, a chemical complex is formed involving the carboxylate group of the product and the extraction solvent system facilitating extraction [31]. For instance, combining an alkylamine and organic solvent has been shown to improve extraction efficiency [14,20,31]. The selection of the solvent is critical to maximize extraction efficiency while minimizing toxic effects to the microorganism.

The objective of the current study was to produce heterogeneous mixtures of primarily BA and HA through anaerobic fermentation from glucose and glucose-rich lignocellulosic hydrolysate. For this purpose, two extractive fermentation strategies (liquid–liquid and pertractive) were investigated. A concentration of 57.2 g/L of a combined BA and HA mixture was obtained in the extracted fraction of the fed-batch, pertractive fermentation at an average productivity of 0.26 g/L/h. Furthermore, the robustness and efficiency of *M. elsdenii* producing BA and HA from biomass hydrolysates was demonstrated, indicating that this organism may eventually be a potential candidate strain for producing mixed VFAs from biomass sugars.

2. Materials and Methods

2.1. Hydrolysate Preparation

In the current work, two different biomass sugar streams generated at the pilot scale at the National Renewable Energy Laboratory (NREL, Golden, CO, USA) were utilized (Table 1). One stream was from deacetylated, dilute-acid (DDA) pretreated, enzymatically hydrolyzed corn stover. DDA material was prepared as previously described [32]. The second stream was obtained from deacetylated, mechanically refined (DMR), enzymatically hydrolyzed corn stover and was prepared as previously reported [33]. Enzymatic hydrolysis of both pretreated materials were run as previously detailed [33] except that 20 mg/g Novozymes Cellic® CTec2 (Franklinton, NC, USA) was used for DDA digestion and 20 mg/g of a 80:20 mixture of Novozymes Cellic® Ctec2 and Novozymes Cellic® HTec2 (Franklinton, NC, USA) was used for DMR hydrolysis. Both enzyme hydrolysates were centrifuged to remove solids, neutralized to pH 6.5 with NaOH (Sigma-Aldrich, St. Louis, MO, USA), and sterile filtered in the anaerobic chamber (Coy Laboratory Products, Grass Lake, MI, USA).

Table 1. Composition of non-diluted deacetylated, dilute-acid (DDA) and deacetylated, mechanically refined (DMR) hydrolysates.

Biomass Stream	Glucose (g/L)	Xylose (g/L)	Galactose (g/L)	Arabinose (g/L)	Acetic Acid (g/L)	Lactic Acid (g/L)	HMF * (g/L)	Furfural (g/L)
DDA	85.9	52.3	3.1	7.8	1.9	0.0	0.1	1.0
DMR	68.0	48.9	0.8	3.6	0.8	2.8	0.0	0.0

* HMF = *5-hydroxymethylfurfural*.

2.2. Microorganism and Growth Media

M. elsdenii NCIMB702410, a strain chosen based on the screening work performed by Choi, et al. [20], was purchased from the National Collection of Industrial, Food and Marine Bacteria (NCIMB) in the United Kingdom. The strain was revived in Peptone Yeast Glucose (PYG) media, adapted from Choi et al. [20] and contained the following components in grams per liter of water: yeast extract (Becton Dickinson, Franklin Lakes, NJ, USA), 10 g; tryptone (Becton Dickinson, Franklin Lakes, NJ, USA), 5 g; peptone (Becton Dickinson, Franklin Lakes, NJ, USA), 5 g; beef extract (Sigma-Aldrich, St. Louis, MO, USA), 5 g; glucose (Caisson Labs, North Logan, UT, USA), 40 g; K_2HPO_4 (Sigma-Aldrich, St. Louis, MO, USA), 2.04 g; KH_2PO_4 (Sigma-Aldrich, St. Louis, MO, USA), 0.04 g; cysteine-HCl (Sigma-Aldrich, St. Louis, MO, USA), 0.5 g; $CaCl_2 \cdot 2H_2O$ (Sigma-Aldrich, St. Louis, MO, USA), 0.01 g; $MgSO_4 \cdot 7H_2O$ (Sigma-Aldrich, St. Louis, MO, USA), 0.02 g; $NaHCO_3$ (Sigma-Aldrich, St. Louis, MO, USA), 0.4 g; NaCl (Fisher Chemicals, Fairlawn, NJ, USA), 0.08 g; resazurin (Sigma-Aldrich, St. Louis, MO, USA), 1 mg; hemin solution, 10 mL; and vitamin K1 solution, 0.2 mL. The hemin solution was prepared by dissolving 50 mg of hemin (Sigma-Aldrich, St. Louis, MO, USA) in 1 mL of 1 M NaOH then adjusted up to 100 mL with water. The vitamin K1 solution was made by dissolving 0.1 mL of vitamin K1 stock (Sigma-Aldrich, St. Louis, MO, USA) in 20 mL 95% ethanol (Fisher Chemicals, Fairlawn, NJ, USA). After combining all ingredients except the cysteine, the media was pH adjusted to 6.5 with NaOH and sparged with nitrogen for 1 h after which the cysteine was added. Then, the media was sterile filtered inside the anaerobic chamber and dispensed into serum bottles, sealed with septa and aluminum crimp seals, and finally autoclaved at 121 °C for 20 min. Media with sugars other than glucose were made as described except the sterile sugars were added after the media was autoclaved. The bacterial frozen stock was prepared from revived bacteria in PYG media during exponential phase which corresponded to an optical density at 600 nm (OD_{600}) of approximately 5.0. A 500-µL aliquot of the culture was injected into a sealed HPLC vial containing 500 µL of 50% anoxic and autoclaved glycerol (Sigma-Aldrich, St. Louis, MO, USA). The vials were gently vortexed and stored at −70 °C.

2.3. Seed Culture Preparation

Inocula were initiated by adding 1 vial of bacterial stock in sealed serum bottles containing 50 mL of PYG and 20 g/L glucose. Cells were grown anaerobically, in a shaker incubator (New Brunswick Scientific, Edison, NJ, USA) at 37 °C and 180 rpm for 8 h. This culture was then used to inoculate the "seed culture" bottle containing the same medium at an initial OD_{600} of 0.05. This culture grew for 18–24 h until an OD_{600} of ~8.0 was obtained. When additional volume (>100 mL) of inoculum was required, the "seed culture" was prepared in a 0.5 L BioStat-Q plus bioreactor (Sartorius AG, Goettingen, Germany) at 37 °C, with 150 rpm agitation, pH controlled at 6.3 and 300 mL PYG media. Inoculation of both serum bottles and fermenters in further experiments was accomplished by approximately 10% (v/v) direct transfer to reach an initial OD_{600} of 0.5.

2.4. Fermentation in Serum Bottles to Test the Effect of Organic Solvents on M. elsdenii Growth

A variety of solvents were tested to evaluate effects on *M. elsdenii* growth. Extractive batch cultures were run using oleyl alcohol (Alpha Aesar, Ward Hill, MA, USA) or octanol (Sigma-Aldrich, St. Louis, MO, USA) as the organic solvent and either trihexylamine (THA) or trioctylaime (TOA) (Sigma-Aldrich, St. Louis, MO, USA) as reactive compounds (at 5% and 10% solvent v/v). Fermentations in serum bottles were run as described above except a 20% volume of anoxic organic extractant was injected into the bottles. This mixture was allowed to equilibrate for at least 1 day. Oxygen was removed from the organic solutions by sparging with nitrogen for 2 h. The inoculum was injected into the aqueous layer with the bottle inverted to minimize contact of the organism with the organic phase. Experiments were performed in duplicate and error bars are presented as the absolute difference between values.

2.5. Batch Extractive Fermentations at Different pHs

Batch extractive fermentations were performed in 0.5 L BioStat-Q plus bioreactors. Fermentors were filled in a sterile hood with 280 mL anoxic PYG and 40 g/L glucose overlaid with 60 mL of 10% trioctylamine diluted in oleyl alcohol and then sparged overnight with nitrogen. The pH was initially adjusted to 5.0, 5.5, 6.0, or 6.5 with NaOH and maintained with either 4 M H_2SO_4 or 4 N NaOH. After inoculation, the sparge was turned off and cultures were maintained at 37 °C and 150 rpm agitation. During sampling, a nitrogen sparge was used to provide positive pressure and to remove any oxygen contamination. Samples (~1 mL) were taken for both aqueous and organic fraction. Acids from the organic phase were extracted against 0.5 N NaOH. Equal volumes (400 μL) of organic solvent and NaOH were mixed, vigorously vortexed, and statically maintained overnight at room temperature. Acid concentrations were measured from both aqueous and NaOH fractions. Glucose and bacterial growth (OD_{600}) were analyzed from the aqueous fraction. Fermentations were performed in duplicate and error bars are presented as the absolute difference between values.

2.6. Fed-Batch Pertractive Fermentations

The pertractive fermentation system utilized in the current study consisted of two Liqui-Cel Extra-Flow 2.5 × 8 membrane contactor units (3M, Charlotte, NC, USA), a 1-L Q-plus, bioreactor (Sartorius AG, Goettingen, Germany), and three FH100 peristaltic pumps (Thermo Scientific, Waltham, MA, USA) (Figure 1). The fermentor contained 1 L of PYG and 40 g/L of glucose and was maintained at pH 6.3, 37 °C, and 180 rpm agitation. The 1-L fermentation broth was continuously circulated through the shell side of the first membrane contactor at 400 mL/min. The organic extractant was continuously sparged with nitrogen and circulated through the lumen side of both membrane contactor units, countercurrent to the aqueous phases at 150 mL/min. The 1.0 N NaOH stripping solution was circulated through the shell side of the second membrane contactor unit at 400 mL/min. Adjustable tubing clamps were used as valves, partially constricting the tubing to apply 3–4 psi of pressure to the membrane contactor units and prevent the organic extractant from bleeding through the membrane. The feed solution was delivered with a model 120U peristaltic pump (Watson Marlow,

Wilmington, MA, USA), continuously sparged with nitrogen, contained 500 g/L glucose, 5 g/L yeast extract, and 2.5 g/L each of tryptone, peptone, and beef extract.

The fermentation portion of the system was sterilized by flushing 100% isopropanol (Fisher Chemicals, Fairlawn, NJ, USA) through the shell to lumen side of the filter unit, followed by circulating 10% H_2O_2 (Sigma-Aldrich, St. Louis, MO, USA) through the shell to lumen side for 6 h. The unit was then drained and flushed with at least 2 L of sterile water from the shell to lumen to remove residual H_2O_2. The filter unit was then dried overnight by flushing sterile filtered air from the shell to lumen. Samples (~1 mL) were taken periodically from the fermentation broth, organic phase, and NaOH. Glucose and bacterial growth were tracked in the fermentor broth. Acids were individually measured in the three solutions. Acids in the organic phase were extracted with 0.5 N NaOH as reported above. Pertractive fermentations were performed in duplicate and error bars are presented as the absolute difference between values.

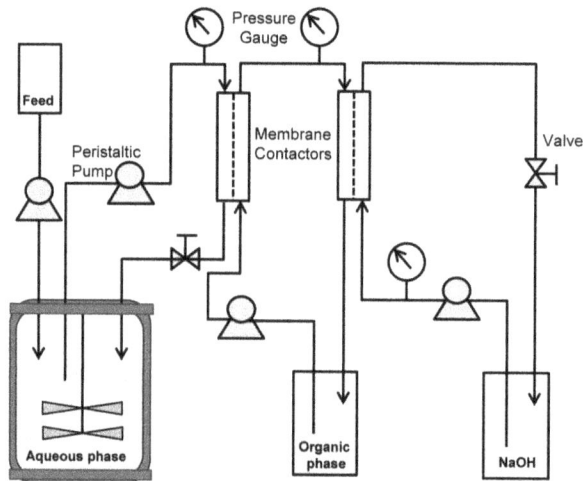

Figure 1. Scheme of the fed-batch pertractive fermentation set up utilized in the current study.

2.7. M. elsdenii Fermentations on Biomass Hydrolysate

Media was prepared in the anaerobic chamber by aseptically adding 5× yeast peptone (PY) (no glucose) media, corn stover hydrolysate (DDA and DMR), and water in sufficient quantities to make 1× PY with a glucose concentration from biomass sugars of 40 g/L. This strategy resulted in differing levels of xylose and other hydrolysate components in the two biomass streams. As a control, media was prepared as detailed above but substituting pure glucose in place of hydrolysate. Media were adjusted to pH 6.5. For extraction, 10% trioctylamine in oleyl alcohol was added at 20% the volume of media. Samples (~1 mL) were taken for both aqueous and organic fraction as reported in Section 2.5. Samples were consistently diluted 20× for OD_{600} measurements to control interference from the hydrolysate. Fermentations were performed in duplicate and error bars are presented as the absolute difference between values.

2.8. Analytical Methods

The optical density of the culture was measured at a wavelength of 600 nm and a 1-cm path length using a Genesys 10uv, UV-vis spectrophotometer (Thermo Scientific, Waltham, MA, USA). The concentrations of carboxylic acids, glucose, and other sugars concentrations were measured via HPLC (Agilent1100 series, Santa Clara, CA, USA)) outfitted with a refractive index detector and using an Aminex HPX-87H (300 × 7.8mm) organic acid column and Cation H$^+$ guard cartridge

(Biorad Laboratories, Hercules, CA, USA). The column was maintained at 65 °C with mobile phase consisting of 0.01 N sulfuric acid with a flow rate of 0.6 mL/min. Mobile phase was prepared from 10.0N sulfuric acid (Ricca Chemical Company, Arlington, TX). Analytes were identified by comparing retention times and spectral profiles with pure standards. The lactic, formic, acetic, butyric, valeric and hexanoic acids, along with 5-hydroxymethylfurfural, furfural, sodium propionate, arabinose and galactose used as standards were obtained from Sigma-Aldrich (St. Louis, MO, USA). Xylose was obtained from (Cascade Biochemicals, Corvallis, OR, USA).

2.9. Yields, Productivities, Carbon Yields, and Mass Balance Calculations

Acid yields were calculated as the coefficient of acetic acid (AA), BA, and HA production (g) and the sugars consumed (g). Yields were not corrected by the dilution factor caused by base addition since base volumes were less than 4% (*v/v*). For yield calculations, the acid sum from all aqueous, organic, and NaOH sources was considered. Lactic and acetic acids present in the biomass sugars were included in yield calculations as carbon sources. Productivity (g/L/h) was calculated as the concentration of total acids (g/L) divided by the time (h) at the end of the fermentation. Carbon yields were expressed as the percentage of moles of carbon in the products (acetic, BA, and HA) divided by the moles of carbon from glucose, lactate, and/or acetate utilized. For mass balance calculations, carbon from cell biomass and CO_2 were considered. Dry cell biomass (g/L) was estimated by multiplying the OD_{600} by the factor 0.261 g/L/OD which we previously determined correlating OD_{600} and cell dry weight (g/L). Carbon content of *M. elsdenii* biomass was 44% as measured with a LECO TruSpec CHN determinator using high temperature combustion in pure oxygen followed by infrared (IR) analysis of H_2O and CO_2. The gas was then reduced and scrubbed to measure nitrogen content via a thermal conductivity detector. CO_2 was estimated to be 33% during butyrate (Equation (1)) and HA production (Equation (2)). These equations are expressed as the maximum carbon from glucose that could lead to product without considering bacterial biomass.

$$C_6H_{12}O_6 \rightarrow C_4H_8O_2 + 2\,CO_2 + 2\,H_2 \tag{1}$$

$$3\,C_6H_{12}O_6 \rightarrow 2\,C_6H_{12}O_2 + 6\,CO_2 + 2\,H_2O + 4\,H_2 \tag{2}$$

3. Results

3.1. Evaluation of Solvent Compatibility for M. elsdenii Growth and Glucose Utilization

A reactive extraction is one where a reactive hydrophobic compound, often an alkylamine, interacts with an acid to facilitate extraction into an organic phase. The reactive compounds are typically diluted in an organic solvent. Tri-alkyl amines have been reported to be effective as reactive compounds for acid extraction [11,20,30,34]. In the current work, we use pure THA and TOA. Oleyl alcohol is one of the most common diluents for in situ extraction [11,20,30] and octanol has also been identified as highly effective for succinic acid extractions [29].

The reactive compounds mentioned above, THA and TOA, and two diluents, octanol and oleyl alcohol, were first evaluated for their compatibility in *M. elsdenii* fermentations. Bacterial growth (OD_{600}) and glucose conversion were employed as indicators of solvent toxicity (Figure 2). In the control case without extraction solvent, bacterial growth peaked at 30 h ($OD_{600} = 14$) and, after 48 h, the OD_{600} dropped to 4. In parallel, the pH declined from 6.5 to ~5, due to the accumulation of VFAs, which stalled the fermentation at ~30 h. In the presence of octanol, cells did not grow and glucose was not utilized, indicating high toxicity. Cultures with oleyl alcohol exhibited enhanced bacterial growth compared to the control and utilized 75% of the glucose. This result suggests that oleyl alcohol has low toxicity and that it extracts carboxylic acids, as previously described by Choi et al. [20]. The reactive compounds, THA and TOA, were also tested at two concentrations diluted in oleyl alcohol. Compared to cultures containing only oleyl alcohol, cultures grown in the presence of 5% and 10% THA exhibited significant decreases in growth and glucose utilization, indicating a moderate level of

THA toxicity. Cultures grown with 5% and 10% TOA showed the highest cell densities and complete glucose utilization. Based on the highest cell density and glucose utilization, 10% TOA in oleyl alcohol was chosen as the extraction solvent for subsequent work.

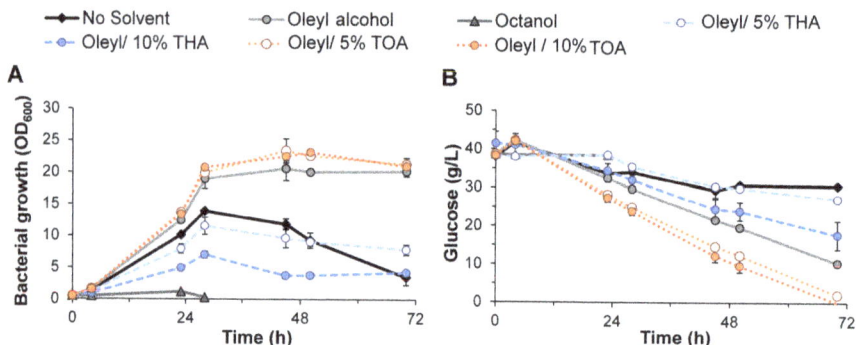

Figure 2. Serum bottle experiments without pH control to test the effect of extraction solvents on *Megasphaera elsdenii* (**A**) growth and (**B**) glucose utilization. Oleyl: oleyl alcohol; THA: trihexylamine; TOA: trioctylamine. Dry cell biomass (g/L) can be calculated by multiplying the optical density at 600 nm (OD_{600}) by the factor 0.261 g/L/OD. Results are duplicate experiments and error bars represent the data range.

3.2. Batch Liquid–Liquid Extractive Fermentation at Different pHs

One of the most important factors for effectively extracting carboxylic acids is pH. In a reactive extraction, only the protonated form of the acid is extracted and consequently, extraction is more efficient at low pH [20]. AA, BA, and HA all have similar pKa values of approximately 4.8. Above this pH, the acids become increasingly deprotonated. However, *M. elsdenii* grows better in the pH range of 5–8. To understand the balance of these requirements, we investigated the effects of different pH (5.0, 5.5, 6.0, and 6.5) in batch fermentations overlaid with 10% TOA in oleyl alcohol.

At pH 5.0, the culture grew very poorly (data not shown). At higher pH, both cell growth and glucose utilization rates were enhanced with increasing pH (Figure 3A,B). The growth rate for pH 6.5 was similar to pH 6.0 and considerably higher than at 5.5. Glucose utilization rates were 0.39, 0.27, and 0.20 g/L/h at pH 6.5, 6.0, and 5.5, respectively. These trends in *M. elsdenii* are similar to those previously reported by Miyazaki et al. [35]. At lower pH, the cells utilize energy to maintain intracellular pH [26] which is reflected as a decrease in growth. Titers of BA and HA (Figure 3C) were similar at pH 6.5 and 6.0 and lower at 5.5, reflecting incomplete glucose utilization at the lower pH. It has been also demonstrated that higher pH enhances titers and productivity due to lower concentrations of protonated acids [21]. Protonated acids in the fermentation broth can diffuse into the cells, decreasing the intracellular pH and causing microbial stress. Although the microbe is equipped to handle excess protons, the cells may not have the capacity to expel the conjugate anion, which can accumulate to toxic levels [36,37]. Carboxylic acid and molar carbon yields were similar for the three cases, ranging between 0.34–0.36 g/g and 0.51–0.55 mol/mol respectively (Table 2). In addition, these fermentations showed a carbon mass balance over 90%, accounting for most of the glucose conversion products.

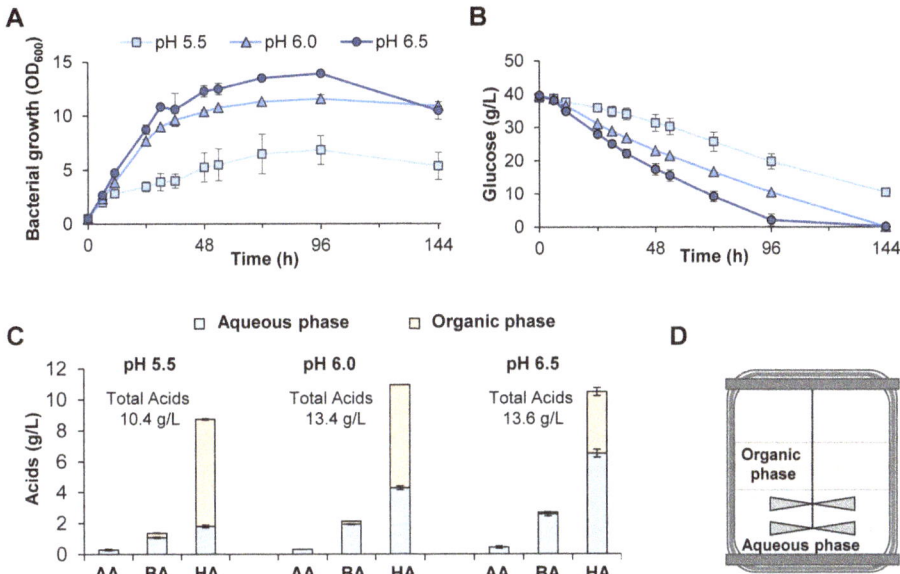

Figure 3. Production of carboxylic acids from glucose by *M. elsdenii* in a liquid–liquid extractive fermentation setup maintained at pH 5.5, 6.0, and 6.5: (**A**) bacterial growth; (**B**) glucose consumption; (**C**) acid concentration in the fermentor (aqueous phase) and extractive solvent (organic phase); and (**D**) representation of the liquid–liquid extractive set up. Results are duplicate experiments and error bars represent the data range. Dry cell biomass (g/L) can be calculated by multiplying the OD_{600} by the factor 0.261 g/L/OD. AA: acetic acid, BA: butyric acid; HA: hexanoic acid.

Table 2. Summary of end-point fermentation parameters for the production of volatile fatty acids (VFAs) by *M. elsdenii* in the different fermentation experiments presented in the current study. Results are duplicate experiments and error represents the data range.

Condition	Total Titer [1] (g/L)	Total Yield [1] (g/g)	AA Yield [2] (g/g)	BA Yield (g/g)	HA Yield (g/g)	Productivity [1,3] (g/L/h)	Carbon Yield [1] (mol/mol)	CMB [4] (%)
Liquid–liquid batch pH 5.5	10.4 ± 0.1	0.36 ± 0.00	0.01 ± 0.00	0.05 ± 0.00	0.31 ± 0.00	0.07 ± 0.00	0.55 ± 0.01	94 ± 1
Liquid–liquid batch pH 6.0	13.4 ± 0.2	0.34 ± 0.01	0.01 ± 0.00	0.05 ± 0.00	0.28 ± 0.01	0.09 ± 0.00	0.51 ± 0.01	92 ± 1
Liquid–liquid batch pH 6.5	13.6 ± 0.2	0.35 ± 0.01	0.01 ± 0.00	0.07 ± 0.01	0.27 ± 0.00	0.09 ± 0.00	0.52 ± 0.01	92 ± 1
Pertractive Fed-batch	61.3 ± 3.6	0.32 ± 0.01	0.02 ± 0.00	0.19 ± 0.01	0.11 ± 0.00	0.26 ± 0.01	0.63 ± 0.02	98 ± 1
Glucose [5]	5.2 ± 0.3	0.44 ± 0.00	0.01 ± 0.01	0.15 ± 0.01	0.28 ± 0.02	0.03 ± 0.00	0.65 ± 0.01	103 ± 1
DDA [5]	6.1 ± 0.0	0.43 ± 0.02	−0.02 ± 0.00	0.16 ± 0.01	0.30 ± 0.01	0.04 ± 0.00	0.65 ± 0.02	104 ± 2
DMR [5]	5.3 ± 0.01	0.37 ± 0.03	−0.02 ± 0.02	0.11 ± 0.00	0.27 ± 0.00	0.04 ± 0.00	0.56 ± 0.03	91 ± 3
Extracted Glucose [5]	16.9 ± 0.1	0.44 ± 0.00	0.01 ± 0.00	0.11 ± 0.01	0.32 ± 0.01	0.10 ± 0.00	0.65 ± 0.00	108 ± 0
Extracted DDA [5]	17.8 ± 0.4	0.42 ± 0.01	−0.01 ± 0.01	0.10 ± 0.00	0.33 ± 0.00	0.10 ± 0.00	0.63 ± 0.01	103 ± 1
Extracted DMR [5]	18.1 ± 0.1	0.44 ± 0.04	0.03 ± 0.00	0.10 ± 0.03	0.32 ± 0.02	0.10 ± 0.00	0.66 ± 0.06	115 ± 7

[1] These parameters account for AA, BA, and HA in aqueous and/or organic fractions; [2] Negative values indicate greater acetate utilization than production; [3] Productivity is calculated at the end of the experiment, irrespective of the time of complete sugar utilization; [4] Carbon mass balance (CMB) accounts for the carbon in acids produced, cell biomass, and a theoretical production of 33% CO_2; [5] These fermentations were performed in serum bottles.

Acid extraction was most efficient at pH 5.5 although still 30% of the carboxylic acids remained in the aqueous phase. Extraction efficiency was progressively worse at higher pH with 49% and 70% carboxylic acids remaining in the broth at pH 6.0 and 6.5, respectively. Furthermore, the high water solubility of BA may have a negative effect on its extraction relative to the less water soluble HA [20]. Although accumulation of BA in the broth would favor chain elongation, limits exist where increasing concentrations would become inhibitory to the organism. These results show that a liquid–liquid extraction process (Figure 3D) is not sufficient to provide high recoveries of extracted acids at an optimal pH for bacterial growth. Therefore, a different fermentation strategy, namely fed-batch pertractive fermentation, was further investigated.

3.3. Fed-Batch Pertractive Fermentation

During liquid–liquid extractive fermentation (Figure 3D), as acid production and extraction proceeds, the organic phase becomes increasingly loaded with acids and the extraction efficiency decreases due to a loss in driving force. Increasing the amount of organic solvent could resolve this issue; however, pertractive fermentation, where the organic solvent is back-stripped with a high pH aqueous phase to remove the acids, will alleviate acid buildup in the organic extractant. The pertractive fermentation system employed in this work (Figure 1) consists of two porous polypropylene membranes that separate the organic (oleyl alcohol + 10% TOA) phase from the aqueous fermentation broth and back stripping (NaOH) phases. Within the pores of the membrane, the two liquid phases are in contact and extraction occurs.

Pertractive fermentation (a means of recovering the product in base as a result of product stripping from the organic solvent) has been used to improve productivity and to mitigate solvent toxicity for the production of BA and propionic acids [11,30] as well as HA from mixed culture fermentations [25,38]. The current work is a demonstration of this technique for the production and extraction of both BA and HA from *M. elsdenii* fermentation. Considering the results presented in Figure 3, the pH selected for these fermentations was 6.3, prioritizing VFA productivity over extraction efficiency. The fermentation was started in a batch mode, containing 40 g/L of glucose. After the glucose concentration dropped below 10 g/L (~36 h), glucose was maintained at 5–10 g/L by continuous feeding. The feed rate was adjusted, as necessary, to maintain this concentration. The glucose utilization rate was 0.42 g/L/h at 23 h. From 23 to 59 h, a period where cells reached a maximal OD_{600}, the glucose utilization rates increased to 1.33 g/L/h. At 59 h, there was an inflection, where cells reached a stationary phase and glucose utilization rates decreased to 0.76 g/L/h (Figure 4).

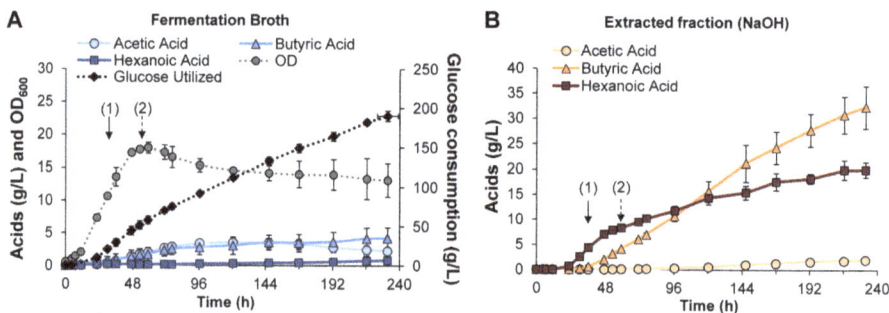

Figure 4. Fed-batch pertractive fermentation by *M. elsdenii* from glucose: (**A**) parameters evaluated in fermentation broth such as bacterial growth, glucose concentration, and acid concentration; (**B**) acid concentration in NaOH from both the organic and NaOH fraction. It is worth noting that only 0.7 and 3.0 g/L of BA and HA, respectively, remain in the organic phase at the end of the fermentation. Experiments were run in duplicate and error bars represent the data range. For samples not taken at the precise same time in the duplicate, time was averaged. Horizontal error bars on the glucose data series show the time data range. Continuous arrows show (1) the end of the batch phase and beginning of the continuous feeding and discontinuous arrow (2) the time point which corresponds to the inflection mentioned above. Dry cell biomass (g/L) can be calculated by multiplying the OD_{600} by the factor 0.261 g/L/OD.

The HA production rate also occurred in three distinct phases corresponding to the glucose utilization rates. During the first 23 h, HA was practically absent, but from 23 to 59 h, the productivity increased to 0.19 g/L/h. From that point to the end of the fermentation, HA production decreased to 0.07 g/L/h. Production of BA lagged about 12 h behind HA production but then remained relatively steady at a production rate of ~0.16 g/L/h. These results suggest that maximum HA production is associated with cell growth and that production declines when the cells enter stationary phase. In contrast, BA production is independent of the phase of the bacteria. Overall, 32 g/L of BA and 20 of g/L HA were recovered in NaOH; in addition, 0.7 and 3.0 g/L of BA and HA, respectively remained in the organic phase. Only 4.5 g/L of BA and 0.7 g/L HA remained in the fermentor broth at the end of the fermentation. AA was only a minor product reaching 2.0 and 0.5 g/L in the NaOH and fermentor broth fractions, respectively. Considering AA, BA, and HA recovered in all fractions, titers of 61.3 g/L and productivities of 0.26 g/L/h were obtained, which in total, are the highest values reported for *M. elsdenii* to our knowledge on glucose.

Although HA was mostly extracted, BA and AA accumulated in the fermentation broth at low levels, requiring neutralization to maintain the pH at 6.3. Other publications noted that the pH would autoregulate around this value due to increased extraction efficiency at lower pH [20]. This was not the case in the current work. However, it is worth noting that a previous single pertractive experiment was run without pH control and the results were similar to those presented here. In this fermentation, the initial pH was 6.5 and dropped to around 5.6 at 80 h before slowly rising to about 6.1 at 200 h. This fermentation produced 3.1, 31.2 and 20.8 g/L of AA, BA and HA respectively for a total of 55.2 g/L acids and a productivity of 0.26 g/L/h.

Carboxylic acid yields were similar to those found in batch experiments (0.32 g/g) as well as the high closure of the carbon mass balance at 98% (Table 2). Carbon yields from carboxylic acids were higher (63%) (mol/mol) for the pertractive experiment than batch experiments, partly due to more glucose being consumed by a similar amount of biomass produced.

3.4. Fermentation of Biomass Sugars with and without Acid Extraction

Subsequently, the performance of *M. elsdenii* on two relevant biomass hydrolysates in the presence and the absence of organic solvent was evaluated to understand if there was any toxicity associated with the hydrolysate. It is worth noting that xylose is a major component of these sugars streams and the *M. elsdenii* strain employed here cannot utilize xylose or other pentose sugars. Regardless, it is important to demonstrate that *M. elsdenii* can tolerate these streams since they have been reported to be toxic for other organisms due to the presence of potential bacterial inhibitors such as acetate, furfural, and phenolic compounds [39]. Considering that xylose is not utilized by this bacterium, a batch experiment in serum bottles was performed; in a fed-batch pertractive fermentation, xylose would accumulate to high concentrations in the fermentor and may present an extra inhibitory effect in the bacterium due to increased osmotic stress.

3.4.1. Non-Extracted Fermentation of Biomass Sugars

Fermentation performance on pure glucose, DDA, and DMR hydrolysates, all at initial glucose concentrations of 40 g/L, was compared. Without extraction, DDA and DMR sugars compared favorably to pure glucose and exhibited similar glucose utilization (11–12 g/L), cell growth, and BA + HA production (4.8, 5.5, and 4.9 g/L for pure glucose, DDA and DMR, respectively) (Figure 5A–C; Table 2). As a result of bacterial contamination during biomass storage and processing, DMR hydrolysates also contained a small amount of lactic acid (1.7 g/L). It is known that *M. elsdenii* can utilize lactic acid to produce propionic, BA, and valeric acids [40]. Lactic acid was included with glucose in the yield calculation, and propionic or valeric acids were not detected. Similar to lactic acid in DMR, both DDA and DMR hydrolysates contain varying levels of AA, which *M. elsdenii* can use for the production of BA and HA [20]. Acetate and lactate utilization were clearly observed during the first 12 h of fermentation in both DDA and DMR as well as a concomitant initial burst of BA production (Figure 5B,C). For clarity, lactate was omitted from the figures. This result suggests rapid lactate utilization (from DMR) and/or chain elongation from AA to BA (from both DDA and DMR). This burst of BA does not occur in pure glucose fermentation, which lacks any initial lactate or acetate (Figure 5A). As expected, the pure glucose fermentation stalled at 72 h due to a combination of product accumulation and decrease in pH, while DDA and DMR fermentations were slower and continued over 100 h. These experiments presented the highest yields in this work, in particular, the fermentation in DMR (0.47 g/g); however, productivities in these conditions were very low (Table 2).

3.4.2. Extracted Fermentation of Biomass Sugars

In contrast to the previous fermentations, extraction resulted in essentially complete glucose utilization for all the conditions (Figure 5D–F). However, with extraction, initial cell growth was delayed compared to the previous set of experiments (Figure 5A–C). Nevertheless, maximum cell densities were 4–6 fold higher than with no extraction. Glucose utilization rates for the pure glucose condition were faster, and all of the glucose was consumed in ~80 h, while the DDA and DMR conditions required more than 100 h. Total BA + HA titers (from both aqueous and organic phases) were 16.45, 16.8, and 17.0 g/L for pure glucose, DDA, and DMR, respectively (Table 2). In this experiment, the accumulation of BA and HA was again observed in the aqueous phase although the extraction was even more efficient than that observed in the batch liquid–liquid fermentations (Figure 3), both starting at pH 6.5. Since no pH control was used in this experiment, this is likely the result of increased extraction efficiency at lower pH. No propionic or valeric acids were detected and the net AA produced was negative for DDA, probably due to the utilization of this compound to produce BA as previously noted. Carboxylic acid yields were again higher than the observed in the fermentors, around 0.44 g/g (Table 2) and carbon mass balances were all slightly over 100%. In this case, BA and HA productivities equaled those observed in the batch liquid–liquid fermentations, reaching 0.10 g/L/h (Table 2).

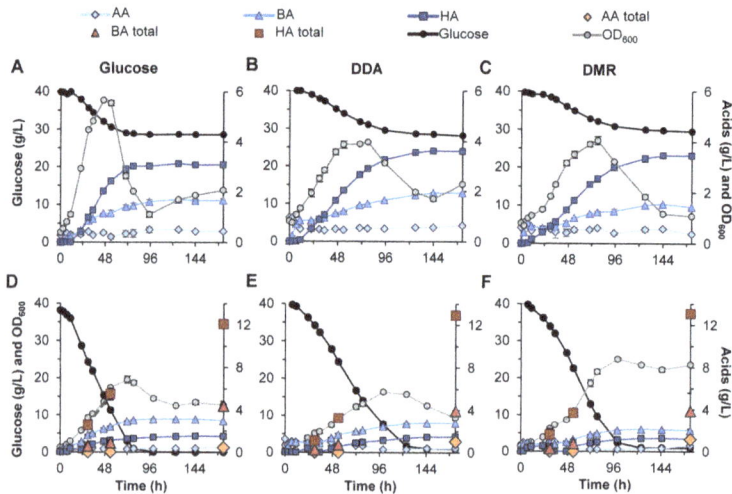

Figure 5. Glucose fermentations by *M. elsdenii* from different carbon sources in serum bottle experiments. Fermentations were performed in (**A–C**) the absence and (**D–F**) the presence of oleyl alcohol plus 10% TOA. Carbon sources consisted of (**A,D**) pure glucose, (**B,E**) DDA hydrolysate, and (**C,F**) DMR hydrolysate. Lines with filled markers present the variables analyzed in the aqueous fraction. Empty markers show AA, BA, and HA total concentration in both aqueous and organic phases. Experiments were run in duplicate and error bars represent the data range. Dry cell biomass (g/L) can be calculated by multiplying the OD_{600} by the factor 0.261 g/L/OD.

4. Discussion

The anaerobic production of VFAs from lignocellulosic hydrolysates is a promising route to generate biofuel precursors. However, due to the toxicity of these compounds, the development of extraction methods integrated to the fermentation system is critical to the production process.

In addition, the use of extractive solvents would considerably reduce the processing cost associated with a neutralization process (including reducing cost and mitigating sustainability issues with salt generation and wastewater treatment) and would also facilitate downstream processing. In the current work, we evaluated different extractive fermentation strategies for the simultaneous production of BA and HA by *M. elsdenii* from glucose and corn stover hydrolysate.

The effects of diverse solvents on the *M. elsdenii* fitness were tested (Figure 2). Octanol was more toxic than oleyl alcohol and THA more toxic than TOA. Similar findings for octanol toxicity have been previously reported for the ethanologen *Zymomonas mobilis* [41]. Although oleyl alcohol plus TOA improved bacterial growth in liquid–liquid extractive fermentations, we cannot definitively state that this mixture does not present any negative, toxic effect on the cells. Since acid extraction clearly provides a positive effect, this could possibly mask a mild toxicity. If bacterial growth is compared between liquid–liquid extraction experiments from serum bottles (Figure 2) and fermentors (Figure 3), one can observe that the maximal OD_{600} are 25 and 15, respectively. It is possible that the increased salinity from the addition of NaOH used to control the pH might osmotically stress the cells. In the current work, the OD_{600} obtained in the pertractive fermentation was close to 20, which also does not achieve the value of 25 observed in the serum bottles. Again, this could be a result of the reduced addition of NaOH used to control pH. It is also worth noting that in the pertractive system, the cells might be subjected to additional stress, possibly due to shear or transient temperature changes as a result of recirculation of the fermentor broth through the system.

Extraction efficiency and bacterial growth are both highly dependent on the pH, but as observed in the current work, these important fermentation metrics are also a function of the fermentation configuration. For instance, BA was hardly extracted in the liquid–liquid system (Figure 3), in contrast to the efficient extraction observed in the pertractive bioreactor configuration (Figure 4). Similarly, the extraction efficiency of HA was also enhanced in the latter system. These results highlight the advantage of some means of backstripping the acids from the organic solvent, which improves extraction efficiency by preventing acid saturation of the organic phase. Apart from this advantage, the possibility of utilizing high pH in the pertractive system also favors higher titers and productivity, which are critical fermentation parameters for industrial use.

The titers reached in the current work are among the highest reported to date when considering mixed BA (36.5 g/L) and HA (20.7 g/L) production by *M. elsdenii*. Roddick et al. [21] reported BA and HA titers of 3.4 and 19 g/L respectively, from glucose by *M. elsdenii* in a fed-batch system, with cells inmobilized and using an anion exchange resin. Choi et al. [20] reached a combined BA and HA titer of 32.0 g/L from sucrose in liquid–liquid extractive fermentation also by *M. elsdenii*. Bacterial co-cultures and mono-culture fermentations of *C. kluyveri* using acetate and ethanol as carbon sources produced HA titers of < 10 and 12.8 g/L, respectively [13,15,18,19] and *Clostridium* sp. BS-1 has been reported to produce up to 32 g/L of HA in 16 days from galactitol. However, as the authors state, *Clostridium* sp. BS-1 displays poor glucose utilization, which would need to be improved if lignocellulosic sugars are to be used [14].

Interestingly, the maximum yield values were obtained in the last experiment from both glucose and biomass hydrolysates (0.42–0.47 g/g), compared to the lowest yield of 0.32 g/g obtained in the pertractive system. Overall, the production of acids was higher in the biomass experiments, including the pure sugar controls. This can be seen in the carbon mass balance values, in most cases, being near or slightly above 100%. Unfortunately, no apparent explanation was found for these acid yield increases but it is noted that one of the main differences in this experiment was the use of serum bottles instead of bioreactors. In the current work, the carbon balance was essentially closed when BA and HA were considered as the major products. It has been also reported that *M. elsdenii* T81 accumulated considerable amounts of glycogen from glucose [40]. Although not measured in this work, this type of intracellular accumulation could be a carbon sink, reducing yields. The maximum HA yield reported in literature is 0.5 g HA/g sucrose, which would be very close to the maximum yields for homo-BA fermentations from glucose (0.49 g BA/g glucose) [12]. However, reported yields in *M. elsdenii* from glucose have varied from 0.27 to 0.39 [21], which are closer to our results.

It has been also demonstrated that productivities were considerably lower in batch extractive fermentations than in the pertractive system (Table 2). This is likely due to a combination of reduced product feedback inhibition, lower salinity from product neutralization, and reduced solvent toxicity in the pertractive system. Comparing glucose utilization rates at maximal OD time points, batch extractive and pertractive fermentations at pH 6.5 yielded values of 0.39 and 0.97 g/L/h respectively. This behavior is reflected in carboxylic acids productivities where ~3-fold higher values were obtained in the pertractive system (0.26 g/L/h) compared to batch extractive (0.09 g/L/h). The pertractive productivity value in this work is the highest reported for the simultaneous production of VFAs using a mono-culture organism from sugars, greater than those obtained from sucrose 0.23 g/L/h [20] and glucose 0.13 g/L/h [21] considering both BA and HA or from acetate and ethanol 0.18 g/L/h [13] and galactitol 0.08 g/L/h [14] for exclusive HA production.

The production of some carboxylic acids from lignocellulosic substrates, such as BA and propionic acid, has been reported [12,42]. However, the production of HA from lignocellulosic sugars by a bacterial mono-culture has not been reported previously to our knowledge. Furthermore, some benefits have been found when using biomass hydrolysates compared to glucose. For instance, the presence of additional acids from biomass deconstruction, such as lactic and AA, which are inherent of some biomass hydrolysates and inhibitors for some organisms, enhance VFA production. BA and HA productivity from hydrolysates decreased compared to pure glucose fermentations since the biomass hydrolysates also contain other inhibitors (e.g., furfural, 5-hydroxymethylfurfural) that the bacterium likely needs to detoxify. The detoxification of furfural to furfuryl-alcohol has been demonstrated in other organisms while growing in anaerobic conditions [43–45]. Another interesting observation was that the HA:BA ratio increased in hydrolysates (1.73 for DDA and 2.0 for DMR, respectively) compared to glucose (1.54). However, as observed in the pertractive fermentation sytem, this ratio might change over the time. Overall, *M. elsdenii* is a robust organism for producing VFAs from biomass hydrolysates.

5. Conclusions

The efficient and simultaneous production of BA and HA by *M. elsdenii* from glucose and corn stover hydrolysates has been demonstrated in the current work. However, presently, this organism cannot utilize xylose, which is one of the main drawbacks for further lignocellulose conversion. Despite the lack of genetic tools for *M. elsdenii*, further efforts to engineer xylose and other minor sugar utilization pathways could elevate this organism as a potential platform host for the production of VFAs from lignocellulosic sugars. In addition, we acknowledge that the media utilized in the current study is rich. Thus, media development will be also critical to decrease costs and improve the economics of the process. Furthermore, cell immobilization may be another means to divert more carbon from cell biomass to product to increase metabolic yields as well as to avoid transient changes in cell recirculation systems.

Acknowledgments: We thank the US Department of Energy BioEnergy Technologies Office (BETO) for funding. The U.S. Government retains and the publisher, by accepting the article for publication, acknowledges that the U.S. Government retains a nonexclusive, paid up, irrevocable, worldwide license to publish or reproduce the published form of this work, or allow others to do so, for U.S. Government purposes. We thank Anne Starache for CHN analysis. We thank Ryan Spiller for his help conducting experiments and Xiaowen Chen, Melvin Tucker, Marykate O'Brian, Nancy Dowe, and Dan Schell for conducting pretreatment and/or providing biomass hydrolysates. We also thank Nancy Dowe and Ed Wolfrum for helpful discussions and the reviewers for providing comments that improved the manuscript.

Author Contributions: Robert S. Nelson performed the experiments and Darren J. Peterson conducted the sugar and carboxylic acid analysis; Robert S. Nelson, Eric M. Karp, Gregg T. Beckham, and Davinia Salvachúa conceived and designed the experiments; Robert S. Nelson and Davinia Salvachúa analyzed the data; Eric M. Karp contributed to the pertractive system design and selection of organic solvents; Robert S. Nelson and Davinia Salvachúa wrote the paper; Darren J. Peterson, Eric M. Karp, and Gregg T. Beckham revised the manuscript.

Conflicts of Interest: The authors declare no conflicts of interest.

References

1. Agler, M.T.; Wrenn, B.A.; Zinder, S.H.; Angenent, L.T. Waste to bioproduct conversion with undefined mixed cultures: The carboxylate platform. *Trends Biotechnol.* **2011**, *29*, 70–78. [CrossRef] [PubMed]
2. Holtzapple, M.T.; Granda, C.B. Carboxylate platform: The MixAlco process Part 1: Comparison of three biomass conversion platforms. *Appl. Biochem. Biotechnol.* **2009**, *156*, 95–106. [CrossRef] [PubMed]
3. Kucek, L.A.; Spirito, C.M.; Angenent, L.T. High *n*-caprylate productivities and specificities from dilute ethanol and acetate: Chain elongation with microbiomes to upgrade products from syngas fermentation. *Energy Environ. Sci.* **2016**, *9*, 3482–3494. [CrossRef]
4. Angenent, L.T.; Richter, H.; Buckel, W.; Spirito, C.M.; Steinbusch, K.J.J.; Plugge, C.M.; Strik, D.P.B.T.B.; Grootscholten, T.I.M.; Buisman, C.J.N.; Hamelers, H.V.M. Chain elongation with reactor microbiomes: Open-culture biotechnology to produce biochemicals. *Environ. Sci. Technol.* **2016**, *50*, 2796–2810. [CrossRef] [PubMed]
5. Renz, M. Ketonization of carboxylic acids by decarboxylation: Mechanism and scope. *Eur. J. Org. Chem.* **2005**, *2005*, 979–988. [CrossRef]
6. Gaertner, C.A.; Serrano-Ruiz, J.C.; Braden, D.J.; Dumesic, J.A. Catalytic coupling of carboxylic acids by ketonization as a processing step in biomass conversion. *J. Catal.* **2009**, *266*, 71–78. [CrossRef]
7. Serrano-Ruiz, J.C.; West, R.M.; Dumesic, J.A. Catalytic conversion of renewable biomass resources to fuels and chemicals. *Annu. Rev. Chem. Biomol. Eng.* **2010**, *1*, 79–100. [CrossRef] [PubMed]
8. Nilges, P.; dos Santos, T.R.; Harnisch, F.; Schroder, U. Electrochemistry for biofuel generation: Electrochemical conversion of levulinic acid to octane. *Energy Environ. Sci.* **2012**, *5*, 5231–5235. [CrossRef]
9. Straathof, A.J.J. Transformation of biomass into commodity chemicals using enzymes or cells. *Chem. Rev.* **2014**, *114*, 1871–1908. [CrossRef] [PubMed]
10. Dwidar, M.; Park, J.-Y.; Mitchell, R.J.; Sang, B.-I. The future of butyric acid in industry. *Sci. World J.* **2012**, *2012*, 471417. [CrossRef]
11. Wu, Z.; Yang, S.T. Extractive fermentation for butyric acid production from glucose by *Clostridium tyrobutyricum*. *Biotechnol. Bioeng.* **2003**, *82*, 93–102. [CrossRef] [PubMed]
12. Yang, S.-T.; Yu, M.; Chang, W.-L.; Tang, I.C. Anaerobic fermentations for the production of acetic and butyric acids. In *Bioprocessing Technologies in Biorefinery for Sustainable Production of Fuels, Chemicals, and Polymers*; John Wiley & Sons, Inc.: Hoboken, NJ, USA, 2013; pp. 351–374.
13. Weimer, P.J.; Stevenson, D.M. Isolation, characterization, and quantification of *Clostridium kluyveri* from the bovine rumen. *Appl. Microbiol. Biotechnol.* **2012**, *94*, 461–466. [CrossRef] [PubMed]
14. Jeon, B.S.; Moon, C.; Kim, B.-C.; Kim, H.; Um, Y.; Sang, B.-I. In situ extractive fermentation for the production of hexanoic acid from galactitol by *Clostridium* sp. BS-1. *Enzyme Microb. Technol.* **2013**, *53*, 143–151. [CrossRef] [PubMed]
15. Kenealy, W.R.; Cao, Y.; Weimer, P.J. Production of caproic acid by cocultures of ruminal cellulolytic bacteria and *Clostridium kluyveri* grown on cellulose and ethanol. *Appl. Microbiol. Biotechnol.* **1995**, *44*, 507–513. [CrossRef] [PubMed]
16. Thanakoses, P.; Black, A.S.; Holtzapple, M.T. Fermentation of corn stover to carboxylic acids. *Biotechnol. Bioeng.* **2003**, *83*, 191–200. [CrossRef] [PubMed]
17. Thanakoses, P.; Mostafa, N.A.A.; Holtzapple, M.T. Conversion of sugarcane bagasse to carboxylic acids using a mixed culture of mesophilic microorganisms. *Appl. Biochem. Biotechnol.* **2003**, *107*, 523–546. [CrossRef]
18. Weimer, P.J.; Nerdahl, M.; Brandl, D.J. Production of medium-chain volatile fatty acids by mixed ruminal microorganisms is enhanced by ethanol in co-culture with *Clostridium kluyveri*. *Bioresour. Technol.* **2015**, *175*, 97–101. [CrossRef] [PubMed]
19. Steinbusch, K.J.J.; Hamelers, H.V.M.; Plugge, C.M.; Buisman, C.J.N. Biological formation of caproate and caprylate from acetate: Fuel and chemical production from low grade biomass. *Energy Environ. Sci.* **2011**, *4*, 216–224. [CrossRef]

20. Choi, K.; Jeon, B.S.; Kim, B.C.; Oh, M.K.; Um, Y.; Sang, B.I. In situ biphasic extractive fermentation for hexanoic acid production from sucrose by *Megasphaera elsdenii* NCIMB 702410. *Appl. Biochem. Biotechnol.* **2013**, *171*, 1094–1107. [CrossRef] [PubMed]

21. Roddick, F.A.; Britz, M.L. Production of hexanoic acid by free and immobilised cells of *Megasphaera elsdenii*: Influence of in-situ product removal using ion exchange resin. *J. Chem. Technol. Biotechnol.* **1997**, *69*, 383–391. [CrossRef]

22. Grootscholten, T.I.; Steinbusch, K.J.; Hamelers, H.V.; Buisman, C.J. Improving medium chain fatty acid productivity using chain elongation by reducing the hydraulic retention time in an upflow anaerobic filter. *Bioresour. Technol.* **2013**, *136*, 735–738. [CrossRef] [PubMed]

23. Elsden, S.R.; Volcani, B.E.; Gilchrist, F.M.C.; Lewis, D. Properties of a fatty acid forming organism isolated from the rumen of sheep. *J. Bacteriol.* **1956**, *72*, 681–689. [PubMed]

24. Jeon, B.S.; Choi, O.; Um, Y.; Sang, B.I. Production of medium-chain carboxylic acids by *Megasphaera* sp. MH with supplemental electron acceptors. *Biotechnol. Biofuels* **2016**, *9*, 129. [CrossRef] [PubMed]

25. Agler, M.T.; Spirito, C.M.; Usack, J.G.; Werner, J.J.; Angenent, L.T. Chain elongation with reactor microbiomes: Upgrading dilute ethanol to medium-chain carboxylates. *Energy Environ. Sci.* **2012**, *5*, 8189–8192. [CrossRef]

26. Jarboe, L.R.; Royce, L.A.; Liu, P. Understanding biocatalyst inhibition by carboxylic acids. *Front. Microbiol.* **2013**, *4*, 272. [CrossRef] [PubMed]

27. Roddick, F.A.; Britz, M.L. Influence of product removal on volatile fatty acid production by *Megasphaera elsdenii*. In Proceedings of the VIIth Australian Biotechnology Conference, Melbourne, Australia, 25–28 August 1986; pp. 386–389.

28. Li, X.; Swan, J.E.; Nair, G.R.; Langdon, A.G. Preparation of volatile fatty acid (VFA) calcium salts by anaerobic digestion of glucose. *Biotechnol. Appl. Biochem.* **2015**, *62*, 476–482. [CrossRef] [PubMed]

29. Kurzrock, T.; Weuster-Botz, D. New reactive extraction systems for separation of bio-succinic acid. *Bioprocess Biosyst. Eng.* **2011**, *34*, 779–787. [CrossRef] [PubMed]

30. Jin, Z.; Yang, S.-T. Extractive fermentation for enhanced propionic acid production from lactose by *Propionibacterium acidipropionici*. *Biotechnol. Prog.* **1998**, *14*, 457–465. [CrossRef] [PubMed]

31. Hong, Y.K.; Hong, W.H.; Han, D.H. Application of reactive extraction to recovery of carboxylic acids. *Biotechnol. Bioprocess Eng.* **2001**, *6*, 386. [CrossRef]

32. Schell, D.J.; Dowe, N.; Chapeaux, A.; Nelson, R.S.; Jennings, E.W. Accounting for all sugars produced during integrated production of ethanol from lignocellulosic biomass. *Bioresour. Technol.* **2016**, *205*, 153–158. [CrossRef] [PubMed]

33. Chen, X.; Kuhn, E.; Jennings, E.W.; Nelson, R.; Tao, L.; Zhang, M.; Tucker, M.P. DMR (deacetylation and mechanical refining) processing of corn stover achieves high monomeric sugar concentrations (230 g·L^{-1}) during enzymatic hydrolysis and high ethanol concentrations (>10% v/v) during fermentation without hydrolysate purification or concentration. *Energy Environ. Sci.* **2016**, *9*, 1237–1245.

34. Zigová, J.; Šturdık, E.; Vandák, D.; Schlosser, Š. Butyric acid production by *Clostridium butyricum* with integrated extraction and pertraction. *Process Biochem.* **1999**, *34*, 835–843. [CrossRef]

35. Miyazaki, K.; Hino, T.; Itabashi, H. Effects of extracellular pH on the intracellular pH, membrane potential, and growth yield of *Megasphaera elsdenii* in relation to the influence of monensin, ethanol, and acetate. *J. Gen. Appl. Microbiol.* **1991**, *37*, 415–422. [CrossRef]

36. Huang, C.B.; Alimova, Y.; Myers, T.M.; Ebersole, J.L. Short- and medium-chain fatty acids exhibit antimicrobial activity for oral microorganisms. *Arch. Oral Biol.* **2011**, *56*, 650–654. [CrossRef] [PubMed]

37. Russell, J.B. Another explanation for the toxicity of fermentation acids at low pH: Anion accumulation versus uncoupling. *J. Appl. Bacteriol.* **1992**, *73*, 363–370. [CrossRef]

38. Ge, S.; Usack, J.G.; Spirito, C.M.; Angenent, L.T. Long-term *n*-caproic acid production from yeast-fermentation beer in an anaerobic bioreactor with continuous product extraction. *Environ. Sci. Technol.* **2015**, *49*, 8012–8021. [CrossRef] [PubMed]

39. Franden, M.A.; Pilath, H.M.; Mohagheghi, A.; Pienkos, P.T.; Zhang, M. Inhibition of growth of *Zymomonas mobilis* by model compounds found in lignocellulosic hydrolysates. *Biotechnol. Biofuels* **2013**, *6*, 99. [CrossRef] [PubMed]

40. Weimer, P.J.; Moen, G.N. Quantitative analysis of growth and volatile fatty acid production by the anaerobic ruminal bacterium *Megasphaera elsdenii* T81. *Appl. Microbiol. Biotechnol.* **2013**, *97*, 4075–4081. [CrossRef] [PubMed]

41. Grzenia, D.L.; Wickramasinghe, S.R.; Schell, D.J. Fermentation of reactive-membrane-extracted and ammonium-hydroxide-conditioned dilute-acid-pretreated corn stover. *Appl. Biochem. Biotechnol.* **2012**, *166*, 470–478. [CrossRef] [PubMed]

42. Liu, Z.; Ma, C.; Gao, C.; Xu, P. Efficient utilization of hemicellulose hydrolysate for propionic acid production using *Propionibacterium acidipropionici*. *Bioresour. Technol.* **2012**, *114*, 711–714. [CrossRef] [PubMed]

43. Taherzadeh, M.J.; Gustafsson, L.; Niklasson, C.; Lidén, G. Conversion of furfural in aerobic and anaerobic batch fermentation of glucose by Saccharomyces cerevisiae. *J. Biosci. Bioeng.* **1999**, *87*, 169–174. [CrossRef]

44. Gutiérrez, T.; Buszko, M.L.; Ingram, L.O.; Preston, J.F. Reduction of furfural to furfuryl alcohol by ethanologenic strains of bacteria and its effect on ethanol production from xylose. *Biotechnol. Fuel Chem.* **2002**, *98*, 327–340.

45. Salvachúa, D.; Mohagheghi, A.; Smith, H.; Bradfield, M.F.A.; Nicol, W.; Black, B.A.; Biddy, M.J.; Dowe, N.; Beckham, G.T. Succinic acid production on xylose-enriched biorefinery streams by *Actinobacillus succinogenes* in batch fermentation. *Biotechnol. Biofuels* **2016**, *9*, 28. [CrossRef] [PubMed]

fermentation

MDPI

Article

Purification of Polymer-Grade Fumaric Acid from Fermented Spent Sulfite Liquor

Diogo Figueira, João Cavalheiro and Bruno Sommer Ferreira *

Biotrend-Inovação e Engenharia em Biotecnologia, S.A. Biocant Park, Núcleo 04 Lote 2,
3060-197 Cantanhede, Portugal; diogo.figueira@biotrend.biz (D.F.); jmcavalheiro@biotrend.biz (J.C.)
* Correspondence: bsf@biotrend.biz; Tel.: +351-231-410-940

Academic Editor: Gunnar Lidén
Received: 28 February 2017; Accepted: 30 March 2017; Published: 1 April 2017

Abstract: Fumaric acid is a chemical building block with many applications, namely in the polymer industry. The fermentative production of fumaric acid from renewable feedstock is a promising and sustainable alternative to petroleum-based chemical synthesis. The use of existing industrial side-streams as raw-materials within biorefineries potentially enables production costs competitive against current chemical processes, while preventing the use of refined sugars competing with food and feed uses and avoiding purposely grown crops requiring large areas of arable land. However, most industrial side streams contain a diversity of molecules that will add complexity to the purification of fumaric acid from the fermentation broth. A process for the recovery and purification of fumaric acid from a complex fermentation medium containing spent sulfite liquor (SSL) as a carbon source was developed and is herein described. A simple two-stage precipitation procedure, involving separation unit operations, pH and temperature manipulation and polishing through the removal of contaminants with activated carbon, allowed for the recovery of fumaric acid with 68.3% recovery yield with specifications meeting the requirements of the polymer industry. Further, process integration opportunities were implemented that allowed minimizing the generation of waste streams containing fumaric acid, which enabled increasing the yield to 81.4% while keeping the product specifications.

Keywords: fumaric acid; purification; spent sulfite liquor; biorefineries

1. Introduction

To ensure sustainable economic growth in a globalized world, resources have to be used in a smarter, more sustainable way. As many natural resources are finite, finding a more environmentally and economically sustainable way of using them has become a priority. In a circular economy, the value of products and materials is maintained for as long as possible. For example, the use of resources is minimized by keeping them within the economy when a product has reached the end of its life, to be used again and again to create further value and reduce the production of waste [1]. Industrial biotechnology provides new processing solutions meeting the requirements for the realization of a circular economy, as it enables the industrial production of added value chemicals from renewable resources and waste streams, including from industrial activity, for example, from the pulp and paper, agricultural and food industries [2]. Indeed, the long-term sustainability of most current industrial processes, which rely on petrochemical sources for raw materials, is compromised. Further, the use of first-generation feedstocks, which compete for raw material and arable land with food and feed uses, is also an issue. Therefore, taking advantage of different, non-conventional carbon sources promotes balance, stability and competitiveness in future processes for the production of chemicals. Concepts of second-generation biorefineries, aiming at utilizing the cellulose in non-food biomass have been put forward for the production of ethanol and other platform chemicals [3], often using existing industrial side-streams, for example, from the pulp and paper industry [4,5]. Blends of C5 and C6 sugars are used today in production of sulfite ethanol, where the cooking liquor from sulfite pulping is used for ethanol production [6]. This cooking liquor, also named spent sulfite liquor (SSL) contains soluble lignosulfonates, with various market applications, and hydrolyzed sugars from the wood hemicelluloses. Yeasts are used for fermentation, and only the C6 sugars (mainly mannose from softwood and glucose from hardwood) are fermented to ethanol while the pentoses are left untouched [6]. Global fumaric acid (FA) market size was estimated at 233.3 kilotons in 2013 and is expected to grow to 346.2 kilotons by 2020 [7]. It is particularly used in food and animal feed for pH adjustment, preservation and flavor enhancement, but increasing demand for unsaturated polyester resins and paints from increasing infrastructure spending, coupled with growing automobile production, particularly in countries such as China, India and Brazil, is likely to surge the demand [7]. Although FA is already available from fermentation on glucose using *Rhizopus* strains, such a production route is only competitive at high oil and gas prices. *Rhizopus* strains are, however, able to utilize C5 sugars such as xylose and arabinose [8]. Recombinant systems based on *Escherichia coli* and *Saccharomyces cerevisiae* may enable the production of FA from sugar-rich streams [9], potentially also using C5 sugars [10], with *S. cerevisiae* providing the extra benefit of allowing the production to be carried out at lower pH, hence obtaining most, if not all, of the product in its fully protonated form instead of a salt [11]. The production of polyesters from bio-based diacid monomers, such as FA, requires the polymerization in solution by direct reaction of diacids and diols at high temperature, mostly using organometallic catalysts [12], and is hindered by the presence of impurities contained in the side and waste streams [13].

This study aims at providing processing options that allow the purification of FA extracted from xylose-metabolising recombinant *S. cerevisiae* fermentation broths that used complex raw materials, such as SSL, as the main carbon source, in order to obtain specifications of the purified product that meet the requirements to enable its use in the manufacture of polymers. Since strains able to produce FA at commercially relevant levels are still under development, in this study a simulated FA-rich fermentation medium was produced as the starting material for the FA purification studies.

The downstream processing of extracellular fermentation products such as FA typically starts by removing biomass from the fermentation broth, producing a clarified broth where the product, sugars, organic acids and other contaminant substances are still present. However, since FA is a dicarboxylic acid with $pK_{a1} = 3.02$ and $pK_{a2} = 4.38$ [14] is actually present in solution as three differently protonated species, the relative amounts of which will depend on the pH of the solution. At low pH, H_2FA is the prevailing form, and if the aqueous solution is saturated, FA can crystallize. At high pH, bivalent fumarate ion FA^{2-} is prevailing and at intermediate pH, the prevailing form is monovalent hydrogenfumarate HFA^-. Varying the pH of the medium will impact on the existing species and on the dissolved/precipitated fractions of FA [15] and on its capability to be adsorbed to sorbent materials. This paper describes the development of an efficient, simple, and cost-effective process of obtaining polymer-grade FA from a fermentation broth based on spent sulfite liquor.

2. Materials and Methods

2.1. Preparation of Spent Sulfite Liquor (SSL)-Based Fermentation Broth

2.1.1. SSL Pretreatment

Crude acidic *Eucalyptus* SSL (Borregaard, Sarpsborg, Norway) was filtered using a Schott Duran® filter funnel G-2 with a 40–100-µm pore size for the removal of solids in suspension and processed by tangential flow filtration (TFF) using a Cogent M1 TFF equipment (Merck Millipore, Molsheim, France), equipped with a Pellicon® 2 filter cassette membrane (Merck Millipore) with a 5 kDa cut-off to remove the larger molecular weight lignosulphonates. The pH of the 5 kDa permeate was adjusted to 5 using 5 N NaOH and the permeate was pasteurized at 70 °C for 1 h in a stirred fermenter equipped with an electric heating blanket (New Brunswick Scientific, Enfield, CT, USA, model BioFlo 110) and stored at 4 °C before further usage. The characterization of the acidic eucalyptus SSL and the 5 kDa filtrate is presented in Table 1.

Table 1. Characterization of acidic eucalyptus SSL.

Parameter	Non-Filtered	5 kDa Filtrate
pH	2.81	2.85
Density (g/mL)	1.07	1.04
Lingnosulphonates (g/L)	109.8	27.0
Mg^{2+}/Ca^{2+} (mM)	144.9	132.1
Carbohydrates (g/L)		
Glucose	1.82	1.29
Xylose	26.9	25.6
Arabinose	0.25	0.15
Galactose	2.61	2.00
Mannose	0.41	0.00
Organic acids (g/L)		
Acetic acid	6.50	5.04
Formic acid	0.00	0.00
Succinic acid	0.00	0.00
Malic acid	0.03	0.00
Fumaric acid	0.00	0.00

2.1.2. Microorganism and Culture

Saccharomyces cerevisiae GSE16 (kindly provided by J. Thevelein, VIB-KU Leuven, Belgium), capable of metabolizing xylose and tolerant to inhibitors [10], was used to produce a fermentation broth to which fumaric acid will be added. The fermentation medium was composed of 30% (*v/v*) of pasteurized 5 kDa SSL permeate, 10.0 g/L of yeast extract (BioSpringer, Maisons-Alfort, France), 20.0 g/L of bacterial peptone (Oxoid), with 40.0 g/L of total sugars (adjusted with addition of xylose (Danisco, Lenzing, Austria)). Cultures were prepared in a 500-mL shake flask with a total cultivation medium volume of 125 mL incubated at 150 rpm and 30 °C for 140 h. The cultures were inoculated from a pre-culture grown a medium with 10.0 g/L of yeast extract, 20.0 g/L of bacterial peptone and 10.0 g/L xylose. The volume of the pre-culture used for inoculation was such as to start the culture with an optical density at 600 nm of 0.1. At the end of the cultivation, cells were removed from the fermented SSL-containing medium by centrifugation (Beckman Coulter, Brea, CA, USA) at 6000 rpm, $9000 \times g$ for 20 min at 20 °C.

2.2. Precipitation of Fumaric Acid

2.2.1. Fumaric Acid Solubility Studies

FA-containing suspensions of concentrations between 5.0 and 50.0 g/L were prepared by adding FA (Merck, ≥99.0%) to 100 mL shake flasks containing 25 mL of reverse osmosis water. The suspensions were incubated at the different temperatures for 1 h in an orbital shaker (New Brunswick Orbital incubator Innova 43). The non-solubilized FA crystals were allowed to settle for 1 h. The crystals were collected by vacuum filtration (VWR 110 mm qualitative filter paper, 5–13 μm particle retention) and oven dried at 60 °C for 24 h until constant weight. FA solubility was determined by the difference between the known amount of FA used for preparing the solutions and the amount of FA recovered in the filtration and confirmed by HPLC analysis of the filtrate.

2.2.2. Fumaric acid Recovery

All downstream processing studies were performed using the SSL-based fermentation broth supplemented with different amounts of FA. In all cases, the pH was adjusted to 4.5 using 10 N NaOH (Applichem Panreac, 98%) to promote complete FA dissolution. A two-stage precipitation protocol was used. In the first stage, 500 g of a 5% (*w/w*) solution was acidified to pH 0.75 by addition of 5 N H_2SO_4 (Fisher Scientific, >95%) and left without agitation at 4 °C overnight, after which the solids were recovered by vacuum filtration as described above. The recovered solids were rinsed with a known amount of ice-cold 0.4 M H_2SO_4 solution. The solids were then placed in 250 g of water to resuspend the material in a shake flask heated at 80 °C. The solution was allowed to cool to room temperature and was then slowly acidified to pH 0.75 by addition of 5 N H_2SO_4. The same experimental conditions were employed in the second stage to obtain the washed FA. At the various steps, the recovered precipitate containing FA was dried and weighed. The percentage of FA in the retained solids was assessed by HPLC after re-suspending the solid material in a known amount of water and, upon heating at 80 °C, achieving the complete dissolution of FA from the recovered precipitate. The FA concentration in the liquid streams was determined by HPLC.

2.3. Use of Activated Carbon

2.3.1. Determination of Fumaric Acid Adsorption Capacity

All downstream processing studies were performed using the fermentation broth produced in Section 2.1.2. Different amounts of activated carbon (Sigma-Aldrich, Saint Louis, MO, USA, >90%), between 0.01 and 0.15 g activated carbon/g FA solution, were added to 25.0 g/L FA solutions at different pH in a shake flask at 150 rpm. The concentration of FA in solution was monitored by HPLC in samples collected after 10, 20, 40, 60 and 80 min of incubation. After vacuum filtration, as described above, FA concentration of the filtrate was assessed by HPLC.

2.3.2. Removal of Contaminants of Fumaric Acid

The precipitate originated from the 2-stage precipitation FA recovery process was dissolved by adding 10 N NaOH until the pH reached 10. The volume of the solution was then adjusted to obtain a 30.0 g/L FA solution to which 0.02 g activated carbon/g FA solution was added. The mixture was incubated at 35 °C, 150 rpm for 60 min. After incubation, the solution was vacuum filtered to remove the activated carbon and any other particulate impurities. The amount of filtrate was determined and FA was analyzed by HPLC. To recover the FA in solution, the filtrate was acidified to pH 0.75 by addition of 5 N H_2SO_4 and then incubated overnight at 4 °C without agitation for the precipitation of FA.

2.3.3. Recovery of Fumaric Acid from Dilute Solutions

One liter of diluted FA solution was placed in a shake flask and 0.1 g activated carbon/g FA solution was added. The suspension was incubated for 60 min at 35 °C, 150 rpm. The suspension was then vacuum filtered, the retained activated carbon was collected and the recovered filtrate was analyzed by HPLC. The thus collected activated carbon was treated with a solution of 1 N NaOH used as desorbent. A 1:1 mixture of wet activated carbon and NaOH solution was incubated for 60 min at 35 °C, 150 rpm. After incubation, the suspension was vacuum filtered and the filtrate was analyzed by HPLC. In order to recover FA from the filtrate, the solution was acidified to pH 0.75 by adding 5 N H_2SO_4 and placed at 4 °C overnight. The solution was then vacuum filtered and the filtrate sampled and analyzed by HPLC. The obtained solids were dried, weighed and the FA content was determined as above.

2.4. Analytical Methods

Optical density of cultures was monitored in a Shimazu UV-1700 (Tokyo, Japan) spectrophotometer at 600 nm. Organic acid and sugar analysis were performed in Shimadzu LC-20AD Prominence HPLC equipped with a Rezex RHM-Monosaccharide H⁺ (8%) LC Column 300 mm × 7.8 mm, from Phenomenex (Torrence, CA, USA), oven temperature set to 35 °C, using as eluent H_2SO_4 5 mM at a flow rate of 0.6 mL/min, 20 µL injection volume. Organic acids were quantified through a Shimadzu SPD-M20A diode array detector at 205 nm and sugars were quantified through a Shimadzu RID-10A detector. Samples were prepared in 50 mM H_2SO_4.

3. Results and Discussion

3.1. Fermentation of SSL Permeate

The time-course of the fermentation of the SSL permeate with xylose-metabolizing *Saccharomyces cerevisiae* GSE16 strain is shown in Figure 1. The culture was able to use all xylose and the residual glucose present in the medium. The residual amounts of galactose, arabinose and mannose were also consumed (data not shown). Additionally, all acetate was also consumed in the first 48 h of fermentation. A slight accumulation of ethanol was detected after 28 h of fermentation, during the stage of maximum growth rate, but all ethanol was consumed afterwards. A very small peak of glycerol was detected, but at concentrations close to or below detection level. The fermentation samples were also analyzed for dicarboxylic acids, such as citric, succinic, fumaric and malic acids, but none were detected. Only traces of pyruvate were observed, but again at concentrations close to or below detection level.

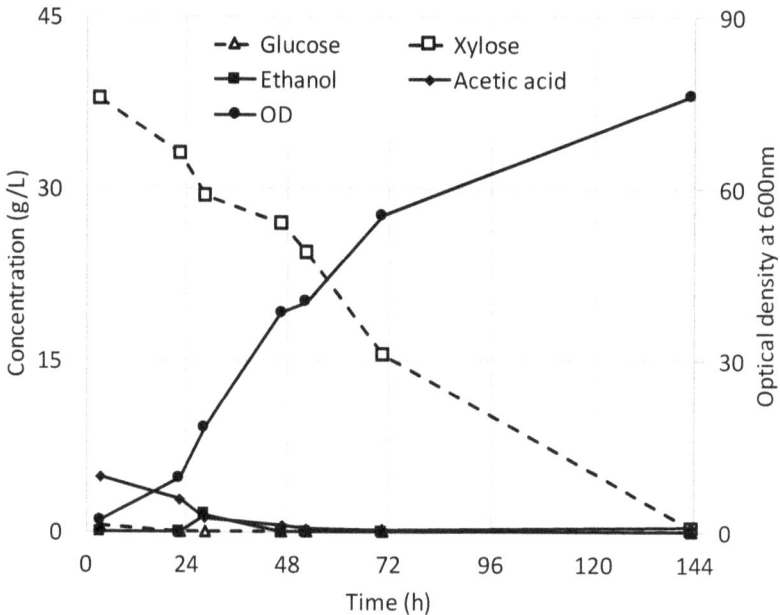

Figure 1. Time course of the fermentation of *Saccharomyces cerevisiae* GSE16 on medium containing 30% spent sulfite liquor (SSL) permeate and xylose added to obtain a total of 40.0 g/L of sugars.

3.2. Influence of pH and Temperature on the Precipitation of Fumaric Acid

Solutions of different total concentration of FA (5.0, 15.0, 35.0 and 50.0 g/L) were prepared and incubated at different temperatures. Upon reaching equilibrium, the amounts of FA that precipitated and that remained in solution were assessed. Total dissolution of the FA in the 0.5% solution and in the 1.5% solution are achieved above 25 °C and at above 45 °C, respectively (Figure 2). In the case of the 35.0 and 50.0 g/L solutions, part of the FA did not dissolve within the temperature ranges tested (5 and 65 °C). The pH of the solutions in equilibrium was measured (Figure 2). At low temperatures, most FA is precipitated and only a small fraction remains in solution, hence the higher pH. When the temperature increases, a higher fraction of FA is solubilized, hence more H^+ is freely in solution and the pH thus decreases. The plateaus on the pH values at temperatures above 25 °C for the experiment with a total 5 g/L FA concentration and above 45 °C for the experiment with a total 15 g/L FA concentration are explained by the total solubilization of FA and no further variation of the concentration of H^+. These results suggest that effective precipitation of FA will occur at temperatures below 15 °C. Further, in order to obtain a high purity of H_2FA with respect to total FA species, the pH should be well below pK_{a1} of FA, for example at or below 2.5.

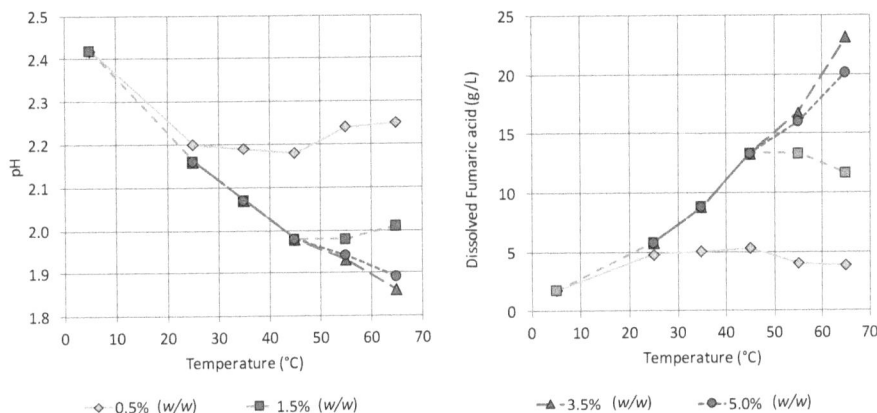

Figure 2. Variation of pH of different fumaric acid (FA) solutions with varying temperature (**left**) and equivalent calculated dissolved FA concentration (**right**).

3.3. Direct Recovery of Fumaric Acid from SSL-Based Fermentation Media

The present work addresses the recovery of FA from two alternative but comparable streams: (i) a ultrafiltered stream of SSL that has been fermented and then centrifuged for cell removal; (ii) a stream of SSL that has been fermented and then centrifuged for cell removal and ultrafiltered. A significant part of the lignosulphonates is removed in a 5-kDa ultrafiltration, but lower molecular weight lignosulphonates and other molecules such as phenolics and aldehydes in addition to metabolites produced by the yeast cells in addition to FA will be present in the stream from which FA is to be recovered. The concentration of the FA produced during the fermentation is not only important to meet the productivity requirements of the process, but may also influence the downstream process. To assess this, two broths with different FA concentrations, 15 g/L and 50 g/L were prepared and a simple one-stage precipitation process was carried out by acidifying the fermentation broth with H_2SO_4 to a pH of 0.75, cooling to 4 °C and filtering out the precipitated material. The thus recovered precipitate was then washed with chilled 0.4 M H_2SO_4 solution to remove impurities while minimizing the loss of FA. Only 22.6% of the FA was recovered in the precipitated material obtained from the broth containing 15.0 g/L FA, while 92.4% FA was recovered in the precipitated material obtained from the broth containing 50 g/L FA, highlighting the impact of the composition of the starting material. The highest fumaric acid concentrations reported obtained through fermentation processes reach 126 g/L [16]; however, these high titers are only possible with relatively low productivities not compatible for commercial deployment. In recent optimization studies, titers of 56.2 g/L were obtained at high productivities [17], hence in all remaining tests, a broth with a concentration of 50.0 g/L FA was used, below which a production process is deemed not feasible.

Another important requirement of the process is the ability to provide a product with a purity meeting the specifications for enabling its use in the production of polymers. After this one-stage purification protocol, a purity of 47.5% was obtained with the broth containing 50.0 g/L FA, clearly showing that the product still contained too many impurities entrained from the fermentation broth. The precipitated material was then used in a second purification stage. The filtered material was washed with water at 80 °C to dissolve the FA while retaining insoluble impurities in the filter. The thus obtained FA-rich stream was then acidified to a pH of 0.75, cooled to 4 °C and the precipitated material is filtered out, washed with a chilled 0.4 M H_2SO_4 solution, dried and recovered. After this two-stage procedure, a recovery yield of 77.8% and a purity of 89.6% were obtained. Despite the fair purity value, the recovered solid was very dark, indicating that a significant amount of impurities from the SSL-based fermentation broth were still present in the product.

3.4. Improving the Purity of Fumaric Acid with Activated Carbon

Since an 89.6% purity is still below specification and the visual inspection clearly showed contamination of the product by compounds from the initial raw material, treatment of the obtained product with activated carbon (AC) was envisaged. AC is a known adsorbent commonly applied in industry, and has been already tested in the purification of FA from fermentation broth using a defined culture medium [18]. The effects of temperature, contact time and pH on the adsorption capacity of FA were tested. It was found that contact time and temperature do not influence the adsorption capacity of AC (data not shown). However, by varying the pH, the capacity of AC to adsorb FA varied markedly (Figure 3).

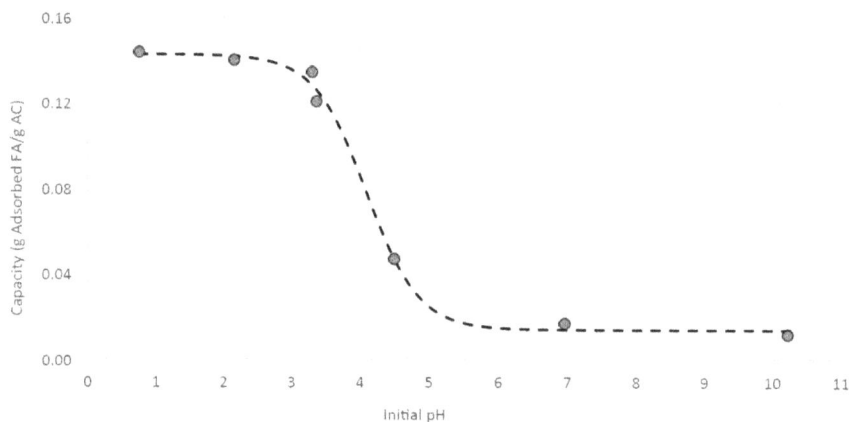

Figure 3. Capacity for FA adsorption of activated carbon (AC) as function of pH. The dashed line is a fit of the experimental data to a sigmoid function.

At the tested conditions, the AC adsorption capacity decreases with increasing pH. Interestingly, a shift occurs at about pH 4, below which the fraction of FA which is fully protonated increases. In order to remove the impurities from the obtained material while minimizing the losses of FA, the dried materials were dissolved in a NaOH solution, to obtain a concentration of around 30 g/L FA and a final pH of 10, at which very low adsorption of FA on the activated carbon occurs. This solution was finally contacted with 0.02 g AC/g solution. After one hour with agitation, the AC was removed by filtration, the obtained solution was acidified to pH 0.75 and cooled at 4 °C to allow for the precipitation of FA. As before, the precipitate was recovered by filtration and washed with a chilled 0.4 M H_2SO_4 solution, dried and recovered. After this AC polishing step, no measurable amounts of impurities were detected in the obtained FA and a white powder was obtained (Figure 4) that could successfully be polymerized with diols (results not shown), thus validating that the FA purity grade obtained meets the requirements. The yield of the AC treatment was 87.8%.

Figure 4. Side-by-side comparison of dry solids obtained without activated carbon treatment (**left**) and with activated carbon treatment (**right**).

3.5. Improving the Yield of the Fumaric Acid Purification Process

The overall yield of the two-stage FA precipitation process and of the AC treatment is 68.3%. The major contributors for losses of product are the liquid streams generated in the chilled washes of FA-rich precipitates, which combined account for the loss of 28.5% of the FA initially present in the broth. However, their FA concentration is fairly low, between two and seven grams/liter, which makes the FA recovery more challenging. We thus envisaged the recovery of FA using again AC, but this time under acidic conditions to retain and concentrate the FA in the AC and desorbing the thus retained FA using a NaOH 1 M solution. This desorption method is more compatible than that described by Zhang et al. [18], which uses acetone, a petrochemically-derived organic solvent, posing safety and environmental concerns and aggressive to most sealing materials used in processing equipment. This procedure allowed recovering 42% of the FA that would have been lost in the water wash streams from the former processing steps, and allowing the overall yield to increase from 68.3% to 80.4%.

3.6. Integration of the Steps

The FA purification process from the fermented SSL-based medium is then effectively achieved by performing a two-stage precipitation process, polishing the thus obtained FA by removing most of the contaminants through AC adsorption under basic conditions and increasing the overall yield by capturing residual FA from liquid waste streams also by AC under acidic conditions. Figure 5 shows a process flow diagram of these operations together with the FA mass balance and concentration in each stream (w/w). By analyzing the processes, a number of optimization opportunities become evident: (i) The FA-rich stream 13 can be sent directly to the polishing step with AC under basic conditions without requiring the energy-consuming drying step; (ii) Stream 37, a slurry with the FA recovered from the chilled washes of FA-rich precipitates (see Section 3.5), contains only about 7.5% of FA, which will not meet purity targets and is too diluted to be dried, but it has a very similar composition to the acidic solution of stream 10. It was thus decided to redirect stream 37 back to the two-stage precipitation process by mixing with stream 10, thus subjecting the contents of stream 37 to a second wash step. Another possibility would be to mix stream 34 with stream 10, since both are acidic solutions generated upon the addition of H_2SO_4, but this would yield a significant dilution of stream 10, thus affecting the efficiency of the subsequent filtration and wash step and hence such possibility was ruled out; (iii) Stream 36 also generates a waste stream containing FA, which actually corresponds to 20.3% of the FA that is to be recovered from the chilled wash of FA-rich precipitates in the process described in Section 3.5. This stream can also be recovered and recycled back to the activated carbon treatment under acidic conditions.

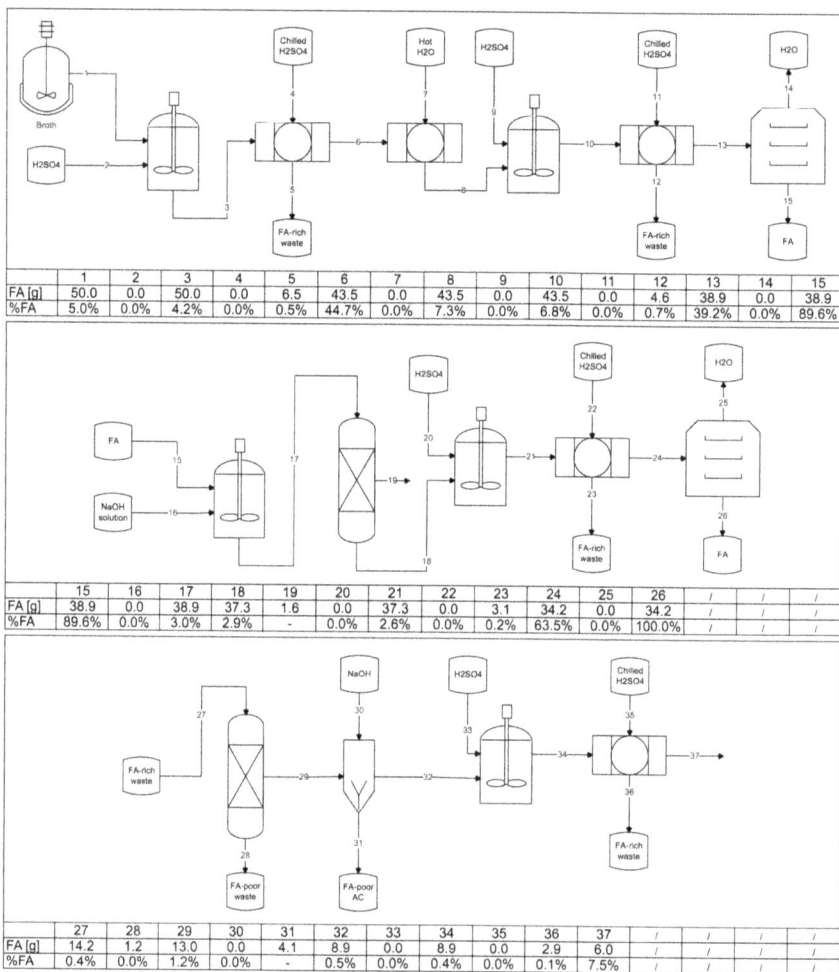

	1	2	3	4	5	6	7	8	9	10	11	12	13	14	15
FA [g]	50.0	0.0	50.0	0.0	6.5	43.5	0.0	43.5	0.0	43.5	0.0	4.6	38.9	0.0	38.9
%FA	5.0%	0.0%	4.2%	0.0%	0.5%	44.7%	0.0%	7.3%	0.0%	6.8%	0.0%	0.7%	39.2%	0.0%	89.6%

	15	16	17	18	19	20	21	22	23	24	25	26			
FA [g]	38.9	0.0	38.9	37.3	1.6	0.0	37.3	0.0	3.1	34.2	0.0	34.2	/	/	/
%FA	89.6%	0.0%	3.0%	2.9%	-	0.0%	2.6%	0.0%	0.2%	63.5%	0.0%	100.0%	/	/	/

	27	28	29	30	31	32	33	34	35	36	37				
FA [g]	14.2	1.2	13.0	0.0	4.1	8.9	0.0	8.9	0.0	2.9	6.0	/	/	/	/
%FA	0.4%	0.0%	1.2%	0.0%	-	0.5%	0.0%	0.4%	0.0%	0.1%	7.5%	/	/	/	/

Figure 5. Process flow diagram and mass balance for FA of the two-stage precipitation process (**top**), polishing the thus obtained FA by removing most of the contaminants through activated carbon adsorption under basic conditions (**middle**) and increasing the overall yield by capturing residual FA from liquid waste streams as well as AC under acidic conditions (**bottom**).

Figure 6 shows a process flow diagram summarizing the process after the operations have been integrated. In addition to reducing the processing and streams, such integration allows further increase of the yield of recovery of FA to 81.4%. Some FA loss occurs in stream 27, but this is considered too diluted to feasibly enable the product recovery. From the mass balances, it can be seen that most FA losses occur in streams 18 and 30, both corresponding to FA adsorbed to AC. In stream 18 most compounds adsorbed together with FA are the contaminating compounds that render the end-product useless, hence the selective separation of FA from these contaminants in this stream does not seem feasible. However, in stream 30, the main compound adsorbed to the AC is actually FA and the losses to the overall process are of FA that has not been desorbed from the activated carbon, so the re-use of at least part of this AC can be envisaged and will be studied in future work. In case 50% of such activated carbon could be reused, the overall process yield could increase to about 87%, similar to the high recovery yield of 93% obtained in the recovery of fumaric acid from a much less complex fermentation broth using refined sugars as a carbon source [18].

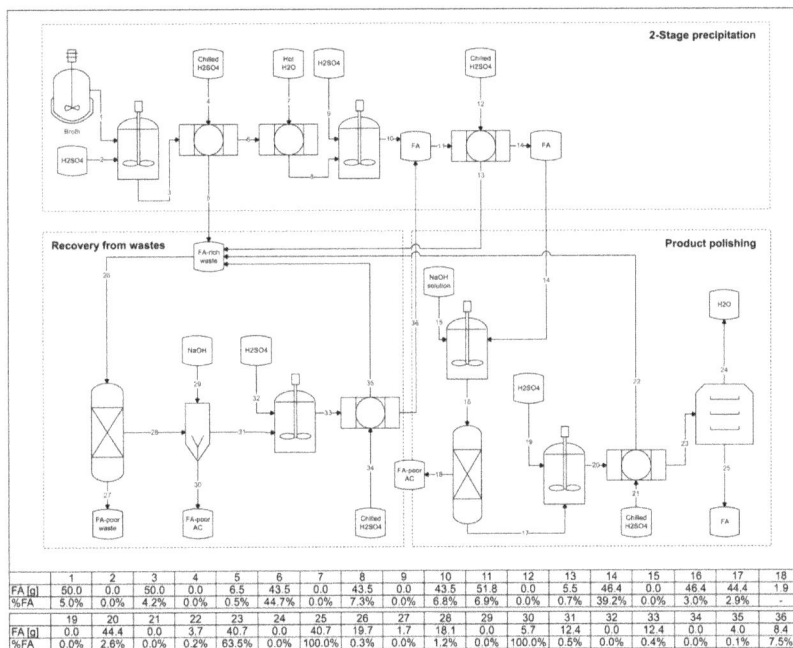

	1	2	3	4	5	6	7	8	9	10	11	12	13	14	15	16	17	18
FA [g]	50.0	0.0	50.0	0.0	6.5	43.5	0.0	43.5	0.0	43.5	51.8	0.0	5.5	46.4	0.0	46.4	44.4	1.9
%FA	5.0%	0.0%	4.2%	0.0%	0.5%	44.7%	0.0%	7.3%	0.0%	6.8%	6.9%	0.0%	0.7%	39.2%	0.0%	3.0%	2.9%	-

	19	20	21	22	23	24	25	26	27	28	29	30	31	32	33	34	35	36
FA [g]	0.0	44.4	0.0	3.7	40.7	0.0	40.7	19.7	1.7	18.1	0.0	5.7	12.4	0.0	12.4	0.0	4.0	8.4
%FA	0.0%	2.6%	0.0%	0.2%	63.5%	0.0%	100.0%	0.3%	0.0%	1.2%	0.0%	100.0%	0.5%	0.0%	0.4%	0.0%	0.1%	7.5%

Figure 6. Process flow diagram and mass balance for the proposed integrated process of purification of FA from SSL-based fermentation broth.

4. Conclusions

The developments described in this paper provide an efficient, simple, and cost-effective process of obtaining polymer-grade FA from complex raw materials, namely spent sulfite liquor. The developed process is based on simple precipitation and adsorption operations and low-cost chemicals with low environmental burden. Further, the dual use of activated carbon for removing contaminants and for capturing FA from dilute streams, tuned by adjusting the pH of the streams, proved to be an innovative and highly effective way to obtain the product with the required specifications while maximizing yields. The capability of the produced FA to be employed as the raw material for polymerization creates an opportunity for the application of this bio-based dicarboxylic acid as a substitute to materials from non-renewable sources, and to the valorization of complex waste streams of industrial origin.

Acknowledgments: This work has received funding from the European Union 7th Framework Programme (FP7/2007–2013) under Grant Agreement number 613771 "BIOREFINE-2G". The authors would like to acknowledge Freddy Tojsås from Borregaard AS (Norway) for providing the SSL, J. Thevelein, from VIB-KU Leuven (Belgium) for providing the *Saccharomyces cerevisiae* GSE16 strain and Amador García from AIMPLAS (Spain) for carrying out the polymerization tests.

Author Contributions: Diogo Figueira, João Cavalheiro and Bruno Sommer Ferreira conceived and designed the experiments; Diogo Figueira and João Cavalheiro performed the experiments; Diogo Figueira and Bruno Sommer Ferreira analyzed the data; Diogo Figueira, João Cavalheiro and Bruno Sommer Ferreira wrote the paper.

Conflicts of Interest: The authors declare no conflict of interest.

Fermentation **2017**, *3*, 13

References

1. *Towards the Circular Economy: Accelerating the Scale-Up across Global Supply Chains*; World Economic Forum: Geneva, Switzerland, 2014.
2. Sheridan, K. Making the bioeconomy circular: The biobased industrie's next goal? *Ind. Biotechnol.* **2016**, *12*, 339–340. [CrossRef]
3. Taha, M.; Foda, M.; Shahasavari, E.; Aburto-Medina, A.; Adetutu, E.; Ball, A. Commercial feasibility of lignocellulose biodegradation: Possibilities and challenges. *Curr. Opin. Biotechnol.* **2016**, *38*, 190–197. [CrossRef] [PubMed]
4. Phillips, R.B.; Jameel, H.; Chang, H.M. Integration of pulp and paper technology with bioethanol production. *Biotechnol. Biofuels* **2013**, *6*, 13. [CrossRef] [PubMed]
5. Palgan, Y.V.; McCormick, K. Biorefineries in Sweden: Perspectives on the opportunities, challenges and future. *Biofuels Bioprod. Biorefin.* **2016**, *10*, 523–533. [CrossRef]
6. Pereira, S.R.; Portugal-Nunes, D.J.; Evtuguin, D.V.; Serafim, L.S.; Xavier, A.M.R.B. Advances in ethanol production from hardwood spent sulphite liquors. *Proc. Biochem.* **2013**, *48*, 272–282. [CrossRef]
7. *Fumaric Acid Market Size by Application (Food & Beverages, Rosin Paper Sizes, Rosin, UPR, Alkyd Resins), Competitive Analysis & Forecast, 2014–2020*; Hexa Research Inc.: Felton, CA, USA, 6 September 2016. Available online: https://www.hexaresearch.com/research-report/fumaric-acid-market (accessed on 15 January 2017).
8. Xu, Q.; Li, S.; Fu, Y.; Tai, C.; Huang, H. Two-stage utilization of corn straw by *Rhizopus oryzae* for fumaric acid production. *Bioresour. Technol.* **2010**, *101*, 6262–6264. [CrossRef] [PubMed]
9. Xu, G.; Zou, W.; Chen, X.; Xu, N.; Liu, L.; Chen, J. Fumaric acid production in *Saccharomyces cerevisiae* by in silico aided metabolic engineering. *PLoS ONE* **2012**, *7*, e52086. [CrossRef] [PubMed]
10. Demeke, M.M.; Dumortier, F.; Li, Y.; Broeckx, T.; Foulquié-Moreno, M.R.; Thevelein, J.M. Combining inhibitor tolerance and D-xylose fermentation in industrial *Saccharomyces cerevisiae* for efficient lignocellulose-based bioethanol production. *Biotechnol. Biofuels* **2013**, *6*, 120. [CrossRef] [PubMed]
11. Abbott, D.A.; Zelle, R.M.; Pronk, J.T.; van Maris, A.J.A. Metabolic engineering of *Saccharomyces cerevisiae* for production of carboxylic acids: Current status and challenges. *FEMS Yeast Res.* **2009**, *9*, 1123–1136. [CrossRef] [PubMed]
12. Díaz, A.; Katsarava, R.; Puiggalí, J. Synthesis, properties and applications of biodegradable polymers derived from diols and dicarboxylic acids: From polyesters to poly(ester amide)s. *Int. J. Mol. Sci.* **2014**, *15*, 7064–7123. [CrossRef] [PubMed]
13. Gallezot, P. Conversion of biomass to selected chemical products. *Chem. Soc. Rev.* **2012**, *41*, 1538–1558. [CrossRef] [PubMed]
14. Lide, D.R. *CRC Handbook of Chemistry and Physics*, 87th ed.; Taylor and Francis: Boca Raton, FL, USA, 2007.
15. Roa-Engel, C.A. Integration of Fermentation and Cooling Crystallisation to Produce Organic Acids. Ph.D. Thesis, Technical University of Delft, Delft, The Netherlands, 2010.
16. Straathof, A.J.; van Gulik, W.M. Production of fumaric acid by fermentation. *Subcell. Biochem.* **2012**, *64*, 225–240. [PubMed]
17. Fu, Y.-Q.; Li, S.; Chen, Y.; Xu, Q.; Huang, H.; Sheng, X.-Y. Enhancement of fumaric acid production by Rhizopus oryzae using a two-stage dissolved oxygen control strategy. *Appl. Biochem. Biotechnol.* **2010**, *162*, 1031–1038. [CrossRef] [PubMed]
18. Zhang, K.; Zhang, L.; Yang, S.-T. Fumaric acid recovery and purification from fermentation broth by activated carbon adsorption followed with desorption by acetone. *Ind. Eng. Chem. Res.* **2014**, *53*, 12802–12808. [CrossRef]

fermentation

MDPI

Review

Microbial Production of Malic Acid from Biofuel-Related Coproducts and Biomass

Thomas P. West

Department of Chemistry, Texas A&M University-Commerce, Commerce, TX 75429, USA;
Thomas.West@tamuc.edu; Tel.: +1-903-886-5399

Academic Editor: Gunnar Lidén
Received: 24 February 2017; Accepted: 6 April 2017; Published: 10 April 2017

Abstract: The dicarboxylic acid malic acid synthesized as part of the tricarboxylic acid cycle can be produced in excess by certain microorganisms. Although malic acid is produced industrially to a lesser extent than citric acid, malic acid has industrial applications in foods and pharmaceuticals as an acidulant among other uses. Only recently has the production of this organic acid from coproducts of industrial bioprocessing been investigated. It has been shown that malic acid can be synthesized by microbes from coproducts generated during biofuel production. More specifically, malic acid has been shown to be synthesized by species of the fungus *Aspergillus* on thin stillage, a coproduct from corn-based ethanol production, and on crude glycerol, a coproduct from biodiesel production. In addition, the fungus *Ustilago trichophora* has also been shown to produce malic acid from crude glycerol. With respect to bacteria, a strain of the thermophilic actinobacterium *Thermobifida fusca* has been shown to produce malic acid from cellulose and treated lignocellulosic biomass. An alternate method of producing malic acid is to use agricultural biomass converted to syngas or biooil as a substrate for fungal bioconversion. Production of poly(β-L-malic acid) by strains of *Aureobasidium pullulans* from agricultural biomass has been reported where the polymalic acid is subsequently hydrolyzed to malic acid. This review examines applications of malic acid, metabolic pathways that synthesize malic acid and microbial malic acid production from biofuel-related coproducts, lignocellulosic biomass and poly(β-L-malic acid).

Keywords: malic acid; biomass; polymalic acid; plant hydrolysates; *Aureobasidium*; *Aspergillus*; *Thermobifida*; *Rhizopus*

1. Introduction

Malic acid is considered as a "building block" chemical for the production of biodegradable polymers [1,2]. Originally, malic acid was extracted from eggshells, fruits and apple juice but the cost of isolating malic acid in that fashion was very expensive since it was labor intensive [2]. The commercial use of malic acid is varied [1–3]. In foods and beverages, it has been used as an acidulant [1,2]. As an acidulant, malic acid has a less bitter taste than citric acid [1,2]. Malic acid can also enhance the flavor of foods [1,2]. Other commercial applications of malic acid are in metal cleaning, textile finishing, pharmaceuticals and agriculture [1,2]. In agriculture, malic acid is used to solubilize aluminum phosphate in soil [4]. Annually, the global market ranges from 40,000–60,000 metric tons of malic acid with the growth rate for malic acid increasing by 4% annually [5]. The value of the market has been estimated as $130 million [5]. It has been reported that the retail price of a pound of malic acid ranges from $1.80 to $2.00 [3]. The current industrial method of malic acid synthesis is petroleum-based. The commercial production of malic acid involves the hydration of maleic anhydride (from the oxidation of benzene or butane) under higher pressure and temperature which synthesizes a racemic mix of D- and L-malic acid where the isomers require chiral resolution [1,2]. An alternate method of synthesizing malic acid is by utilizing the enzyme fumarate hydratase or *Saccharomyces cerevisiae* cells containing the overexpressed gene for fumarate hydratase to catalyze the conversion of fumarate to malate [2,3,6]. The most recent method of synthesizing malic acid is by the acid hydrolysis of poly(β-L-malic acid) (PMA) that is metabolically synthesized by microbes such as *Aureobasidium pullulans* [2]. PMA can also be synthesized from malic acid or it can be hydrolyzed to malic acid. PMA may be useful in the production of biodegradable polymers for use in pharmaceuticals and agriculture since it is a water-soluble biopolymer [7–10]. Also, low molecular weight PMA has been reported to be a protease inhibitor [11]. Green chemistry approaches to produce malic acid from low value biomass or processing coproducts have begun being investigated. Such low value products could help to reduce the price of malic acid compared to its production using petroleum-based products. By lowering its price/pound, biobased malic acid production could more effectively compete with petroleum-based production for use in various food or non-food products. This review of biobased malic acid production first explores its metabolic synthesis in various microbes. Next, possible production from biofuel-related coproducts, lignocellulosic biomass or polymalic acid is examined.

2. Pathways of Malic Acid Biosynthesis

Possible pathways of malic acid biosynthesis have been explored and it has been concluded that there are likely three pathways by which microorganisms synthesize malic acid (Figure 1). The first pathway is characterized as a reductive pathway (Figure 1). This pathway is thought to involve pyruvate carboxylase and malate dehydrogenase where pyruvate initially undergoes carboxylation to oxaloacetate and oxaloacetate is reduced to malate [2]. This pathway is thought to require carbon dioxide derived from the addition of a carbonate salt to the culture medium or sparging the medium with CO_2. Therefore, this pathway requires fixation of CO_2 but is neutral relative to ATP. There is evidence that this pathway is occurring in eukaryotes such as *Aspergillus flavus*, *Aspergillus oryzae*, *Candida glabrata*, *Penicillium* spp. and *Saccharomyces cerevisiae* and in the prokaryotes *Bacillus subtilis* and a metabolically engineered strain of *Escherichia coli* [2,12–14]. In *S. cerevisiae*, it was shown that pyruvate carboxylase and malate dehydrogenase activities were elevated in a strain exhibiting elevated malic acid production [13]. In a strain of the thermophilic soil bacterium, *Thermobifida fusca* muC, phosphoenolpyruvate is converted to oxaloacetate by phosphoenolpyruvate carboxylase and the oxaloacetate reduced to malate by malate dehydrogenase [5]. Thus far, the microorganisms utilizing the reductive pathway produced the highest malic acid levels and yields [2,12–14]. The second pathway of malate biosynthesis by microbes involves the glyoxylate pathway where two molecules of acetyl-CoA are used (Figure 1). The glyoxylate pathway enzyme isocitrate lyase converts isocitrate into succinic acid and glyoxylate. The second enzyme of the pathway malate synthetase catalyzes a reaction involving acetyl-CoA, glyoxylate and water to form malate and Co-A. Carboxylation of malate is necessary for oxaloacetate production to occur and the glyoxylate cycle to continue. It has been shown that *S. cerevisiae* utilizes this pathway to produce malate [2,15]. As shown in Figure 1, the oxidative pathway of malic acid synthesis by microbes involves the tricarboxylic acid cycle [2]. Acetyl-CoA undergoes condensation with oxaloacetate and enters the tricarboxylic acid cycle as citrate until malate is formed with the release of two CO_2 [2]. It is thought that *S. cerevisiae* can use this pathway to produce excess malate particularly if fumarase is overexpressed in the yeast cells [2,6]. Of the three pathways to synthesize malic acid, the reductive pathway appears to be the primary pathway utilized by most organisms.

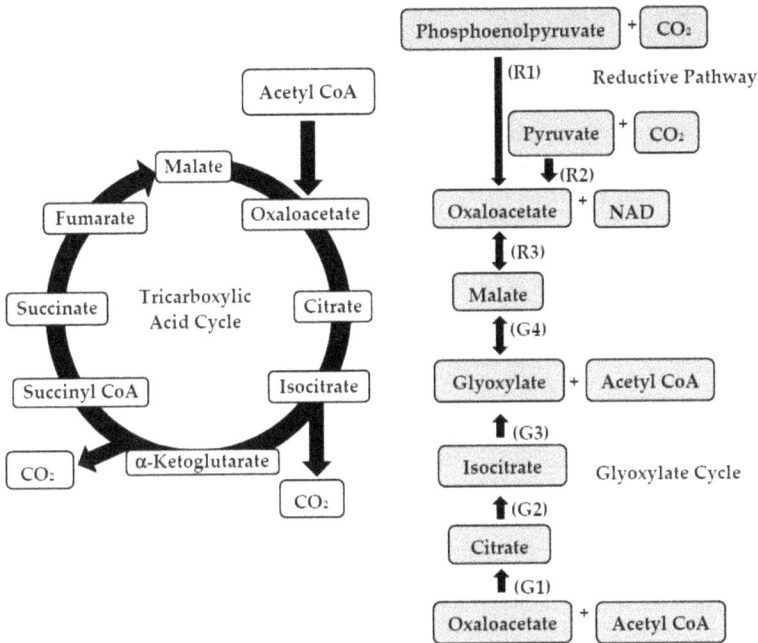

Figure 1. Three metabolic pathways in microorganisms utilized to synthesize malic acid. The tricarboxylic acid cycle which forms malate oxidatively is indicated on the right. The reductive pathway (R) of malic acid synthesis involves phosphoenolpyruvate carboxykinase (R1), pyruvate carboxylase (R2) and malate dehydrogenase (R3). The glyoxylate cycle (G) pathway enzymes include citrate synthase (G1), aconitase (G2), isocitrate lyase (G3) and malate synthase (G4). The tricarboxylic acid pathway is mitochondrial (no shading) while the reductive and glyoxylate cycle pathways are cytosolic (shaded in gray).

3. Microbial Malic Acid Production from Sugars

The microbial production of malic acid from sugars has been investigated and it has been concluded that glucose is the optimum fermentable sugar in the presence of a carbonate salt when the nitrogen concentration is limiting in the medium [16–18]. The carbonate salt serves a dual function as a neutralizing agent and a source of CO_2 [16]. Fermentable sugars have been shown to support malic acid production by species of the fungus *Aspergillus* [16–18]. Prior studies demonstrated malic acid production by *A. flavus* ATCC 13697. This strain grown in a 10 L fermentor (0.5 volume of air/volume of culture/min with a stirring speed of 300 rpm) for 192 h at 25 °C from 10% (*w/v*) glucose as the carbon source and 80 g/L calcium carbonate produced 36.4 g/L malic acid [16]. Malic acid was synthesized from pyruvate with pyruvate carboxylase being involved [17,19]. In a 16 L fermentor, ATCC 13697 produced 113 g/L malic acid using 12% (*w/v*) glucose as a carbon source and 60 g/L calcium carbonate after 190 h (0.5 volume of air/volume of culture/min with a stirring speed of 400 rpm producing a 20% dissolved oxygen concentration) at 32 °C [18]. The problem associated with large-scale malic acid production by *A. flavus* is that the fungus can produce aflatoxins and it would not be granted a "generally regarded as safe status" [2]. It was determined that nitrogen-limiting conditions the *A. oryzae* strains stimulated malic acid production and that peptone as a nitrogen source increased malic acid production from glucose [20]. When the incubation temperature for *A. oryzae* DSM 1863 growth was increased from 30 to 35 °C, its malic acid production in a 2 L fermentor (0.5 volume of air/volume of culture/min with a stirring speed of 300 rpm) after 168 h on 114 g/L (*w/v*) glucose and 90 g/L (*w/v*) calcium carbonate was increased by 1.7-fold [21]. The five carbon sugar xylose also supported malic acid production by *A. oryzae* DSM 1863 but its molar yield (0.56) was lower than on glucose (0.82) as a carbon source [21]. In the fungus *Rhizopus delemar*, production of 60 g/L malic acid from corn straw hydrolysate (100 g/L glucose and 25 g/L xylose) occurred in 60 h at 30 °C in a 3 L fermentor (aeration provided by magnetic stirring) [22]. For the fungus *Penicillium sclerotium* K301, a level of 88.6 or 92 g/L calcium malate was produced from 142 g/L (*w/v*) glucose as a carbon source and 50 g/L (*w/v*) calcium carbonate in shake flasks or in a 10 L fermentor (aeration rate of 8 L/min with stirring speed of 300 rpm) after 96 or 72 h at 28 °C, respectively [23]. Another species of the fungus, namely *Penicillium viticola* 152 (isolated from a marine environment) was shown to produce 168 g/L calcium malate in a medium containing 140 g/L (*w/v*) glucose, 40 g/L (*w/v*) calcium carbonate and 0.5% (*v/v*) corn steep liquor in a 10 L fermentor (aeration rate of 8 L/min with stirring speed of 300 rpm) for 96 h at 28 °C [4]. Metabolic engineering of various species is also being explored as a way to produce high levels of malic acid under controlled conditions in a fermentor [2,5,12,14,15,24,25]. It was reported that a metabolically engineered *A. oryzae* ATCC 56747 produced approximately four-fold higher levels of malic acid than did the wild type strain [12]. In the genetically engineered strain, pyruvate carboxylase (Figure 1), malate dehydrogenase (Figure 1) and the C4-dicarboxylate transporters were overexpressed [12]. The metabolically engineered ATCC 56747 produced 154 g/L malic acid after 164 h at 34 °C (1 volume of air/volume of culture/min with an agitation rate set at 500 rpm) when the dissolved oxygen level was maintained above 50% [12]. Genetic manipulation of strains of *S. cerevisiae* to increase yield (0.42 mol malic acid/mol glucose) and *E. coli* (1.42 mol malic acid/mol glucose) have also been used to improve their malic acid production [14,15]. Although the metabolically engineered *S. cerevisiae* strain was grown on 188 g/L glucose at 30 °C in the presence of 500 mM calcium carbonate at pH 6.0, the genetically modified *E. coli* strain produced a higher yield on only 50 g/L glucose at 37 °C in the presence of 100 mM potassium bicarbonate at pH 7.0 [14,15]. Similarly, the metabolic engineering of *C. glabrata* CCTC M202019 to overproduce the activities of pyruvate carboxylase, malate dehydrogenase and a malate transporter caused a several fold increase in malate production compared to its parent strain [25]. In a metabolically engineered strain of *A. oryzae*, malic acid overproduction was observed when pyruvate carboxylase and malate dehydrogenase were increased by several-fold compared to its parent strain [24].

4. Malic acid Production from Biofuel-Related Coproducts

It has been shown that biofuel-related coproducts can be used to support microbial malic acid production (Table 1) but the level of malic acid produced will depend on the type of coproduct and the ability of the microbe to utilize or tolerate it. One coproduct that can be used to support microbial malic acid production is thin stillage which is formed during the dry milling of corn to produce ethanol (Table 1). Thin stillage is the coproduct that is recovered from whole stillage after removal of yeast cells by centrifugation and its composition includes a high percentage of glycerol and about 1% nitrogen [26]. It is usually mixed with wet distillers' grains and dried to produce dried distillers' grains with solubles [26]. To improve the economics of ethanol production, thin stillage may not be mixed with wet distillers' grains and dried [26]. Instead, wet distillers' grains is marketed alone which has reduced the value of thin stillage as a coproduct. This stillage does contain glycerol which is suitable for microbial fermentation [27]. A prior study has examined the utilization of thin stillage to support malic acid production by species of *Aspergillus* [28]. A number of strains were tested for their ability to synthesize malic acid from thin stillage including *Aspergillus niger* ATCC 9029, ATCC 9142, ATCC 10577 and *A. flavus* ATCC 13697 [28]. Of the strains screened, the yields of malic acid from thin stillage by *A. niger* ATCC 9142 and ATCC 10577 were highest (0.79 g malic acid/g glucose and glycerol in thin stillage) [28]. It was concluded that the ethanol production coproduct thin stillage could be utilized for fungal malic acid production [28].

Crude glycerol is another coproduct that could be utilized for the synthesis of malic acid and is produced during the production of biodiesel (Table 1). The production of biodiesel involves the esterification of a vegetable oil using methanol at an alkaline pH using heat [29]. This process results in the coproduct crude glycerol being formed [30]. The composition of crude glycerol includes glycerol, fatty acids, and methyl esters of fatty acids [29]. Approximately 10% of the coproduct formed during the esterification process is characterized as the crude glycerol fraction. It is expected that substantial volumes of crude glycerol will have to be processed as waste with the current global annual production of biodiesel being more than 30 million tons [31]. With crude glycerol being valued at a current price of $0.05/pound [32], there will be large volumes of crude glycerol available as low-cost feedstock for microbial fermentation. Crude glycerol has been reported to be a substrate for microbial production of malic acid [33]. It has been shown that crude glycerol derived from the production of biodiesel was capable of supporting fungal malic acid synthesis [1,2,30]. In a recent study, *A. niger* ATCC 9142, ATCC 10577 and ATCC 12846 were found to be capable of producing malic acid on 10% (v/v) crude glycerol after 192 h at 25 °C [34]. Of the three strains, *A. niger* ATCC 12846 produced the highest malic acid concentration from crude glycerol [34]. The ability to produce malic acid from crude glycerol did not seem to be highly correlated with cellular biomass since ATCC 10577 produced a higher biomass level than did ATCC 12846 after 192 h at 25 °C [34]. Although *A. oryzae* DSM 1863 was also capable of producing 45 g/L malic acid from 84 g/L (w/v) glycerol after 353 h at 35 °C, its ability to utilize crude glycerol as a substrate for malic acid production has not been investigated [19]. The fungus *Ustilago trichophora* TZ1 was investigated for its ability to produce malic acid from glycerol and crude glycerol (Table 1) in recent studies [35,36]. It was noted that high concentrations of glycerol or crude glycerol had an inhibitory effect on the fungal growth rate. The medium was supplemented with calcium carbonate to provide carbon dioxide as well as to maintain the pH above 5.4 to allow fungal malic acid production to occur [35]. As the level of crude glycerol in the culture was increased from 100 to 200 g/L (w/v), malic acid production by *U. trichophora* was diminished [36]. It was concluded that for the fungus to effectively grow on crude glycerol that it will likely require adaptive evolution of strain TZ1.

Table 1. Malic acid production by microorganisms grown on biofuel-processing coproducts or hydrolyzed lignocellulosic biomass.

Coproduct/Hydrolyzed Lignocellulosic Biomass	Microorganism	Growth Conditions	Malic Acid (g/L)	Yield (g/g)	Reference
Corn stover	*T. fusca* muC-16	55 °C	21.5	0.43	[4]
Corn straw	*R. delemar* HF-119	30 °C	60.0	0.48	[22]
	R. delemar HF-121	30 °C	121.8	0.97	[22]
Thin stillage	*A. flavus* ATCC 13697	25 °C	10.2	0.48	[28]
	A. niger ATCC 9029	25 °C	1.0	0.05	[28]
	A. niger ATCC 9142	25 °C	16.9	0.79	[28]
	A. niger ATCC 10577	25 °C	16.4	0.79	[28]
Crude glycerol	*A. niger* ATCC 9142	25 °C	16.5	0.17[1]	[34]
	A. niger ATCC 10577	25 °C	20.3	0.20[1]	[34]
	A. niger ATCC 12846	25 °C	23.5	0.24[1]	[34]
	U. trichophora TZ1	30 °C	108.0	0.26	[35]
Syngas (plant biomass)	*C. ljungdahli* DSM 13528/ *A. oryzae* DSM 1863	25 °C	1.1	0.17	[37]
Biooil (plant biomass)	*A. oryzae* DSM 1863	32 °C	0.0	0.0	[38]

[1] The malic acid yields produced by *A. niger* ATCC 9142, ATCC 10577 and ATCC 12846 on 10% (v/v) crude glycerol represent unpublished data.

5. Lignocellulosic Biomass-Based Malic Acid Production

An alternative to using coproducts from biofuel production could be the utilization of plant biomass hydrolysates or biooils to synthesize malic acid (Table 1). The challenge of using lignocellulosic biomass instead of a processing coproduct to microbially produce malic acid is the cost of enzymes (cellulases and xylanases) to degrade the biomass to fermentable sugars (glucose or xylose). The hydrolysis of lignocellulosic biomass by physical, chemical and enzymatic treatments to produce glucose from cellulose and xylose from xylan represents a green chemistry approach since plant biomass is biorenewable [39–41]. Hydrolyzed corn straw has been used as a substrate for the fungus *R. delemar* strain HF-119 to synthesize malic acid (Table 1). The corn straw was heat at 160 °C for 20 s and treated with 0.4% sulfuric acid [22]. The resultant solid residue was treated with cellulase and β-glucosidase at 120 rpm for 72 h at 52 °C to produce glucose and xylose [22]. A fluoroacetate mutant strain HF-121 was isolated (Table 1) and shown to be capable of producing 120.5 g/L malic acid from the corn straw hydrolysate (125 g/L mixed sugars) at 30 °C within 60 h [22]. This strain was noted to utilize glucose and xylose as carbon sources more efficiently than strain HF-119 [22]. Xylose could also be utilized by other fungi such as *A. oryzae* DSM 1863 to synthesize malic acid [21]. Syngas can be converted to malic acid using a sequential approach [37]. Initially, a medium sparged with a syngas produced from the gasification of straw is used to anaerobically grow *Clostridium ljungdahli* at 37 °C to synthesize acetate. The acetate is aerobically converted by *A. oryzae* to 4.34 g/L malic acid in the absence of nitrogen (Table 1). It was also shown that bioreactor production of malic acid was possible using this two-step process [37]. A pyrolysis oil produced from the fast hydrolysis of wheat straw has been shown to support malic acid production by *A. oryzae* DSM 1863 (Table 1) depending on the growth conditions [38]. It was noted that the strain could tolerate 1%–2% biooil but the biooil alone did not support fungal malic acid synthesis [38]. The addition of pyrolysis oil (0.5%) to a glucose-containing medium supported malic aid production by the strain but inclusion of higher biooil concentrations (1%–2%) decreased malic acid levels in the same medium after 7 days at 32 °C [38]. Inhibitors within the biooil were found to be responsible for the diminution in malic acid production by the strain [38]. It has been shown that a strain of the filamentous soil bacterium *T. fusca* muC could produce malic acid from cellulose at 55 °C. The bacterium produces a cellulase allowing it to degrade cellulose over a wide pH range [5]. The presence of yeast extract in the medium stimulated cell growth and malate production at 55 °C at 250 rpm [5]. It was also determined that the histidine phosphocarrier protein was the repressor of malic acid production in *T. fusca* muC [42]. A metabolic engineered strain *T. fusca* muC-16 that contained the pyruvate carboxylase gene from *Corynebacterium glutamicum* ATCC 13032 produced a malic acid yield that was 48% higher than its parent strain [5]. The metabolically engineered strain could ferment cellulose and corn stover into malic acid [5]. The strain muC-16 produced about 63 g/L malic acid from 100 g/L cellulose after 124 h at 55 °C. This strain could also convert corn stover (26.9% glucan and 19.3% xylan) to 21.5 g/L malic acid after 120 h at 55 °C (Table 1). It is clear that malic acid can be synthesized by microorganisms from lignocellulosic biomass but additional research on the utilization of other types of hydrolyzed biomass, such as wood, for microbial malic production is necessary.

6. Malic Acid Derived from poly(β-L-malic acid) (PMA) Production

The fungal production of PMA for eventual conversion to malic acid may be the most promising and economical method to synthesize malic acid. It has been shown that malic acid can also be produced from the acid hydrolysis of PMA synthesized by *A. pullulans* [7–9]. PMA is hydrolyzed at 85 °C to pure malic acid using 2 M sulfuric acid [8]. When calcium carbonate was used as a neutralizing agent during the fungal production of PMA, the molecular weight of the PMA was increased compared to using sodium carbonate [11]. It was demonstrated that alkaline peroxide-treated corn straw or wheat straw supported PMA by four strains of *A. pullulans* [7]. If 3% (*w/v*) calcium carbonate and hydrolytic enzymes (such as cellulase, xylanase and glucosidase) were added to medium, PMA production by the strains was enhanced within 7 days at 25 °C [7]. *A. pullulans* NRRL 50383 produced greater than 20 g/L PMA within 7 days at 25 °C under optimal conditions [7]. The mutant strain ZX-10, isolated from *A. pullulans* NRRL Y-2311-1, produced PMA (76.2 g/L) and malic acid (87.6 g/L) from 180 g/L glucose after 140 h at 25 °C in a stirred-tank 5 L bioreactor [8]. The yield of malic acid from glucose was 0.49 g malic acid/g glucose [7]. Another strain of *A. pullulans*, namely YJ6-11, produced 28.6 g/L PMA or 32.4

g/L malic acid from a corncob hydrolysate (produced by treatment with 1% sulfuric acid and treatment with cellulase and xylanase) after 72 h at 25 °C in a 5 L fermentor [43]. The corncob hydrolysate contained 90 g/L mixed sugars including glucose, xylose and arabinose [43]. The advantage of using strain YJ6-11 is that it is able to utilize glucose or xylose at a comparable rate to produce PMA [43]. Using an aerobic fibrous bed bioreactor, this strain was further adapted to corncob hydrolysate by fermentation on the hydrolysate (96 g/L glucose, 54% xylose, 2.5% furfurals and 1.8% acetic acid) for 864 h with a new isolate (CCTCC M2012223) of the original strain being isolated [10]. The strain CCTCC M2012223 was isolated by adaption to increasing concentrations of corncob hydrolysate that contained the growth inhibitors furfural, 5-hydroxymethyfurfural, formic acid and acetic acid [10]. The resultant evolved strain was more resistant to growth inhibition by the known inhibitors [10]. In another study, PMA production by strain CCTCC M2012223 was found to be influenced by nitrogen availability [44]. PMA production was maximum at 36 g/L when the isolated strain was grown on the hydrolysate (110 g/L sugar mix) and ammonium sulfate (2 g/L) in the fibrous bed reactor at 25 °C with aeration for 120 h [44]. Free cells (stirred tank bioreactor) or immobilized cells (in an aerobic fibrous bed bioreactor) of *A. pullulans* CCTCC M2012223 could produce PMA in a medium containing raw sweet potato hydrolysate (120 g/L carbohydrate), yeast extract and citrate [44]. Using free cells of the strain, 29.6 g/L PMA was produced after 120 h at 25 °C [45]. In the immobilized cell system, the strain produced 57.5 g/L PMA after 156 h at 25 °C [45]. PMA can also be produced by *A. pullulans* ZX-10 on sugarcane molasses (44% sucrose, 6% glucose and 5% fructose) [46]. The strain produced 52.6 g/L PMA after 187.5 h at 25 °C during batch fermentation [46]. In a 5 L fermentor, the strain produced 116.3 g/L PMA after 340 h at 25 °C [46]. Under fed-batch conditions, *A. pullulans* ZX-10 was also capable of utilizing a soybean hull hydrolysate (26.8 g/L carbohydrate) supplemented with corn steep liquor (10 g/L, *w/v*) at 25 °C for 168 h to produce 31.3 g/L malic acid from 27.2 g/L PMA [47]. In the same study, *A. pullulans* ZX-10 was capable of growing on soybean molasses (26.8 g/L carbohydrate) under fed-batch conditions for about 260 h to produce 71.9 g/L malic acid from 62.6 g/L PMA [47]. Microbial production of PMA from agricultural biomass would seem to be as effective as direct microbial malic acid fermentation to synthesize this industrially-valuable organic acid.

7. Conclusions

In summary, an opportunity to use low value biomass or processing coproducts such as thin stillage, crude glycerol or straw, exists to synthesize the platform chemical malic acid. From a "green chemistry" perspective, it could help reduce the reliance on petroleum-based chemicals to synthesize malic acid by substituting biobased processes. The most pressing issues that have to be addressed for the biobased production of malic acid include strain development and process development. Both factors are critical for the cost of biobased malic acid to be competitive with petroleum-based malic acid. New environmentally-friendly, microbial-based approaches can only be achieved with microbes capable of elevated malic acid production in the presence of growth inhibitors in plant biomass hydrolysates or processing coproducts. Using these microbes, process development will be essential for the efficient bioconversion of biomass hydrolysates or coproducts into malic acid to effectively compete with the current petroleum-based production methods from an economic standpoint.

Acknowledgments: Financial support of this work was provided by the South Dakota Agricultural Experiment Station Grant SD00H434-12 and Welch Foundation Grant T-0014.

Author Contributions: The author wrote and edited this review article.

Conflicts of Interest: The author declares no conflict of interest.

References

1. Mondala, A.H. Direct fungal fermentation of lignocellulosic biomass into itaconic, fumaric and malic acids: Current and future prospects. *J. Ind. Microbiol. Biotechnol.* **2015**, *42*, 487–506. [CrossRef] [PubMed]

2. Chi, Z.; Wang, Z.-P.; Wang, G.-Y.; Khan, I.; Chi, Z.-M. Microbial biosynthesis and secretion of L-malic acid and its applications. *Crit. Rev. Biotechnol.* **2016**, *36*, 99–107. [CrossRef] [PubMed]
3. West, T.P. Microbial malic acid production: Exploring new avenues of synthesizing a commercially-valuable chemical. *J. Microb. Biochem. Technol.* **2016**, *8*, 321. [CrossRef]
4. Khan, I.; Nazir, K.; Wang, Z.-P.; Liu, G.-L.; Chi, Z.-M. Calcium malate overproduction by *Penicillium viticola* 152 using the medium containing corn steep liquor. *Appl. Microbiol. Biotechnol.* **2014**, *98*, 1539–1546. [CrossRef] [PubMed]
5. Deng, Y.; Mao, Y.; Zhang, X. Metabolic engineering of a laboratory-evolved *Thermobifida fusca* muC strain for malic acid production on cellulose and minimal treated lignocellulose. *Biotechnol. Prog.* **2016**, *32*, 14–20. [CrossRef] [PubMed]
6. Neufeld, R.J.; Peleg, Y.; Rokem, J.S.; Pines, O.; Goldberg, I. L-Malic acid formation by immobilized *Saccharomyces cerevisiae* amplified for fumarase. *Enzym. Microb. Technol.* **1991**, *13*, 991–996. [CrossRef]
7. Leathers, T.D.; Manitchotpisit, P. Production of poly(β-L-malic acid) (PMA) from agricultural biomass substrates by *Aureobasidium pullulans*. *Biotechnol. Lett.* **2013**, *35*, 83–89. [CrossRef] [PubMed]
8. Zou, X.; Zhou, Y.; Yang, S.T. Production of polymalic acid and malic acid by *Aureobasidium pullulans*. *Biotechnol. Bioeng.* **2013**, *110*, 2105–2113. [CrossRef] [PubMed]
9. Chi, Z.; Liu, G.L.; Liu, C.-G.; Chi, Z.-M. Poly(β-L-malic acid) (PMA) from *Aureobasidium* spp. and its current proceedings. *Appl. Microbiol. Biotechnol.* **2016**, *100*, 3841–3851. [CrossRef] [PubMed]
10. Zou, X.; Wang, Y.; Tu, G.; Zan, Z.; Wu, X. Adaption and transcriptome analysis of *Aureobasidium pullulans* in corncob hydrolysate for increased inhibitor tolerance to malic acid production. *PLoS ONE* **2015**. [CrossRef]
11. Cao, W.; Chen, X.; Luo, J.; Yin, J.; Qiao, C.; Wan, Y. High molecular weight poly(β-L-malic acid) produced by *A. pullulans* with Ca^{2+} added repeated batch culture. *Int. J. Biol. Macromol.* **2016**, *85*, 192–199. [CrossRef] [PubMed]
12. Brown, S.H.; Bashkirova, L.; Berka, R.; Chandler, T.; Doty, T.; McCall, K.; McCulloch, M.; McFarland, S.; Thompson, S.; Yaver, D.; Berry, A. Metabolic engineering of *Aspergillus oryzae* NRRL 3488 for increased production of L-malic acid. *Appl. Microbiol. Biotechnol.* **2013**, *97*, 8903–8912. [CrossRef] [PubMed]
13. Oba, T.; Kusomoto, K.; Kichise, Y.; Izumoto, E.; Nakayama, S.; Tashiro, K.; Kuhara, S.; Kitagaki, H. Variations in mitochondrial membrane potential correlate with malic acid production by natural isolates of *Saccharomyces cerevisiae* sake strains. *FEMS Yeast Res.* **2014**, *14*, 789–796. [CrossRef] [PubMed]
14. Zhang, X.; Wang, X.; Shanmugam, K.T.; Ingram, L.O. L-Malate production by metabolically engineered *Escherichia coli*. *Appl. Environ. Microbiol.* **2011**, *77*, 427–434. [CrossRef] [PubMed]
15. Zelle, R.M.; Hulster, E.; van Winden, W.A.; de Waard, P.; Dijkema, C.; Winkler, A.A.; Geertman, J.M.A.; van Dijken, J.P.; Pronk, J.T.; van Maris, A.J. Malic acid production by *Saccharomyces cerevisiae*: Engineering of pyruvate carboxylation, oxaloacetate reduction, and malate export. *Appl. Environ. Microbiol.* **2008**, *74*, 2766–2777. [CrossRef] [PubMed]
16. Peleg, Y.; Stieglitz, B.; Goldberg, I. Malic acid accumulation by *Aspergillus flavus*. I. Biochemical aspects of acid biosynthesis. *Appl. Microbiol. Biotechnol.* **1988**, *28*, 69–75.
17. Peleg, Y.; Barak, A.; Scrutton, M.C.; Goldberg, I. Malic acid accumulation by *Aspergillus flavus*. III. ^{13}C NMR and isoenzymes analyses. *Appl. Microbiol. Biotechnol.* **1989**, *30*, 176–183. [CrossRef]
18. Battat, E.; Peleg, Y.; Bercowitz, A.; Rokem, J.S.; Goldberg, I. Optimization of L-malic acid production by *Aspergillus flavus* in a stirred fermentor. *Biotechnol. Bioeng.* **1991**, *37*, 1108–1116. [CrossRef] [PubMed]
19. Bercovitz, A.; Peleg, Y; Battat, E.; Rokem, J.S.; Goldberg, I. Localization of pyruvate carboxylase in organic acid producing *Aspergillus* strains. *Appl. Environ. Microbiol.* **1990**, *56*, 1594–1597. [PubMed]
20. Knuf, C.; Nookaew, I.; Brown, S.; McCulloch, M.; Berry, A.; Nielsen, J. Investigation of malic acid production in *Aspergillus oryzae* under nitrogen starvation conditions. *Appl. Environ. Microbiol.* **2013**, *79*, 6050–6058. [CrossRef] [PubMed]
21. Ochsenreither, K.; Fischer, C.; Neumann, A.; Syldatk, C. Process characterization and influence of alternative carbon source and carbon-to-nitrogen ratio on organic acid production by *Aspergillus oryzae* DSM1863. *Appl. Microbiol. Biotechnol.* **2014**, *98*, 5449–5460. [CrossRef] [PubMed]
22. Li, X.; Liu, Y.; Yang, Y.; Zhang, H.; Wang, H.; Wu, Y.; Zhang, M.; Sun, T.; Cheng, J.; Wu, X.; Pan, L.; Jiang, S.; Wu, H. High levels of malic acid production by the conversion of corn straw hydrolyte using an isolated Rhizopus Delemar strain. *Biotechnol. Bioprocess Eng.* **2014**, *19*, 478–492. [CrossRef]

23. Wang, Z.-P.; Wang, G.-Y.; Khan, I.; Chi, Z.-M. High-level production of calcium malate from glucose by *Penicillium sclerotium* K302. *Bioresour. Technol.* **2013**, *143*, 674–677. [CrossRef] [PubMed]

24. Knuf, C.; Nookaew, I.; Remmers, I.; Khoomrung, S.; Brown, S.; Berry, A.; Nielsen, J. Physiological characterization of the high malic acid-producing *Aspergillus oryzae* strain 21031-68. *Appl. Microbiol. Biotechnol.* **2014**, *98*, 3517–3527. [CrossRef] [PubMed]

25. Chen, X.; Xu, G.; Xu, N.; Zou, W.; Zhu, P.; Liu, Liming; Chen, J. Metabolic engineering of *Torulopsis glabrata* for malate production. *Metab. Eng.* **2013**, *19*, 10–16. [CrossRef] [PubMed]

26. Kim, Y.; Mosier, N.S.; Hendrickson, R.; Ezeji, T.; Blaschek, H.; Dien, B.; Cotta, M.; Dale, B.; Ladisch, M.R. Composition of corn dry-grind ethanol by-products: DDGS, wet cake, and thin stillage. *Bioresour. Technol.* **2008**, *99*, 5165–5176. [CrossRef] [PubMed]

27. Gonzalez, R.; Campbell, P.; Wong, M. Production of ethanol from thin stillage by metabolically engineered *Escherichia coli*. *Biotechnol. Lett.* **2010**, *32*, 405–411. [CrossRef] [PubMed]

28. West, T.P. Malic acid production from thin stillage by *Aspergillus* species. *Biotechnol. Lett.* **2011**, *33*, 2463–2467. [CrossRef] [PubMed]

29. Tapasvi, D.; Wiesenborn, D.; Gustafson, C. Process model for biodiesel production from various feedstocks. *Trans. ASAE* **2005**, *48*, 2215–2221. [CrossRef]

30. West, T.P. Crude glycerol: A feedstock for organic acid production by microbial bioconversion. *J. Microb. Biochem. Technol.* **2012**. [CrossRef]

31. Quispe, C.A.G; Coronado, C.J.R.; Carvalho, J.A., Jr. Glycerol: Production, consumption, prices, characterization and new trends in combustion. *Renew. Sustain. Energy Rev.* **2013**, *27*, 475–493. [CrossRef]

32. Yang, F.; Hanna, M.A.; Sun, R. Value-added uses for crude glycerol—A byproduct of biodiesel production. *Biotechnol. Biofuels* **2012**, *5*, 13–22. [CrossRef] [PubMed]

33. Dobson, R.; Gray, V.; Rumbold, K. Microbial utilization of crude glycerol for the production of value-added products. *J. Ind. Microbiol. Biotechnol.* **2012**, *39*, 217–226. [CrossRef] [PubMed]

34. West, T.P. Fungal biotransformation of crude glycerol into malic acid. *Z. Naturforsch. C* **2015**, *70*, 165–167. [CrossRef] [PubMed]

35. Zambanini, T.; Kleineberg, W.; Sarikaya, E.; Buescher, J.M.; Meurer, G.; Wierckx, N.; Blank, L.B. Enhanced malic acid production from glycerol with high-density *Ustilago trichophora* TZ1 cultivations. *Biotechnol. Biofuels* **2016**, *9*, 135. [CrossRef] [PubMed]

36. Zambanini, T.; Sarikaya, E.; Kleineberg, W.; Buescher, J.M.; Meurer, G.; Wierckx, N.; Blank, L.B. Efficient malic acid production from glycerol with *Ustilago trichophora* TZ1. *Biotechnol. Biofuels* **2016**, *9*, 67. [CrossRef] [PubMed]

37. Oswald, F.; Dorsam, S.; Veith, N.; Zwick, M.; Neumann, A.; Ochsenreither, K.; Syldatk, C. Sequential mixed cultures: From syngas to malic acid. *Front. Microbiol.* **2016**. [CrossRef] [PubMed]

38. Dorsam, S.; Kirchhoff, J.; Bigalke, M.; Dahmen, N.; Syldatk, C.; Ochsenreither, K. Evaluation of pyrolysis oil as carbon source for fungal fermentation. *Front. Microbiol.* **2016**. [CrossRef] [PubMed]

39. Zhang, Y.H.P. Reviving the carbohydrate economy via multi-product lignocellulose biorefineries. *J. Ind. Microbiol. Biotechnol.* **2008**, *35*, 367–375. [CrossRef] [PubMed]

40. West, T.P. Xylitol production by *Candida* species grown on a grass hydrolysate. *World J. Microbiol. Biotechnol.* **2009**, *25*, 913–916. [CrossRef]

41. West, T.P. Effect of nitrogen source concentration on curdlan production by *Agrobacterium* sp. ATCC 31749 grown on prairie cordgrass hydrolysates. *Prep. Biochem. Biotechnol.* **2016**, *46*, 85–90. [CrossRef] [PubMed]

42. Deng, Y.; Lin, J.; Mao, Y.; Zhang, X. Systematic analysis of an evolved *Thermobifida fusca muC* producing malic acid on organic and inorganic nitrogen sources. *Sci. Rep.* **2016**, *6*, 30025. [CrossRef] [PubMed]

43. Zou, X.; Yang, J.; Tian, X.; Guo, M.; Li, Z.; Li, Y. Production of polymalic acid and malic acid from xylose and corncob hydrolysate by a novel *Aureobasidium pullulans* YJ 6-11 strain. *Process Biochem.* **2016**, *51*, 16–23. [CrossRef]

44. Wang, Y.; Song, X.; Zhang, Y.; Wang, B.; Zou, X. Effects of nitrogen availability on polymalic acid biosynthesis in the yeast-like fungus *Aureobasidium pullulans*. *Microb. Cell Fact.* **2016**, *15*, 146. [CrossRef] [PubMed]

45. Zan, Z.; Zou, X. Efficient production of polymalic acid from raw sweet potato hydrolysate with immobilized cells of *Aureobasidium pullulans* CCTC M2012223 in aerobic fibrous bed bioreactor. *J. Chem. Technol. Biotechnol.* **2013**, *88*, 1822–1827. [CrossRef]

46. Wei, P.; Cheng, C.; Lin, M.; Zhou, Y.; Yang, S.-T. Production of poly(malic acid) from sugarcane juice in fermentation by *Aureobasidium pullulans*: Kinetics and process economics. *Bioresour. Technol.* **2017**, *224*, 581–589. [CrossRef] [PubMed]

47. Cheng, C.; Zhou, Y.; Lin, M.; Wei, P.; Yang, S.-T. Polymalic acid fermentation from *Aureobasidium pullulans* for malic acid production from soybean hull and soy molasses: Fermentation kinetics and economic analysis. *Bioresour. Technol.* **2017**, *223*, 166–174. [CrossRef] [PubMed]

fermentation

MDPI

Article

Valorization of a Pulp Industry By-Product through the Production of Short-Chain Organic Acids

Diogo Queirós, Rita Sousa, Susana Pereira and Luísa S. Serafim *

CICECO-Aveiro Institute of Materials, Chemistry Department, University of Aveiro,
Campus Universitário de Santiago, 3810-193 Aveiro, Portugal; dc.queyros@gmail.com (D.Q.);
sousa.rita@ua.pt (R.S.); susana.raquel@ua.pt (S.P.)
* Correspondence: luisa.serafim@ua.pt; Tel.: +351-234-370-360

Academic Editor: Gunnar Lidén
Received: 8 February 2017; Accepted: 8 May 2017; Published: 12 May 2017

Abstract: In this work, hardwood sulfite spent liquor (HSSL)—a by-product from a pulp and paper industry—was used as substrate to produce short-chain organic acids (SCOAs) through acidogenic fermentation. SCOAs have a broad range of applications, including the production of biopolymers, bioenergy, and biological removal of nutrients from wastewaters. A continuous stirred tank reactor (CSTR) configuration was chosen to impose selective pressure conditions. The CSTR was operated for 88 days at 30 °C, without pH control, and 1.76 days of hydraulic and sludge retention times were imposed. The culture required 46 days to adapt to the conditions imposed, reaching a pseudo-steady state after this period. The maximum concentration of SCOAs produced occurred on day 71—7.0 g carbon oxygen demand (COD)/L that corresponded to a degree of acidification of 36%. Acetate, propionate, butyrate, valerate, and lactate were the SCOAs produced throughout the 88 days, with an average proportion of 59:17:19:1.0:4.0%, respectively.

Keywords: short-chain organic acids; hardwood sulfite spent liquor; activated sludge; acidogenic fermentation; mixed microbial cultures

1. Introduction

The increasing interest in new renewable sources of energy and materials is a consequence of several factors, including the rapid depletion of petroleum and the colossal generation of residues and wastes—both direct consequences of human population growth and its activities. To exploit the potential of using wastes as feedstock, waste management needs to move from treatment-oriented processes to the integration of technologies able to valorize organic waste streams for the production of value-added products [1,2].

Currently, a considerable effort is being made to develop technologies that are able to produce value-added products using residues and by-products. Koutinas et al. [3] enlisted several compounds to be produced from wastes: building-blocks such as succinic acid, 2,3-butanediol, and 1,3-propanediol; biofuels like bioethanol and biohydrogen; and polymers such as bacterial cellulose and polyhydroxyalkanoates (PHAs); and short-chain organic acids (SCOAs) [3].

SCOA production is currently achieved by chemical synthesis, followed by distillation at atmospheric pressure [2]. These compounds have a maximum of six carbon atoms, and present a broad range of applications in the production of biopolymers [4,5], bioenergy [6], and the biological removal of nutrients from wastewaters [4,7].

Organic industrial and urban wastes can act as substrate for SCOA production through acidogenic fermentation (AF). This is a stage of anaerobic digestion (AnD) where several organic compounds present in wastes are transformed into SCOAs including acetate, propionate, butyrate, or lactate, and alcohols such as ethanol [8]. AnD is a complex process which can be divided into four individual stages: hydrolysis, acidogenic fermentation, acetogenesis, and methanogenesis. If AF is the objective, the process should be interrupted at acidogenesis or acetogenesis by inhibiting methanogenesis. AnD is a mature technology that is usually employed worldwide within full-scale facilities for the treatment of industrial and urban wastewaters and organic solid wastes. AnD is advantageous over aerobic activated sludge systems because of its high organic content removal, low energy input requirements, energy production, and low sludge production [9]. Usually, the final goal of AnD is the production of methane and carbon dioxide (biogas), and only recently is being driven by the formation of SCOAs. For SCOA production, it is essential to establish selective pressure conditions to inhibit methanogenic microorganisms and select those which are capable of synthesizing SCOAs, maximizing their production [10]. This selection is possible because methanogenic microorganisms are quite sensitive to operational conditions like temperature, or to the presence of inhibitors, and usually present low growth rates compared to the acidogenic population [11]. Furthermore, the thermodynamics of AF is more favorable than methanogenesis [8,12].

AF has the potential to produce value-added compounds from low-cost waste-based materials. This aspect is of vital importance to the integration of an AF stage into other processes, once the substrate costs often contribute significantly to the overall process economy. Therefore, there are critical parameters to be ascertained to select the most suitable waste streams for acidogenic fermentation, and thus to establish a more cost-effective process. These include the biochemical acidogenic potential of the waste stream, which is the composition and amount of SCOAs that can be generated from the fermentation of the organic wastes [2]. Generated SCOAs can be used further in other processes; namely, the production of PHAs by mixed microbial cultures (MMCs) [13,14]. Parameters such as pH [15,16], temperature [15], hydraulic retention time (HRT), solids retention time (SRT) [17,18], and organic loading rate (OLR) [19] have already been studied in the production of SCOA from industrial or municipal wastes.

The objective of this work was to evaluate the possibility of the application of AF to hardwood sulfite spent liquor (HSSL)—an industrial by-product from the pulp and paper industry—to produce SCOAs. Sulfite spent liquors have the potential to be valorized through a lignocellulosic biorefinery approach [20]. Due to the considerable amount of sugars such as glucose, xylose, and in smaller amounts, galactose, mannose, and arabinose, SCOA production is a strong opportunity to be explored. Instead of using the traditional anaerobic sludge as the inoculum of the process, an aerobic activated sludge was chosen and submitted to anaerobic conditions without pH control and temperature-controlled at 30 °C. This procedure was expected to inhibit the presence of methanogenic bacteria and increase the presence of the acidogenic population [11,21].

2. Materials and Methods

2.1. Microbial Culture

The reactor was inoculated with an MMC collected from an aerobic tank of the wastewater treatment plant (WWTP) Aveiro Sul, SIMRia. The inoculum concentration was 10.5 g carbon oxygen demand (COD)/L.

2.2. Experimental Setup

A continuous stirred tank reactor (CSTR) configuration was chosen to perform the AF of chemically pre-treated HSSL under anaerobic conditions. The working volume of the reactor was 1.55 L, and the flow rate of the feeding solution was 0.85 L/day, resulting in a hydraulic retention time (HRT) of 1.76 days. The reactor had no retention system for the biomass, resulting in a SRT similar to the HRT. The effluent was collected at the outlet of the reactor by overflow. Reactor stirring was performed by a magnetic stirrer and kept constant at 100 rpm. Nitrogen was sparged regularly to assure anaerobic conditions. Oxidation–reduction potential (ORP) was monitored with a transmitter M300 2-channel ORP meter (Mettler-Toledo Thornton, Inc., Greifensee, Switzerland). The system worked with temperature control at 30.1 ± 1.0 °C and without pH control.

2.3. Substrate

HSSL from magnesium-based acidic sulfite pulping of *Eucalyptus globulus* was supplied by Caima–Indústria de Celulose S.A. (Constância, Portugal). Pre-evaporated HSSL was collected from an inlet evaporator from a set of multiple-effect evaporators to avoid the presence of free SO_2. To remove part of the most recalcitrant compounds, HSSL was submitted to a preliminary pretreatment [22]. The pretreatment started with a pH adjustment to 7.0 with 6 M KOH, followed by aeration with compressed air (2880 L of compressed air per liter of HSSL in 6 h). Then, the liquor was centrifuged for 1 h at 5000 rpm. The precipitated colloids were filtered using a 1 µm glass microfiber filter. Finally, the pre-treated HSSL was stored at 4 °C. The total carbon oxygen demand (COD) of pretreated HSSL was determined ($\approx$267 g COD/L). Lignosulphonates (LS) and phenolic components were still the main constituents (ca. 190 g/L), along with xylose, acetate, and glucose (43.5, 14.4, and 7.9 g/L, respectively). No phosphates or ammonia were detected in the HSSL.

2.4. Fermentation Medium

To achieve an organic load rate (OLR) of 11.8 g COD/L day in the CSTR, HSSL was diluted with a mineral solution at a ratio of 1:12.8 (*v:v*). The mineral solution was composed by (per liter of distilled water): 80 mg of $CaSO_4 \cdot 2H_2O$, 160 mg of $FeSO_4 \cdot 7H_2O$, 160 mg of $MgSO_4 \cdot 7H_2O$, 80 mg of $Na_2MoO_4 \cdot 2H_2O$, 160 mg of NH_4Cl. The pH of the medium was adjusted to 7.0, and the medium was autoclaved for 20 min at 121 °C. KH_2PO_4 (160 mg/L) and K_2HPO_4 (80 mg/L) were added under sterile conditions.

2.5. Sampling

Samples were collected every 4 days, three times a day, at intervals of 3 h (sample volume of 5 mL). Samples were further centrifuged at 13,000 rpm for 10 min, the pellet discarded, and the pH of the supernatant measured before storage under −16 °C for later determination of glucose, xylose, SCOAs, COD, and LS concentrations. Five milliliter samples were regularly collected for the determination of total suspended solids (TSS) and volatile suspended solids (VSS).

2.6. Analytical Methods

Biomass concentration was determined using total suspended solids (TSS) and volatile suspended solids (VSS) procedure described in *Standard Methods* [23]. Microfiber filters with 1.0 µm pore size were calcined for 30 min at 550 °C to remove all organic matter. After cooling to room temperature, filters were weighed, and 5.0 mL samples were filtered. Then, filters with the biomass were dried in the oven for 24 h at 105 °C to remove water and were then weighed at room temperature to achieve the TSS concentrations. Finally, the filters were calcined at the same conditions as earlier, and finally weighed at room temperature to achieve the VSS concentrations.

COD was measured accordingly to *Standard Methods* [23]. Replicates were prepared for each sample, which were diluted according to the detection range of the method (100–900 mg/L). Diluted sample (2.0 mL) was mixed with 2.8 mL of digestion solution (20.43 g of $K_2Cr_2O_7$, 66.6 g of $HgSO_4$, 334 mL of H_2SO_4 in 2 L of distilled water) and 1.2 mL of acidic solution (23.3 g of $AgSO_4$ in 2 L of H_2SO_4). In the preparation of blank, 2.0 mL of distilled water was added to the test tubes instead of sample. Tubes suffered a vigorous agitation and were placed on a pre-warmed incubator for 2 h at 150 °C. After the digestion, tubes were taken from the incubator and placed in the dark to cool down to room temperature. Lastly, the absorbance of the tubes was read at 600 nm with a colorimeter. COD concentrations were then calculated based on a calibration curve performed with glucose standards.

Acetate, propionate, butyrate, valerate and lactate, xylose, and glucose were measured by HPLC. For this, 650 µL of each sample were filtered using centrifuge tube filters with a cellulose acetate membrane, 0.2 µm pore size at 8000 rpm for 20 min. Samples were then injected (Auto-sampler HITACHI L-2200, Hitachi, Ltd., Chiyoda, Japan) in an ion exchange column Aminex HPX-87H

(Bio-Rad, Hercules, CA, USA) at 40 °C (Oven Gecko-2000, CIL Cluzeau, Sainte-Foy-la-Grande, France), and analyzed by a refractive index detector (HITACHI L-2490, Hitachi, Ltd., Chiyoda, Japan). The eluent 0.01 N H_2SO_4 was pumped at a flow rate of 0.6 mL/min (HITACHI L-2130 pump, Hitachi, Ltd., Chiyoda, Japan) at room temperature. The eluent was prepared with milli-Q water and filtered with a cellulose acetate membrane, 0.22 μm pore size. The concentrations of sugars and SCOAs in g/L were determined by comparison with the calibration curves of each analyzed compound obtained using standards of known concentrations. The standards concentrations were within the range of the expected concentrations of the analytes: 0.15 to 3.00 g/L for lactate and valerate; 0.20 to 4.00 g/L for propionate and butyrate; 0.25 to 5.00 g/L for glucose, xylose, and acetate.

LS were measured according to Restolho et al. [24]. The absorbance of samples was measured in a spectrophotometer (Shimadzu UVmini-1240, Shimadzu Corp., Kyoto, Japan) at 273 nm, after a dilution of 1:400. LS concentration was calculated using the Beer–Lambert law with a molar attenuation coefficient of 7.41 g/cm [25].

2.7. Calculations

The concentrations of SCOAs, sugars, and biomass were determined by volume of mixed liquor and were converted from g/L to g COD/L using conversion factors that represent the mass (g) of oxygen required to oxidize 1 g of compound based on the oxidation reactions for each compound. The overall oxidation equation is represented below.

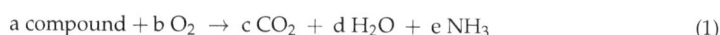

$$a \text{ compound} + b \text{ } O_2 \rightarrow c \text{ } CO_2 + d \text{ } H_2O + e \text{ } NH_3 \tag{1}$$

In which a, b, c, d, and e represent the stoichiometric coefficients of the equation. Therefore, the conversion factor (cf) was calculated according with the following Equation:

$$cf \text{ } (gO_2/g) = \frac{b \times M(O_2)}{a \times M(\text{compound})} \tag{2}$$

The conversion factors were 1.07 g O_2/g for glucose, xylose, lactate, and acetate; 1.51 g O_2/g for propionate; 1.82 g O_2/g for butyrate; and 2.04 g O_2/g for valerate. For biomass, an empirical molecular formula of $C_5H_7NO_2$ that corresponded to a conversion factor of 1.42 g O_2/g for biomass was assumed [19].

Additionally, acidification degrees (ADs) were calculated for the fermentative process. The total acidification degree (AD_{Total}) represents the amount of substrate consumed to produce SCOAs considering all the organic matter entering the reactor Equation (3). The sugars acidification degree (AD_{Sugars}) represents the amount of sugars consumed to produce SCOAs, considering the xylose and glucose fed to the reactor Equation (4). These calculations were performed as percentages.

$$AD_{Total} \text{ } (\%COD) = \frac{[SCOAs]_{out} - [SCOAs]_{in}}{COD_{In}} \tag{3}$$

$$AD_{Sugars} \text{ } (\%COD) = \frac{[SCOAs]_{out} - [SCOAs]_{in}}{COD_{Sugars\text{ }in}} \times 100 \tag{4}$$

3. Results and Discussion

The CSTR inoculated with an aerobic consortium was operated with a HRT of 1.76 days. A short SRT was chosen since short SRTs were already observed to promote the growth of acidogenic organisms comparatively to methanogens, usually presenting low growth rates [2]. Moreover, the use of an aerobic population as inoculum was also based on the fact that methanogens are strict anaerobes and acidogenic bacteria are facultative anaerobes [11,26]. Consequently, the conditions imposed were expected to favor the acidogenic population. The initial COD concentration of the CSTR feed (20.8 g COD/L) was chosen considering preliminary assays of acidogenic fermentation of HSSL (data not shown) and taking into account the COD and LS concentrations of HSSL obtained after the chemical pre-treatment (266 g COD/L and 190 g/L, respectively). This corresponded to an organic load rate of 11.8 g COD/L day. Temperature was kept in the mesophilic range (30 °C), because at this value the process can occur efficiently without major energy requirements [2]. Finally, pH was monitored but not controlled along the operational period of the CSTR. This can be advantageous at an industrial level, considering that lower amounts of chemicals are required and no extra equipment for pH control is needed, thus reducing operation costs. This would also be beneficial for the process scale-up.

3.1. Acidogenic Fermentation of HSSL

The variation of SCOA composition during the CSTR operational period, which lasted about 88 days, as well as of the main sugars, glucose and xylose, is shown in Figure 1.

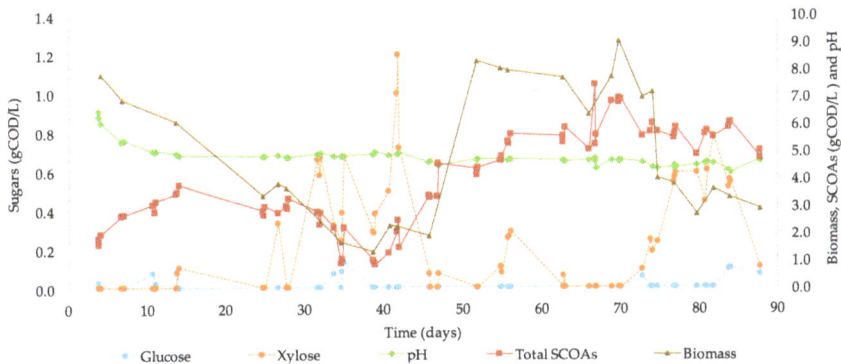

Figure 1. Evolution of short-chain organic acids (SCOAs), sugars, biomass, and pH variation during the fermentative process. COD: carbon oxygen demand.

At the beginning of the operational period, the CSTR lost a significant amount of biomass, reaching the lowest value (1.3 g COD/L) on the 39th day. The observed loss of biomass probably resulted from the operational conditions imposed. The short SRT would not allow for the survival of slow-growing bacteria. Additionally, the obligate aerobes initially present in the inoculum were probably washed out from the system due to the selective pressure that resulted from the anaerobic conditions imposed. Then, from the 46th day, a quick increase was observed in the following five days, and then it stabilized between 7.0 and 9.0 g COD/L until the 74th day, when it started to decrease again. A same pattern of biomass concentration was observed by Fernández-Morales et al. [26] during the acclimatization of conventional aerobic activated sludge to obtain an enriched acidogenic culture.

Most days, glucose was exhausted while xylose was not. This could show a preference for glucose over xylose by the culture. It is known that the metabolic pathway for glucose conversion to SCOAs is simpler than for xylose, with fewer enzymatic steps. Glucose enters directly in the glycolysis step, producing pyruvic acid, from which SCOAs are synthetized [10]. On the other hand, xylose needs to be converted to the intermediary D-xylulose-5-phosphate and go through pentose phosphate pathway (most commonly) to finally arrive in glycolysis and then be converted to pyruvate, and consequently, into SCOAs [10,27,28]. Moreover, Temudo et al. [10] verified that xylose catabolism was more efficient than glucose under anaerobic conditions, since more ATP was generated per mole of substrate. However, when authors compared the ATP-utilization for biomass production, biomass growth from xylose was less efficient than from glucose [10].

During the first 15 days of operation, a gradual increase of SCOAs production was observed, and then it stabilized until the 30th day. A drastic decrease in SCOA production was observed between days 30 and 40, which could be a consequence of the increase of glucose and xylose concentrations observed in the reactor. The minimum value of total SCOA concentration (0.86 g COD/L) was obtained on the 39th day (Figures 1 and 2). The decrease in SCOA concentration could be a consequence of the lowest biomass concentration observed on this day in the CSTR (Figure 1), suggesting that the culture was still unstable. The pH also fluctuated during this period, probably due to the verified instability. Generally, throughout the CSTR operational period, a higher concentration of SCOAs in the reactor effluent corresponded to a lower amount of sugars. An increase in the amount of sugars in the reactor effluent coincided with lower SCOAs production.

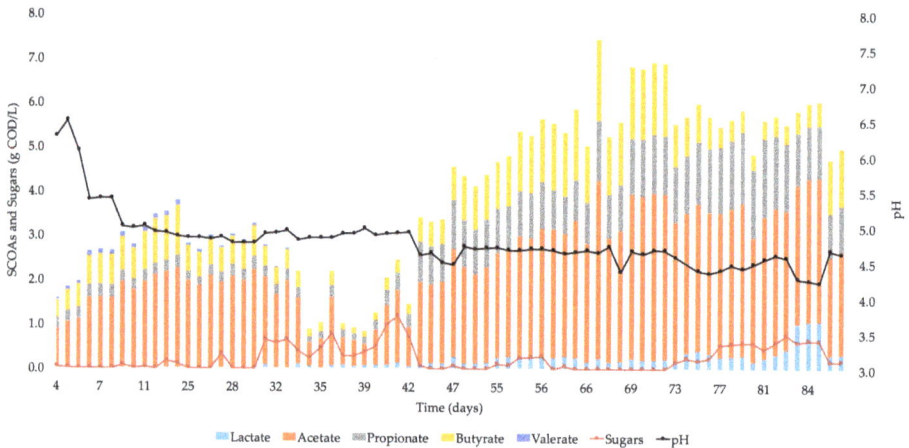

Figure 2. SCOA production profile, sugars, and pH evolution throughout fermentation time.

Total SCOA maximum concentration was achieved on day 71, and corresponded to an average value of 7.0 g COD/L, which was comparable to the values reported in the literature. For instance, Bengtsson et al. [29] tested the AF of four industrial wastes in batch mode with pH controlled at 6.00. The maximum SCOA concentration obtained was 3.96 g COD/L using a waste derived from a paper mill [29]. Additionally, Silva et al. [8] studied the AF of eight organic streams in batch experiments, and the maximum SCOA concentration obtained was 3.37 g COD/L for cheese whey. Lastly, Jie et al. [30] tested different pH values and their impact on AF of excess sludge, and observed the highest SCOA concentration for pH 10.0, 3.16 g COD/L. Moreover, the SCOA concentration obtained seemed to be related not only to the conditions imposed during the AF process, but also to the AF potential of the waste used. In this way, despite the presence of inhibitory components, and due to the sugar content, HSSL bears a significant AF potential, comparable to the wastes tested in the literature [8,29,30].

The concentrations of sugars and SCOAs were the parameters chosen to evaluate the stability of the microbial population. More stable sugars and SCOA concentrations—3.8 ± 0.25 and 5.5 ± 0.84 g COD/L, respectively—were obtained from day 45 onward. A pseudo-stationary state was considered to have been reached after this day, meaning that after inoculation, the system required 44 days to adapt to substrate and conditions imposed (Table 1). The reduction in the biomass concentration at the beginning of the process could be responsible for the long adaptation time. Moreover, the presence of known bacterial inhibitors in HSSL such as gallic acid, pyrogallol, and furfural, as reviewed by Pereira et al. [22], could affect the SCOAs and biomass concentrations obtained, and consequently contribute to the long adaptation time [26,31]. An introduction of a pre-adaptation step in batch mode could accelerate the adaptation of the MMC to HSSL.

Table 1. Main results from the acidogenic fermentation of hardwood sulfite spent liquor (HSSL).

Parameters	Time (Day)	Sugars Consumed *		SCOAs * (g COD/L)	SCOAs Profile (% COD) *					AD * (%)
		g COD/L	(%)		Lactate	Acetate	Propionate	Butyrate	Valerate	
Operation	0–88	3.8 ± 0.29	94 ± 7.2	4.0 ± 1.76	3.9 ± 3.91	59 ± 6.8	17 ± 6.8	19 ± 6.5	1.0 ± 1.3	20 ± 9.0
PSS	46–88	3.8 ± 0.25	94 ± 6.3	5.5 ± 0.84	5.7 ± 3.91	53 ± 3.2	22 ± 4.2	19 ± 7.5	0.0 ± 0.00	28 ± 4.3

* mean values ± standard deviation. AD: acidification degree. PSS: Pseudo-steady state

3.2. Short-chain Organic Acids (SCOAs) Production

As stated previously, pH was not controlled during the process, and its variation is shown in Figure 2. The SCOA concentration was maximal during a period with pH values lower than 5.0. Previous studies indicated that this range of pH values generally results in very low yields of SCOA production [2,29]. By not controlling the pH in the CSTR and letting it remain under 5.0, methanogens would be washed out from the system, resulting in a possible way to improve SCOA concentration. Consequently, this could be a way to control the methanogenic population without adding inhibitors for methanogenic bacteria.

Acetate, propionate, and butyrate were the SCOAs obtained in higher concentrations during the operational period. During the PSS, 3.0 ± 0.53 g COD/L of acetate, 1.2 ± 0.24 g COD/L of propionate, and 1.0 ± 0.45 g COD/L of butyrate were produced. Acetate was the SCOA with highest concentration most of the time, followed by propionate (Figure 2). The presence of these SCOAs in the mixed liquor was already expected since they are usually the main products of AF [2]. Chang et al. [32] verified that those were the most common SCOAs generated after the AF of diverse complex substrates that included food wastes, pig and chicken manure, rice straw, and corn stover [32].

Propionate and butyrate concentrations appeared to be somehow inversely related, and dependent on the pH (Figure 3). For pH lower than 4.6, propionate seemed to dominate over butyrate, and the opposite could be observed for higher pH values. This is in agreement with the literature that reports different metabolic pathways for the production of the two acids, suggesting the existence of two different types of populations competing for the carbon source [29,30,33].

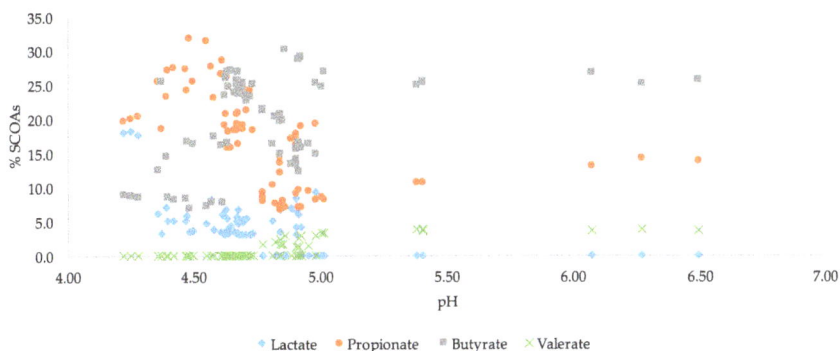

Figure 3. Short-chain Organic Acids (SCOAs) concentrations versus pH values obtained during the continuous stirred tank reactor (CSTR) operation.

The concentrations of valerate and lactate also seemed to be related to pH values (Figure 3). During the first 32 days, valerate concentration reached its maximum value of 0.11 g COD/L. Experimental data showed that valerate production occurred when pH values were higher than 4.8, especially at the beginning of the CSTR operational period. Then, pH values decreased over time (probably as a result of the increase of SCOAs concentration over time), and the concentration of valerate decreased. This fact is consistent with the results obtained by Lim et al. [34] for AF of food waste after testing three pH values (5.0, 5.5, 6.0) in a semi-continuous reactor with once-a-day feeding and draw-off. At pH 5.0, valerate production was not detected [34]. Regarding lactate, it was only detected when pH values lower than 5.0 (Figure 3). These findings were in agreement with some AF studies that observed lactate production under similar conditions [34–36]. Thus, lactate achieved a maximum concentration at pH 4.3 after 84 days of fermentation. The maximum lactate obtained corresponded to 18% of the total SCOAs (Figure 2).

The pH seemed to be a crucial parameter in the definition of the SCOAs profile during AF, despite being a direct consequence of SCOAs production. It probably influenced the selection of different acidogenic populations in the MMC and their ability to produce different SCOAs. The influence of pH on the type of SCOAs obtained has already been shown by Albuquerque et al. [37]. Moreover, a tight relationship between pH and type of SCOAs produced should be achieved in future works for HSSL, since tailoring the distribution of SCOAs produced based on pH is essential when using the SCOAs mixtures as substrates in bioprocesses such as PHAs production by MMC [37].

3.3. Acidification Degree

Acidification degrees (ADs) relatively to the total COD, AD_{total}, and to the main sugars xylose and glucose, AD_{sugars}, were determined along the CSTR operational period. AD_{total} had a maximum of 36% on day 71, which corresponded to the maximum value of SCOA concentration obtained. The reason for the relatively low AD_{total} could be explained by the majority of the COD present in the feed being constituted by phenolic compounds, LS, and other recalcitrant compounds, which are more difficult to biodegrade than sugars by the microbial population [38].

The maximum AD_{sugars} obtained was 175% on day 71. This value shows that besides xylose and glucose, other compounds present in the HSSL were probably used for SCOAs production. These compounds probably included other monomeric sugars such as rhamnose, arabinose, mannose, or galactose [39], usually below the detection limit of HPLC due to the dilution applied to samples before analysis. Assuming the data provided by Caima–Indústria de Celulose SA, the HSSL after the chemical pre-treatment would have 1.36, 1.14, 2.28, 2.28 g COD/L of rhamnose, arabinose, mannose, and galactose, respectively. Taking these values into consideration and assuming their total consumption, the $AD_{TotalSugars}$ would be 154%. This result means that besides sugars, some phenolic components or even LS would be used for SCOAs production. Hence, it would be interesting to analyze the evolution of the different components of HSSL in more detail to understand which compounds were being consumed to produce SCOAs.

In order to improve the AD_{Total}, higher HRT values should be imposed to the CSTR to assess whether the most recalcitrant compounds of HSSL—LS and phenolics—could be acidified. The possibility of maximizing the amount of components of HSSL acidified not only increases the concentration of SCOAs obtained, but also decreases the amount of possible microbial inhibitors, which favors the utilization of the fermented stream in further bioprocesses.

4. Conclusions

A CSTR was operated for 88 days in order to acidify part of the organic components of HSSL. The system took about 45 days to reach a pseudo-steady state under the operational conditions imposed. On day 71, a maximum SCOA concentration of 7.0 g COD/L was obtained, which corresponded to an AD_{Total} of 36.0%. The experimental results showed that it was possible to perform the acidification of HSSL and obtain SCOA by inoculating the CSTR with an aerobic MMC. The use of an aerobic inoculum and leaving the pH uncontrolled could be responsible for the significant SCOA production. Future studies should focus on testing higher HRTs, different reactor conformation (e.g., moving bed biofilm reactor), and controlling the pH to different values in order to improve the concentration of SCOAs and the amount of HSSL components consumed. Finally, microbiological studies should be performed in order to understand how the microbial community changed over time.

Acknowledgments: This work was developed within the scope of the project CICECO-Aveiro Institute of Materials (Ref. FCT UID/CTM/50011/2013), financed by national funds through the FCT/MEC and when appropriate co-financed by FEDER under the PT2020 Partnership Agreement. Authors acknowledge Eng. A. Prates from CAIMA-Indústria de Celulose S.A., Constância, Portugal for HSSL. Diogo Queirós thanks FCT for his PhD grant (SFRH/BD/87758/2012).

Author Contributions: Diogo Queirós and Luísa S. Serafim conceived and designed the experiments; Diogo Queirós, Rita Sousa and Susana Pereira performed the experiments; Diogo Queirós, Rita Sousa and Luísa S. Serafim analyzed the data. Diogo Queirós and Rita Sousa wrote the paper.

Conflicts of Interest: The authors declare no conflict of interest.

References

1. Fava, F.; Totaro, G.; Diels, L.; Reis, M.; Duarte, J.; Carioca, O.B.; Poggi-Varaldo, H.M.; Ferreira, B.S. Biowaste biorefinery in Europe: Opportunities and research & development needs. *New Biotechnol.* **2015**, *32*, 100–108.
2. Lee, W.S.; Chua, A.S.M.; Yeoh, H.K.; Ngoh, G.C. A review of the production and applications of waste-derived volatile fatty acids. *Chem. Eng. J.* **2014**, *235*, 83–99. [CrossRef]
3. Koutinas, A.A.; Vlysidis, A.; Pleissner, D.; Kopsahelis, N.; Lopez Garcia, I.; Kookos, I.K.; Papanikolaou, S.; Kwan, T.H.; Lin, C.S.K. Valorization of industrial waste and by-product streams via fermentation for the production of chemicals and biopolymers. *Chem. Soc. Rev.* **2014**, *43*, 2587–2627. [CrossRef] [PubMed]
4. Frison, N.; Katsou, E.; Malamis, S.; Oehmen, A.; Fatone, F. Development of a novel process integrating the treatment of sludge reject water and the production of polyhydroxyalkanoates (PHAs). *Environ. Sci. Technol.* **2015**, *49*, 10877–10885. [CrossRef] [PubMed]
5. Shen, L.; Hu, H.; Ji, H.; Cai, J.; He, N.; Li, Q.; Wang, Y. Production of poly(hydroxybutyrate-hydroxyvalerate) from waste organics by the two-stage process: Focus on the intermediate volatile fatty acids. *Bioresour. Technol.* **2014**, *166*, 194–200. [CrossRef] [PubMed]
6. Woo Park, G.; Fei, Q.; Jung, K.; Chang, H.N.; Kim, Y.-C.; Kim, N.; Choi, J.; Kim, S.; Cho, J. Volatile fatty acids derived from waste organics provide an economical carbon source for microbial lipids/biodiesel production. *Biotechnol. J.* **2014**, *9*, 1536–1546. [CrossRef] [PubMed]
7. Elefsiniotis, P.; Wareham, D.G.; Smith, M.O. Use of volatile fatty acids from an acid-phase digester for denitrification. *J. Biotechnol.* **2004**, *114*, 289–297. [CrossRef] [PubMed]
8. Silva, F.C.; Serafim, L.S.; Nadais, H.; Arroja, L.; Capela, I. Acidogenic fermentation towards valorisation of organic waste streams into volatile fatty acids. *Chem. Biochem. Eng. Q.* **2013**, *27*, 467–476.
9. Ke, S.; Shi, Z.; Fang, H.H.P. Applications of two-phase anaerobic degradation in industrial wastewater treatment. *Int. J. Environ. Pollut.* **2005**, *23*, 65–80. [CrossRef]
10. Temudo, M.F.; Mato, T.; Kleerebezem, R.; van Loosdrecht, M.C.M. Xylose anaerobic conversion by open-mixed cultures. *Appl. Microbiol. Biotechnol.* **2009**, *82*, 231–239. [CrossRef] [PubMed]
11. Visvanathan, C.; Abeynayaka, A. Developments and future potentials of anaerobic membrane bioreactors (AnMBRs). *Membr. Water Treat.* **2012**, *3*, 1–23. [CrossRef]
12. De Aquino, S.F.; Chernicharo, C.A.L. Acúmulo de ácidos graxos voláteis (AGVs) em reatores anaeróbios sob estresse: causas e estratégias de controle. *Eng. Sanit. E Ambient.* **2005**, *10*, 152–161. [CrossRef]
13. Queirós, D.; Rossetti, S.; Serafim, L.S. PHA production by mixed cultures: A way to valorize wastes from pulp industry. *Bioresour. Technol.* **2014**, *157*, 197–205. [CrossRef] [PubMed]
14. Queirós, D.; Fonseca, A.; Lemos, P.C.; Serafim, L.S. Long-term operation of a two-stage polyhydroxyalkanoates production process from hardwood sulphite spent liquor. *J. Chem. Technol. Biotechnol.* **2016**, *91*, 2480–2487. [CrossRef]
15. Jiang, J.; Zhang, Y.; Li, K.; Wang, Q.; Gong, C.; Li, M. Volatile fatty acids production from food waste: Effects of pH, temperature, and organic loading rate. *Bioresour. Technol.* **2013**, *143*, 525–530. [CrossRef] [PubMed]
16. Jankowska, E.; Chwiałkowska, J.; Stodolny, M.; Oleskowicz-Popiel, P. Effect of pH and retention time on volatile fatty acids production during mixed culture fermentation. *Bioresour. Technol.* **2015**, *190*, 274–280. [CrossRef] [PubMed]
17. Scoma, A.; Bertin, L.; Fava, F. Effect of hydraulic retention time on biohydrogen and volatile fatty acids production during acidogenic digestion of dephenolized olive mill wastewaters. *Biomass Bioenergy* **2013**, *48*, 51–58. [CrossRef]

18. Miron, Y.; Zeeman, G.; van Lier, J.B.; Lettinga, G. The role of sludge retention time in the hydrolysis and acidification of lipids, carbohydrates and proteins during digestion of primary sludge in CSTR systems. *Water Res.* **2000**, *34*, 1705–1713. [CrossRef]
19. Dogan, E.; Demirer, G.N. Volatile fatty acid production from organic fraction of municipal solid waste through anaerobic acidogenic digestion. *Environ. Eng. Sci.* **2009**, *26*, 1443–1450. [CrossRef]
20. Rueda, C.; Calvo, P.A.; Moncalián, G.; Ruiza, G.; Coz, A. Biorefinery options to valorize the spent liquor from sulfite pulping. *J. Chem. Technol. Biotechnol.* **2015**, *90*, 2218–2226. [CrossRef]
21. Zygmunt, B.; Banel, A. Formation, occurrence and determination of volatile fatty acids in environmental and related samples. In Proceedings of the 3rd WSEAS International Conference on Energy Planning, Energy Saving, Environmental Education, Renewable Energy Sources, Waste Management, Tenerife, Spain, 2009; pp. 476–481.
22. Pereira, S.R.; Ivanuša, S.; Evtuguin, D.V; Serafim, L.S.; Xavier, A.M.R.B. Biological treatment of eucalypt spent sulphite liquors: A way to boost the production of second generation bioethanol. *Bioresour. Technol.* **2012**, *103*, 131–135. [CrossRef] [PubMed]
23. Clesceri, L.S.; Greenberg, A.E.; Eaton, A.D. *Standard Methods for the Examination of Water and Wastewater*, 20th ed.; American Public Health Association, 1998. Available online: https://www.standardmethods.org/ (accessed on 1 March 2015).
24. Restolho, J.A.; Prates, A.; de Pinho, M.N.; Afonso, M.D. Sugars and lignosulphonates recovery from eucalyptus spent sulphite liquor by membrane processes. *Biomass Bioenergy* **2009**, *33*, 1558–1566. [CrossRef]
25. Xavier, A.M.R.B.; Correia, M.F.; Pereira, S.R.; Evtuguin, D.V. Second-generation bioethanol from eucalypt sulphite spent liquor. *Bioresour. Technol.* **2010**, *101*, 2755–2761. [CrossRef] [PubMed]
26. Fernández-Morales, F.J.; Villaseñor, J.; Infantes, D. Modeling and monitoring of the acclimatization of conventional activated sludge to a biohydrogen producing culture by biokinetic control. *Int. J. Hydrog. Energy* **2010**, *35*, 10927–10933. [CrossRef]
27. Jeffries, T.W. Advances in biochemical engineering/biotechnology. In *Pentoses and Lignin*; Fieehter, A., Ed.; Springer: Berlin/Heidelberg, Germany, 1983; pp. 1–32.
28. Prakasham, R.S.; Brahmaiah, P.; Sathish, T.; Sambasiva Rao, K.R.S. Fermentative biohydrogen production by mixed anaerobic consortia: Impact of glucose to xylose ratio. *Int. J. Hydrog. Energy* **2009**, *34*, 9354–9361. [CrossRef]
29. Bengtsson, S.; Hallquist, J.; Werker, A.; Welander, T. Acidogenic fermentation of industrial wastewaters: Effects of chemostat retention time and pH on volatile fatty acids production. *Biochem. Eng. J.* **2008**, *40*, 492–499. [CrossRef]
30. Jie, W.; Peng, Y.; Ren, N.; Li, B. Volatile fatty acids (VFAs) accumulation and microbial community structure of excess sludge (ES) at different pHs. *Bioresour. Technol.* **2014**, *152*, 124–129. [CrossRef] [PubMed]
31. Cohen, A.; Gemert, J.M.V.; Zoetemeyer, R.J.; Breure, A.M. Main characteristics and stoichiometric aspects of acidogenesis of soluble carbohydrate containing wastewater. *Process Biochem.* **1984**, *19*, 282–286.
32. Chang, H.N.; Kim, N.-J.; Kang, J.; Jeong, C.M. Biomass-derived volatile fatty acid platform for fuels and chemicals. *Biotechnol. Bioprocess Eng.* **2010**, *15*, 1–10. [CrossRef]
33. Horiuchi, J.-I.; Shimizu, T.; Tada, K.; Kanno, T.; Kobayashi, M. Selective production of organic acids in anaerobic acid reactor by pH control. *Bioresour. Technol.* **2002**, *82*, 209–213. [CrossRef]
34. Lim, S.J.; Kim, B.J.; Jeong, C.M.; Choi, J.D.; Ahn, Y.H.; Chang, H.N. Anaerobic organic acid production of food waste in once-a-day feeding and drawing-off bioreactor. *Bioresour. Technol.* **2008**, *99*, 7866–7874. [CrossRef] [PubMed]
35. Gouveia, A.R.; Freitas, E.B.; Galinha, C.F.; Carvalho, G.; Duque, A.F.; Reis, M.A. Dynamic change of pH in acidogenic fermentation of cheese whey towards polyhydroxyalkanoates production: Impact on performance and microbial population. *New Biotechnol.* **2017**, *37*, 108–116. [CrossRef] [PubMed]
36. Itoh, Y.; Tada, K.; Kanno, T.; Horiuchi, J.I. Selective production of lactic acid in continuous anaerobic acidogenesis by extremely low pH operation. *J. Biosci. Bioeng.* **2012**, *114*, 537–539. [CrossRef] [PubMed]
37. Albuquerque, M.G.E.; Eiroa, M.; Torres, C.; Nunes, B.R.; Reis, M.A.M. Strategies for the development of a side stream process for polyhydroxyalkanoate (PHA) production from sugar cane molasses. *J. Biotechnol.* **2007**, *130*, 411–421. [CrossRef] [PubMed]

38. Jantsch, T.G.; Angelidaki, I.; Schmidt, J.E.; Braña de Hvidsten, B.E.; Ahring, B.K. Anaerobic biodegradation of spent sulphite liquor in a UASB reactor. *Biores. Technol.* **2002**, *84*, 15–20. [CrossRef]
39. Pereira, S.R.; Portugal-Nunes, D.J.; Evtuguin, D.V.; Serafim, L.S.; Xavier, A.M.R.B. Advances in ethanol production from hardwood spent sulphite liquors. *Process Biochem.* **2013**, *48*, 272–282. [CrossRef]

fermentation

MDPI

Review

Microbial Propionic Acid Production

R. Axayacatl Gonzalez-Garcia [1,†], Tim McCubbin [1,†], Laura Navone [1,2], Chris Stowers [3],
Lars K. Nielsen [1] and Esteban Marcellin [1,*]

[1]   Australian Institute for Bioengineering and Nanotechnology, The University of Queensland,
      Brisbane, QLD 4072, Australia; r.gonzalezgarcia@uq.edu.au (R.A.G.-G.);
      timothy.mccubbin@uqconnect.edu.au (T.M.); laura.navone@qut.edu.au (L.N.);
      lars.nielsen@uq.edu.au (L.K.N.)
[2]   Present address: Science and Engineering Faculty, Queensland University of Technology,
      Brisbane, QLD 4000, Australia
[3]   Dow Agrosciences LLC, Indianapolis, 46268 IN, USA; CCStowers@dow.com
*   Correspondence: e.marcellin@uq.edu.au; Tel.: +61-733-464-298
†   These authors contributed equally to this work.

Academic Editor: Gunnar Lidén
Received: 27 March 2017; Accepted: 7 May 2017; Published: 15 May 2017

Abstract: Propionic acid (propionate) is a commercially valuable carboxylic acid produced through microbial fermentation. Propionic acid is mainly used in the food industry but has recently found applications in the cosmetic, plastics and pharmaceutical industries. Propionate can be produced via various metabolic pathways, which can be classified into three major groups: fermentative pathways, biosynthetic pathways, and amino acid catabolic pathways. The current review provides an in-depth description of the major metabolic routes for propionate production from an energy optimization perspective. Biological propionate production is limited by high downstream purification costs which can be addressed if the target yield, productivity and titre can be achieved. Genome shuffling combined with high throughput omics and metabolic engineering is providing new opportunities, and biological propionate production is likely to enter the market in the not so distant future. In order to realise the full potential of metabolic engineering and heterologous expression, however, a greater understanding of metabolic capabilities of the native producers, the fittest producers, is required.

Keywords: propionic acid; propionibacteria; fermentation

1. Introduction

Propionic acid is an FDA approved, generally regarded as safe (GRAS) three-carbon chemical with applications in a wide variety of industries. Propionate is primarily used for its antimicrobial properties with major markets as a food preservative or herbicide [1]. The antimicrobial properties of propionate result in its increasing use in construction and cleaning products as sodium, calcium or potassium salt. Propionate salts are effective suppressors of mould growth on surfaces and, when combined with lactic and acetic acids, can inhibit the growth of *Listeria monocytogenes* [2]. In the plastic industry, it is used in the manufacture of cellulose derived plastics such as textiles, membranes for reverse osmosis, air filters, and as a component of lacquer formulations and moulding plastics [3]. In the pharmaceutical industry, sodium propionate is used primarily in animal therapy for the treatment of wound infections and as a component of conjunctivitis and anti-arthritic drugs [4]. In the cosmetics industry, propionate salts are used as perfume bases together with butyl rubber to improve the consistency and shelf life of products. Propionate finds additional application as a flavour enhancer in the form of citronellyl or geranyl propionate.

Global production of propionic acid is estimated at ~450,000 tonnes per year with a 2.7% annual growth [5] and a price ranging between $2–$3 USD/kg. Four manufacturers supply 90% of the global propionate market: BASF covers approximately 31% of the market with plants in Germany and

China; The Dow Chemical Company supplies 25% of the global market with production in the USA; Eastman Chemical provides 20% of the market with production in the USA; and Perstorp in Sweden supplies 14% of the global market. At present, propionate is industrially synthesized by petrochemical processes, predominantly through the Reppe process, which converts ethylene, carbon monoxide and steam into propionate, and the Larson process, which converts ethanol and carbon monoxide into propionate in the presence of boron trifluoride. Other less common synthesis techniques include oxidation of propionaldehyde, the Fischer-Tropsch process and pyrolysis of wood.

However, recent market needs demand biological propionic acid biosynthesis as a sustainable alternative [6]. First described by Albert Fitz in 1878 [7], *Propionibacterium* species can ferment sugars into propionic acid as their main fermentation product. Later on, Swick and Wood [8] described the set of reactions involved in the process of propionate production currently known as the Wood-Werkman cycle (Figure 1(AIII)). After many years of development, the gap between production costs of propionate via petrochemical processes and by fermentation by propionibacteria is narrowing [6,9]. Fermentation economics of low-value products such as propionate depend on the ability to convert carbon sources at a high yield and productivity to a high titre. Recent advances in metabolic engineering and fermentation have significantly improved the economic viability of the propionic acid fermentation process, however, separation of the various organic acids produced alongside propionic acid remains a problem. This problem is compounded by the fact that most product specifications for propionic acid require >99% purity which is reasonably achievable through the petrochemical manufacturing routes, but very challenging to achieve biologically. To overcome the downstream purification challenges, improvement of the biocatalyst remains the most viable and economical option. As a result of advances in systems and synthetic biology, new strains are being developed that meet the fermentation yield target while maintaining a high productivity; hereby overcoming one of, if not the, most technically challenging hurdles to commercialization of bio-propionic acid synthesis [10].

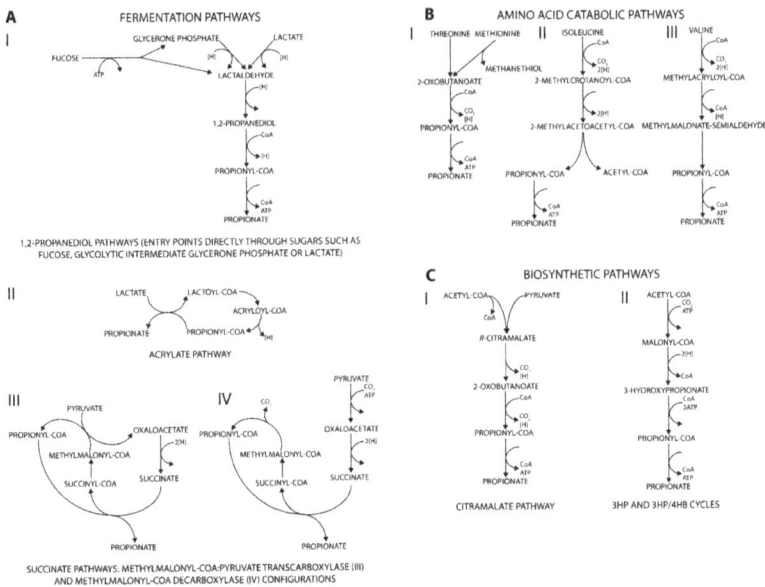

Figure 1. Metabolic pathways allowing the synthesis of propionate. **A**: Fermentation pathways (I-propanediol pathways, II-acrylate pathway, III-Wood-Werkman cycle, IV-sodium-pumping); **B**: Amino acid catabolic pathways (I-threonine and methionine catabolism, II-isoleucine catabolism, III-valine catabolism); **C**: Biosynthetic pathways (I-citramalate pathway, II-3-hydroxypropanoate and 3-hydroxypropanoate/4-hydroxybutanoate pathways).

In this review, we look at the current stage, progress and perspective of propionic acid biosynthesis, in particular at the improvements over the past decade as a result of the omics revolution. We present an in-depth analysis of different metabolic pathways for propionic acid production from an energy optimization perspective. This contribution provides a clear path to numerous development opportunities, including metabolic engineering of current native producers as well as future heterologous production via a deep understanding of propionibacteria metabolism. We demonstrate that *Propionibacterium* is the best natural producer of propionate and remains the most suitable candidate for industrial-scale production.

2. Overview of Developments in the Fermentation Process

The fermentation process for propionate production has been studied for more than 150 years (Table 1). In recent years, many studies have aimed at improving the fermentation process as a means of improving process economics by enhancing the productivity, yield and final titre.

Native propionate producers, propionibacteria, have been the primary candidates for the development of a biological process due to their unique metabolism. As will be subsequently shown, the Wood-Werkman cycle is energetically the most efficient propionate fermentation route currently known. Propionibacteria are Gram-positive, non-motile, non-spore forming, rod-like, facultative anaerobic bacteria. Dairy species, especially *P. acidipropionici* strains, have been explored as potential propionic acid producers due to their tendency to produce propionate at a higher titre and yield [9]. However, even the best performing native producers are limited by low growth rates and correspondingly low productivities, acid stress and, most importantly, the co-generation of by-products which increase downstream processing costs and limit the economic feasibility of the fermentation process. The benchmark for an economically feasible fermentation process by 2020 from sugar has been set at a yield of 0.6 g/g (1.52 mol propionate/mol glucose) and a productivity of 1 g/L/h [9]. Various fermentation technologies including batch, fed-batch, continuous fermentation and immobilised cells to prevent product inhibition [11–15] have been explored, but still fermentations fall short on simultaneously meeting both criteria (Table 2). While a single study simultaneously met both requirements, an exceptionally high initial biomass of ~56g/L was used and a low final propionate titre of 34.5 g/L was achieved [15]. For economic feasibility, the final titre should be closer to 100 g/L, and the requirement for the high initial biomass may present additional economic hurdles. Anaerobic processes are preferred for the production of propionate as they do not require additional aeration, thereby reducing the costs related to pneumatic power and fermenter scale-up. While biomass immobilisation has been used as a leading strategy to overcome acid stress limitations as previously reviewed [16], the use of these reactors is yet to be realised on an industrial scale due two key challenges. Firstly, it is necessary to adapt the reactors from a three-dimensional system into a two-dimensional system, which would limit the amount of biomass within the reactors. Secondly, the cleaning process presents a major challenge. In addition, most of the immobilization techniques also increase cost as there is a membrane/cloth/gel bead that wears out after a few fermentation cycles.

Table 1. Overall historical development of the propionic acid fermentation process.

Year	Event
1854	Adolph Strecker observed the formation of propionic acid from sugar in a mixture of calcium carbonate-sugar [1].
1861–1879	Pasteur showed that fermentation occurs due to the activity of microbes.
1878	Fist work on propionic acid production by Propionibacteria. Albert Fitz predicted that 3 moles of lactic acid would lead to the production of 2 moles of propionic acid, 1 mole acetic acid, 1 mole CO_2 and 1 mole H_2O [7].
1906	11 species of propionibacteria were identified as propionic acid producers during cheese making [1].

Table 1. *Cont.*

Year	Event
1928	First mention of glycerol as carbon source for propionic acid production [1].
1937	First complete study on propionibacteria metabolism during propionic acid fermentation by Wood [1].
1949	A complete review of the factors affecting propionic acid fermentation was published [1].
1920–1953	17 patents for propionic acid production by different *Propionibacterium* strains were approved [1].
1961	Immobilized cells are first used to reach higher production yields [1].
1962	The Wayman process was developed. It consisted of a continuous system with immobilised cells of *P. acidipropionici* [1].
1960–2010	Selection of overproducer strains and new production strategies.
2011–2013	Complete genome of *P. shermanii* [17] and *P. acidipropionici* [18] were sequenced and published.
2013–2014	Techno economic studies suggest the fermentation of sugar to propionate can be profitable if productivity reaches 1–2 g/L/h, yield reaches 0.6 g/g and final titre reaches ≈100 g/L [9,19].

Improving fermentation yield is the most critical step to achieve an economically viable fermentation, given productivity is readily improved by densifying the inoculation medium or by cell recycle [9]. The maximum theoretical yield of 0.7 g propionate/g glucose (1.71 mol/mol) can be achieved through the Wood-Werkman cycle if reduced cofactors are supplemented through an alternative pathway, such as the redirection of carbon from glycolysis to the pentose phosphate pathway, making an apparent yield of 0.6 g propionate/g glucose in complex media achievable. One technique to improve yield is to restrict biomass production [20]. This has been done by restricting nutrient availability [21] and by reducing the pH which also favours propionate production over acetate [22,23].

Table 2. Comparison of some propionic acid (PA) fermentation approaches from the literature dating 2010 to 2015.

Strain	Fermentation Approach	Substrate (s)	Titre (g/L)	PA Yield (g/g)	Productivity (g/L/h)	References
P. acidipropionici	Batch	Glucose/Glycerol	22	0.57	0.152	[24]
	Fed-batch	Glucose/Glycerol	30	0.54	0.152	[24]
	Sequential batch	Glucose	35	0.62	1.28	[15]
	Fed-batch	Glucose	56	0.43	2.23	[15]
	Continuous	Lactose	19	0.4	0.9	[25]
	Fed-batch	Glucose	71	-	-	[26]
	Fed-batch	Glycerol	48	0.59	0.2	[27]
	Batch	Glucose	45	0.45	2	[6]
	Batch	Corn mash	24	0.6	0.5	[9]
	Sequential batch (with cell recycle)	Glycerol	27	0.78	0.22	[28]
	Fed-batch	Xylose	53	0.35	0.23	[29]
	Fed-batch	Corncob molasses	72	-	0.28	[29]
	PEI-Poraver bioreactor (Continuous)	Glycerol	14	0.86	1.4	[30]
	Fibrous-bed bioreactor (Fed-batch)	Glucose	51	0.43	0.71	[31]
	Fibrous-bed bioreactor (Fed-batch)	Sugarcane bagasse hydrolysate	59	0.37	0.38	[31]
P. shermanii	Fibrous-bed bioreactor (Repeated-batch)	Glucose/Glycerol	75	0.57	0.25	[32]

Table 2. *Cont.*

Strain	Fermentation Approach	Substrate (s)	Titre (g/L)	PA Yield (g/g)	Productivity (g/L/h)	References
P. freudenreichii	Multi-point fibrous-bed bioreactor (Fed-batch)	Glucose	67	0.43	0.14	[23]
	Plant fibrous-bed bioreactor (Fed-batch)	Hydrolysed cane molasses	92	0.46	0.36	[33]
	Plant fibrous-bed bioreactor (Fed-batch)	Hydrolysate of cane molasses & waste *Propionibacterium* cells	80	0.4	0.26	[33]

In these cases, productivity typically suffers and increased stress lowers final titres. Recently, carbon dioxide sparging was identified as a viable alternative, where propionate yield was improved without loss of productivity and a higher titre of both propionate and biomass was obtained through increased substrate catabolism (unpublished data [34]). Inoculating at a higher cell density can also improve yield while enhancing productivity, achieving yields of up to 0.62 g propionate/g glucose [9]. The use of alternative carbon sources has also been explored. Glycerol has been studied in particular, as it has a similar redox state to propionate and consequently can achieve much higher maximum theoretical yields (Table 2). Given the relatively high cost of media compared to propionate, a greater emphasis has been placed on finding cheaper feedstocks. From this perspective, a fermentation using enzymatically treated corn mash has achieved the yield target of 0.6 g propionate/g glucose, where additional nutrients in the mash are responsible for the increase in apparent yield [9].

While progress has been made in optimizing fermentation conditions, Table 2 demonstrates few sugar fermentations are able to exceed the yield targets, particularly in an industrially scalable fermentation. This has resulted in an increasing reliance on genetic engineering to further enhance fermentation yields and restrict by-product fermentation to improve economics of the process.

3. Biological Propionic Acid Biosynthesis

Propionic acid is a metabolic by-product of many organisms, ranging from bacteria to humans, although few organisms produce it as a primary fermentation product (Table 3). Metabolic pathways leading to the production of propionic acid can be classified into three classes. The primary fermentation pathways catabolize different carbon sources to propionate and include the well-known acrylate and Wood-Werkman cycle pathways of native propionate producers (Figure 1A). Catabolic pathways can degrade a number of amino acids to propionic acid (Figure 1B). Finally, anabolic pathways associated with the production of biomass precursors from pyruvate or carbon dioxide can be harnessed to produce propionate (Figure 1C). Redox balancing is the main limitation to achieving higher yields from glucose. More favourable yields can be achieved by using more reduced substrates such as glycerol with a similar degree of reduction to propionate, although these are not likely to be as economically favourable as sugars [6]. We have therefore calculated all pathway yields referenced to glucose catabolism.

Pathways leading to propionate are typically linked to substrate level phosphorylation via the promiscuous activity of enzymes associated with acetate metabolism and can act as either electron sinks or sources. Because glycolysis results in the net production of reduced cofactors, pathways that act as electron sinks achieve more favourable maximum yields. In this work, we have analysed the feasibility of propionate production with a focus on energetic optimality; studying two separate scenarios. The first is the maximal propionate yield achievable while metabolism operates to capture as much potential energy in the form of ATP as possible. Energy maximisation is consistent with the evolutionary drive to maximise growth, which can approximate the behavior of microorganisms [35]. The second scenario is the maximum energy yield when propionate is produced at the maximum theoretical yield. This represents maximum energy available to the organism for growth and stress tolerance when metabolism is artificially perturbed away from optimality, such as by gene overexpression and deletion, to achieve maximum propionate yield. All calculations are performed under anaerobic conditions because this is the most practical approach for propionate production, in terms of both process costs and the ability to growth-couple propionate production; a requirement given the high target yield. An underlying model of central carbon metabolic pathways common to *E. coli* (glycolysis, pentose phosphate pathway, acetate and ethanol metabolism and the dicarboxylic branch of the tricarboxylic acid cycle (TCA) leading to succinate production) was used. For simplicity, all reduced cofactors are treated as NADH. An optimal pathway will not only have a high maximum theoretical yield for propionate, but a high energetic yield; driving the organism to produce propionate as a primary fermentation product while providing sufficient energy to overcome the inhibitory effects of propionate accumulation. We also compare our in silico calculations to performances obtained by these pathways in vivo where data are available.

3.1. Fermentation Routes for Propionate Production

Propionate, as a primary fermentation product, is produced via pathways that contain 1,2-propanediol (PDO) as an intermediate, the acryloyl-CoA pathway and the methylmalonyl-CoA or succinate pathways. Compared to the amino acid degradation and biosynthetic pathways (see Sections 3.2 and 3.3), fermentative pathways provide energy and help consume reduced cofactors that result from the catabolism of sugars. Both their role in energy generation and maintaining a redox balance permits these pathways to be growth coupled.

Table 3. Microbial species able to generate propionic acid during fermentation.

Microorganism	Substrates	Products	Pathway
Propionibacteria acidipropionici *P. freudenreichii* [1] *P. shermanii* [2]	Glucose, sucrose, lactate, glycerol	Propionate, acetate, succinate, CO_2	Wood-Werkman cycle (Figure 1(AIII))
Clostridia propionicum	Glycerol, lactate, alanine, serine, threonine	Propionate, succinate, formate, acetate, n-propanol	Acrylate pathway (Figure 1(AII))
Bacteroides fragilis B. ruminicola	Glucose	Acetate, lactate propionate, succinate, formate, CO_2	Succinate pathway (Figure 1(AIV))
Veillonella parvula V. alcalescens	Lactate, succinate	Propionate, acetate, CO_2, H_2	Succinate pathway (Figure 1(AIV))
Propionigenum modestum	Succinate	Propionate, CO_2	Succinate pathway (Figure 1(AIV))
Selenomonas ruminantium *S. sputigena*	Lactate Glucose	Propionate, lactate, acetate, CO_2	Succinate pathway (Figure 1(AIV))
Megasphaera elsdenii	Lactate	Acetate, propionate, butyrate	Acrylate pathway (Figure 1(AII))
Salmonella typhimurium	Deoxy sugars, glucose, 1,2-propanediol	1,2-propanediol, propanol, propionate, acetate, formate, lactate, CO_2	1,2-propanediol pathway (Figure 1(AI))

[1] *P. freudenreichii* subsp. *Freudenreichii*; [2] *P. freudenreichii* subsp. *Shermanii*.

3.1.1. 1,2-propanediol Associated Pathways

Propionic acid production occurs in the rumen as a net result of the microbial consortia consisting of PDO fermenters and PDO consumers. Some organisms including *Salmonella typhimurium* [36] and *Roseburia inulinivorans* [21] are known to perform both processes, demonstrating a novel fermentative pathway for propionate production. PDO can be generated from the catabolism of deoxy sugars via lactate or from the glycolytic intermediate glycerone phosphate (DHAP) [37]. PDO is catabolized stoichiometrically to propionate yielding one ATP and one reduced cofactor by the combined actions of diol dehydratase and two promiscuous enzymes commonly associated with acetate metabolism; the CoA-dependent aldehyde dehydrogenase phosphotransacylase and acetate kinase (Figure 1(AI)). Table 4 contains the yield calculations for each pathway. While the biological production of PDO is an area of active research, its further conversion to propionic acid production has not been explored because of the higher value of PDO, despite the fact that the conversion of PDO to propionate is energetically favourable. Because of the liberation of an additional reduced cofactor through PDO catabolism; the maximum molar yield of propionate from glucose is 30% higher; accompanied by a 70% increase in ATP generated as compared to PDO. In the case of the lactate pathway, this additional energy increases the net ATP yield from glucose to 4 and growth couples the production of propionate

(Table 4). While the lactate pathway appears extremely promising, almost all reported strategies to produce PDO to date utilise the less efficient DHAP pathway [38]. This focus on the DHAP pathway is due to a lack of biochemical evidence to support the existence of the lactate pathway [39] which has been postulated to exist in just a single source [40]. The high cost for deoxy sugars also limits their potential application since the cheapest sells for over $300/kg [41]. Still, promising progress has been made constructing an artificial pathway for the conversion of lactate to propanediol in *E. coli* which consumes 1 ATP equivalent [39,42]. While this strategy could benefit similarly if PDO were subsequently degraded to propionate, propanediol production from glucose in an engineered mutant only just exceeded 1 g/L [39].

Table 4. Estimated yields for propionate and ATP production in propanediol fermentative pathways.

PDO pathway	Maximum Yields (mol/mol Glc)		Expected Yields (mol/mol Glc)	
	PA	ATP	PA	ATP
Deoxy sugar [1]	1	2.5	1	3 [2]
DHAP	1.71	0	0	3
Lactate	1.71	3.43	1.33	4
Engineered lactate pathway	1.71	0 [3]	1	3
All	1.71	3.43	0	3

[1] This pathway utilises fucose instead of glucose; [2] Without the pyruvate formate lyase, the pathway is expected to be active producing 1 propionate and 2.5 ATP and also conferring a phenotypic advantage. [3] Based on a cost of 2 ATP consumed to reassimilate acetate, 1.71 ATP can be extracted if the reassimilation cost is reduced to 1 ATP.

3.1.2. Acrylate Pathway

The acrylate pathway enables ATP neutral conversion of lactate to propionate with the consumption of NADH. The pathway is found in several distantly related bacteria including *Clostridium propionicum*, *Megasphaera elsdenii* and *Prevotella ruminicola*. Though a variety of substrates can be catabolised to propionate and acetate, including lactate, serine, alanine and ethanol, the fermentation of glucose does not appear to result in the production of propionate in any native producer, presumably because glucose fermentation does not trigger expression of the lactate racemase required to initiate the cycle [43].

The metabolic advantage of the acrylate cycle is not immediately clear since a pyruvate formate lyase (PFL) is present in *C. propionicum*. The combination of these metabolic functionalities enables the energetically equivalent production of either propionate or ethanol with acetate in a 1:1 molar ratio. However, it does enable the consumption of acrylate to propionate [44]. Propionate production could be growth coupled to glucose catabolism if strains were engineered to utilise this pathway. In the presence of glucose, net yields of 3 ATP/glucose and about 0.4 g propionate/g glucose (1 mol propionate/mol glucose) could be achieved, while higher yields can be achieved through the use of the pentose phosphate pathway with a concomitant energetic penalty (Table 5, Table 6). While this pathway has been successfully engineered into *E. coli* [45], initial yields were 2 orders of magnitude lower than the 0.4 g/g that would be expected.

Table 5. Maximum propionate yield analysis of all metabolic pathways leading to propionate production.

Products	Catabolic Pathways			Biosynthetic Pathways		Fermentation Routes				Overall [1]
	Val/Iso	Thr	Met	Citramalate	3HP/4HB	Propanediol	Acrylate	Na⁺ Pumping	Wood-Werkman	
ATP	2.29	0	1	2.4	0	3.43	1.71	2.57	3.43	3.43
Propionate	0.29	1.33	1	0.4	1.33	1.71	1.71	1.71	1.71	1.71
Acetate	0	0.67	1	0	0.67	0	0	0	0	0
Ethanol	1.43	0	0	1.2	0	0	0	0	0	0
Formate	1.43	0	1	2	0	0	0	0	0	0
CO_2	0.86	0.67	0	0.4	0.67	0.86	0.86	0.86	0.86	0.86

Pathway Yields (mol/mol Glc)

[1] To test whether synergistic interactions could occur between pathways to improve propionate production, all glucose catabolising pathways leading to propionate production were allowed to carry flux, although performance did not improve over the Wood-Werkman cycle.

Table 6. Maximum energy yield analysis of all metabolic pathways leading to propionate production.

Products	Catabolic Pathways			Biosynthetic Pathways		Fermentation Routes				Overall [1]
	Val/Iso	Thr	Met	Citramalate	3HP/4HB	Propanediol	Acrylate	Na⁺ Pumping	Wood-Werkman	
ATP	3	3	3	3	3	4	3	3.25 [2]	4	4
Propionate	0	0	0	0	0	1.33	1	1 [2]	1.33	1.33
Acetate	1	1	1	0	1	0.67	1	1 [2]	0.67	0.67
Ethanol	1	1	1	1	1	0	0	0	0	0
Formate	2	2	2	2	2	0	1	1 [2]	0	0
CO_2	0	0	0	0	0	0	0.67	0 [2]	0	0.67

Pathway Yields (mol/mol Glc)

[1] To test whether synergistic interactions could occur between pathways to improve energy production, all glucose catabolising pathways leading to propionate production were allowed to carry flux, although performance did not improve over the Wood-Werkman cycle. [2] In the absence of a pyruvate formate lyase, propionate is expected to be produced in a 2:1 ratio with acetate giving a yield of 1.33 propionate, 0.67 acetate and CO_2 and 3 ATP per glucose.

3.1.3. Succinate Pathway

The catabolism of pyruvate to succinate via the dicarboxylic branch of the TCA cycle offers an alternative electron sink to ethanol. While an ATP is typically consumed or lost to fix carbon dioxide and pyruvate or phosphoenolpyruvate into oxaloacetate, this is at least partially compensated by an anaerobic electron transport chain consisting of the NADH dehydrogenase and fumarate reductase. In the absence of the PFL, the cell suffers an energetic penalty; less energy is gained through substrate level phosphorylation via production of succinate and acetate in a 2:1 ratio as compared to the dissimilation of glucose to acetate and ethanol in a 1:1 ratio. Some organisms have evolved energy conservation strategies that allow this fermentation strategy to equal or even better the energy available via the mixed acid fermentation strategy by the further decarboxylation of succinate to propionate. Two separate mechanisms have evolved to facilitate this; the sodium pumping methylmalonyl-CoA decarboxylase (Figure 1(AIV)) and the methylmalonyl-CoA:pyruvate transcarboxylase (Figure 1(AIII)).

The sodium pumping pathway, found in organisms such as *Propionigenium modestum*, couples the decarboxylation of methylmalonyl-CoA derived from succinate to propionyl-CoA with the pumping of two sodium ions across the cell membrane [46]. The mechanism of this reaction is likely to be identical to the well-studied oxaloacetate decarboxylase [47] and is probably linked to the consumption of a periplasmic proton [48], leading to a net energy gain of roughly 0.25 ATP.

While a modest conservation of energy, the 2:1 production of propionate to acetate becomes energetically equivalent to the mixed acid fermentation mode yielding 3 ATP/glucose; although this pathway is typically associated with *Veilonella* and *Parvula* species which do not catabolise sugars. Further energy can be extracted in the presence of the PFL enabling the production of 3.25 ATP/glucose and 1:1 production of propionate and acetate.

The second pathway, the Wood-Werkman cycle, is found predominantly in *Propionibacterium* and produces propionate in a similar way to the sodium pumping pathway, except the decarboxylation step is replaced by the methylmalonyl-CoA:pyruvate transcarboxylase which transfers a carboxyl group from methylmalonyl-CoA to pyruvate to generate propionyl-CoA in an ATP-independent manner (Figure 1(AIII)).

As opposed to the sodium pumping pathway, this bypasses the loss of ATP required to fix carbon dioxide to oxaloacetate and therefore conserves an entire ATP. This additional energy growth couples the 2:1 production of propionate and acetate, regardless of the presence of the PFL, while enabling the generation of 4 ATP per glucose.

While both of these pathways can improve energy yields from catabolising glucose and favour propionate production, the Wood-Werkman cycle is far superior to any other pathway in terms of its ability to promote propionate production corresponding to the metabolic objective of energy maximisation (Table 5, Table 6). Only the hypothetical lactate pathway for PDO production combined with PDO catabolism could equal the Wood-Werkman cycle in terms of energetic efficiency. Given propionibacteria both contain the Wood-Werkman cycle and naturally ferment sugars to achieve high propionate yields; it is of little surprise that they have been the focus of the bulk of the effort to design and scale a biological process for the industrial-scale synthesis of propionate.

3.2. Degradation of Amino Acids to Produce Propionate

The degradation of valine, threonine, isoleucine and methionine (Figure 1(BI-III)) can lead to the production of propionate and ATP via propionyl-CoA. As such, fermentations using complex media can result in the production of propionate in many organisms. However, the low market value of propionate compared to amino acids is restrictive. Alternatively, since pathways for the synthesis and subsequent catabolism of amino acids are present in a broad range of microorganisms, the combination of amino acid anabolic and catabolic pathways can be used to produce propionate from glucose. For example, Table 7 shows the metabolic costs of the synthesis, degradation, and combination of the two pathways for the production of 1 mole of amino acid from pyruvate and subsequent degradation to 1 mole of propionate. The maximum theoretical yields for propionic acid production from amino acid fermentations are shown in Table 8. In all cases, the synthesis and subsequent degradation of amino acids result in a net reduction in the energy yield from glucose. Therefore, these pathways are inconsistent with the metabolic objective of maximizing energy for growth and, consequently, rely on the over-expression of each enzyme.

Table 7. Theoretical molar yields [1] for propionic acid production using via amino acids synthesis and degradation.

Amino Acid	Valine				Isoleucine				Threonine				Methionine			
Substrate	Pyr	ATP	NADH	NADPH	Pyr	ATP	NADH	NADPH	Pyr	ATP	NADH	NADPH	Pyr	ATP	NADH	NADPH
Degradation	0	1	5	0	0	2	3	1	0	1	0	1(0)	0	1	0	1(0)
Biosynthesis	−2	0	−1	−1	−3	0	1	−2(−1)	−1	−3	−1	−3	−2	−2	−1	−2(−1)
Combined	−2	1	4	−1	−2	1	4	−1(0)	−1	−2	−1	−1(−2)	−2	−1	−1	−1

[1] Calculations were performed by lumping ferredoxin and $FADH_2$ with NADPH and NADH, respectively. Values in parentheses indicate when formate is produced through the pyruvate formate lyase. These calculations consider the costs of regenerating all substrates of the metabolic pathways.

Table 8. Estimated yields for propionic acid production via degradation of amino acids.

Amino Acid Pathway	Maximum Yields (mol/mol Glc)		Expected Yields [1] (mol/mol Glc)	
	PA	ATP	PA	ATP
Valine/Isoleucine	0.29	2.29	0	3
Threonine	1.33	0	0	3
Methionine	1 [2]	1 [2]	0	3
All	1.45	0	0	3

[1] Expected yields assume cells operate to maximise energy. [2] The yield for propionate production through the methionine pathway allows a stoichiometric yield of 1.125 propionate and 0 ATP per glucose to be obtained if acetate produced by methionine degradation is allowed to be re-consumed to pyruvate. If the energetic cost of this step can be reduced to 1 ATP, the pathway will perform similarly to the threonine pathway.

The high redox generation associated with propionate production via branched chain amino acids ultimately limits the yield by requiring the generation of an oxidized product such as ethanol. In these cases, targeting propanol instead can improve the maximum theoretical yields by 33%. While both methionine and threonine have a net consumption of reduced cofactors which can be balanced through the use of the pentose phosphate pathway, the threonine pathway is both more energetically efficient and much shorter, making it the most feasible pathway for propionate production. Threonine production and catabolism have been explored extensively in *E. coli* as a possible source of propionyl-CoA for the production of various chemicals including propanol [49,50], erythromycin (6-DB) [51] and 3-hydroxyvalerate [52]. *E. coli* has been engineered to co-produce propanol and butanol previously using just threonine synthesis and catabolism with some success. The final concentration reached was 1 g/L of propanol (apparent yield of 0.09 g/g glucose, 29% of the maximum theoretical yield for the citramalate pathway) in a semi-defined medium [53]. Also, the combination of the valine/isoleucine and threonine pathways are synergistic given the redox and energy balances and can, together, slightly improve the propionate yield.

3.3. Biosynthetic Routes via Propionyl-CoA

Three pathways associated with anabolic metabolism that lead to the synthesis of propionyl-CoA have been explored: the citramalate pathway (Figure 1(CI)) associated with isoleucine biosynthesis, the 3-hydroxypropanoate (3HP) cycle and the 3-hydroxypropanoate/4-hydroxybutanoate (4HB) cycles related to carbon fixation (Figure 1(CII)).

3.3.1. Citramalate Pathway

Many organisms contain the citramalate pathway, which condenses pyruvate and acetyl-CoA to generate 2-oxobutanoate. This compound is a precursor for isoleucine biosynthesis and is also an intermediate of methionine and threonine degradation pathways to propionate. The direct synthesis of 2-oxobutanoate from pyruvate results in the net production of just one reduced cofactor when the PFL is used to generate acetyl-CoA. Overall this pathway can yield 0.4 mol propionate per mol glucose and 2.4 ATP, which is again less than the 3 ATP that would otherwise be extracted from glucose through central carbon metabolism. The potential for the citramalate pathway to supply propionyl-CoA for the production of its derivatives has been analyzed for three separate products. When propanol was targeted, a yield of 0.11 g propanol/g glucose (~0.28 mol/mol) has been achieved [31], about half of the maximum theoretical yield for the citramalate pathway. Additionally, it has been utilised for the production of erythromycin [51] and determined to be the most significant source of propionyl-CoA in the native production of poly(3-hydroxybutyrate-*co*-3-hydroxyvalerate) in *Haloferax mediterranei* [54]. The synergistic interaction of the threonine and citramalate pathways improves the maximal propionate yield to 1.6 mol propionate per mol glucose without ATP production; this interaction was observed experimentally when combining these pathways further improved propanol yield to 0.15 g/g glucose in a semi-defined medium [53].

3.3.2. 3HP/4HB Cycles

The 3HP and 4HB cycles are particularly attractive pathways for the production of propionate due to their capacity to fix carbon dioxide as a sole carbon source. These pathways have been identified in the phototrophic *Chloroflexacae* and Archaea, respectively. Both cycles enable carbon fixation for biomass generation through an acetyl-CoA/propionyl-CoA carboxylase [55] and only differ in the final steps; the 3HP cycle fixes carbon dioxide to glyoxylate whereas the 4HB cycle generates acetyl-CoA. Because the shared initial steps of the pathways enable acetyl-CoA to be converted to propionyl-CoA, it could be possible to source propionate from acetyl-CoA resulting from glucose dissimilation (Figure 1(CII)). Indeed, the 3HP cycle has been reconstituted successfully in *E. coli* in sub-pathways including that for the production of propionyl-CoA [56], but to date there are no reports in the literature attempting to utilise this pathway for the production of propionic acid.

A key limitation of this strategy is the high ATP requirements of these reactions; consuming a net two ATP and two NADPH per acetyl-CoA consumed, resulting in a maximum theoretical yield of 1.33 mol propionate/mol glucose with no ATP generation (Table 5).

Although energy demanding and highly reducing, these cycles may offer new opportunities to explore the production of propionate from carbon dioxide as a substrate. The 3HP pathway is likely the second-greatest contributor to propionyl-CoA production for poly(3-hydroxybutyrate-*co*-3-hydroxyvalerate) in *Haloferax mediterranei* [54]. The production of propionate through the 4HB cycle is possible with one and a half turns of the cycle; leading to the overall consumption of 8 ATP and 7 reducing equivalents (8 NADPH consumed, 1 NADH generated). Natively, the 4HB pathway operates in aerobic conditions to supply the high ATP requirements and depends on hydrogen gas to feed the reducing equivalents [55]. The 3HP cycle is slightly more energetically demanding; leading to a net consumption of 9 ATP and 7 reducing equivalents (8 NADPH consumed and one quinol produced) in the production of propionate. Despite this, the properties of organisms that contain this pathway are even more unusual, such as the phototrophic *Chloroflexus aurantiacus* that performs anoxygenic photosynthesis with hydrogen or hydrogen sulfide as electron donors [57].

4. Genetic Engineering to Overcome the Current Challenges for Propionate Production

Targeted genetic engineering of propionibacteria remains challenging for several reasons: propionibacteria have a high GC content, which complicates genetic manipulation and contributes to poor gene annotations [58]; relatively few closed genomes are available [17,18]; a small number of cloning vectors are available [22,59,60]; the ability of strains to readily develop spontaneous antibiotic resistance; thick cell walls; and the presence of strong restriction modification systems which contribute to the low transformation efficiency of propionibacteria [61]. A number of recent studies have reported the modification of *P. freudenreichii* subsp. *shermanii* (*P. shermanii*) and *P. jensenii* while only a couple of contentious studies have reported modification of the high-producing, genetic modification resistant strain, *P. acidipropionici* (Table 9). The expression of methylation components of restriction modification systems in host organisms has resulted in large improvements in the transformation efficiencies of non-model organisms [62,63] and may be a critical step to improving the transformation efficiency in propionibacteria, particularly *P. acidipropionici*. Despite promising progress in the rational design of *P. shermanii* and *P. jensenii*, these modified strains still fall short of the natively high-producing *P. acidipropionici*. Therefore, a second line of research has focussed on random-mutagenesis strategies to enhance propionic acid production in *P. acidipropionici*.

Table 9. Genetic engineering strategies performed in propionibacteria to improve propionic acid production [1].

Aim	Strategy	Strain	Results	Reference
Decrease by-products	Genome editing	*P. acidipropionici* ACK-Tet strain	Acetate production reduced ~14%. ~13% improvement of propionate production.	[64]
	Genome editing, overexpression	*P. jensenii poxB* or *ldh* knock-out and *ppc* overexpression	Maximum 30% improvement in titre and 24% improvement productivity	[65]
Improve acid tolerance	Overexpression	*P. acidipropionici otsA* overexpression strain	Propionic acid yield 11% higher.	[66]
	Overexpression	*P. jensenii* strains overexpressing *gadB, arcA, arc, gdh* or *ybaS*	Up to a 1.5-fold increase in yield and 5.4-fold increase in titre, in shake flasks	[67]
Increase of metabolic flux towards propionate production	Overexpression	*P. shermanii CoAT* overexpression strain	Increase yield and productivity, maximum 10% and 46%, respectively.	[68]
	Overexpression of heterologous enzymes from *P. acidipropionici*	*P. shermanii* overexpressing *mmc, pyc* or *mmd*	Strongest phenotype observed with *mmc* overexpressing strain with 14% increase in yield from glucose and 17% increase in productivity from glucose/glycerol co-fermentation. Performed in serum bottles.	[69]
	Overexpression of heterologous enzymes from *E. coli*	*P. shermanii* overexpressing *ppc* strain	Improved productivity on glycerol only, no improvement in yield.	[70]
	Overexpression of heterologous enzymes from *E. coli* and *Klebsiella pneumoniae*	*P. jensenii* co-expression of *gdh* and *mdh*	Increase in propionate synthesis, but slow growth of the mutant strain.	[71]

[1] Results correspond to experiments performed in reactors unless otherwise stated.

4.1. Empirical Strain Design

Non-rational engineering approaches, such as random mutagenesis and genome shuffling have been extensively used to optimise *P. acidipropionici* strains [6,9,64,72–74]. Propionate production serves as an electron sink and generates energy through oxidative phosphorylation through the Wood-Werkman cycle, thus from a metabolic engineering perspective it is growth coupled. One option is relying on high-throughput HPLC to identify high producers [6]. Alternatively, growth rate may be used to identify mutants that are more acid tolerant and exhibit improved propionate yields [10,75,76].

Genome-shuffling has by far been the most successful empirical approach. Developed in the late 1970s, genome shuffling is routinely used to improve strains in industry. Compared to classical strain improvement methods such as chemical or UV mutagenesis, genome shuffling accelerates the evolutionary process by using multiple genotypes to provide an initial pool of genetic diversity, which can be refined for genomes that display desirable and diverse phenotypes. Genome shuffling combines the advantages of multi-parental crossing facilitated by DNA exchange, thus allowing the incorporation of foreign DNA [77]. Recursive genomic recombination combines classical breeding (asexual recursive mutagenesis), DNA shuffling (sexual recursive recombination) and screening for the desired phenotype and provides a feasible strategy to improve strains.

Genome shuffling in propionibacteria has been used to improve vitamin B12 production in *P. shermanii* [78] and propionic acid in *P. acidipropionici* [71,75,76]. Recently, multiple propionibacteria species were genome shuffled resulting in a strain which achieved a yield of 0.55 g propionate/g glucose [75,76]. Next-generation sequencing was used to analyse recombination events and identify novel/unique regions from each strain leading to the improved phenotype, including changes linked to acid tolerance mechanisms and possibly to a new transcriptional mechanism through mutation in ribosomal RNAs. Further rounds of genome-shuffling produced a strain that exceeded the 0.6 g propionate/g glucose and 1 g/L/h yield and productivity targets for an economically viable fermentation [10,76].

4.2. Rational Strain Engineering

Research into the rational design of propionibacteria metabolism to enhance propionic acid production remains in its infancy. Necessarily, the strategies implemented to date have been relatively simple; in fact, the first study to co-express two genes in propionibacteria was published in 2015 [79].

Recently, genome-scale models for a number of closed *Propionibacterium* genomes [58] were developed and these were used to design rational engineering strategies in *P. shermanii* [34]. Two strategies were explored to enhance propionate production. The first relied on increasing the availability of reduced cofactors by overexpressing the pentose phosphate pathway to favour propionate production, where a 4-fold improvement in the propionate to acetate ratio was observed at the end of the exponential growth phase. The second strategy introduced an alternative, high-energy linear pathway for propionate production that includes the phosphoenolpyruvate carboxykinase and the sodium pumping methylmalonyl-CoA decarboxylase [34]. This work demonstrated the power of genome-scale models to rationally design propionibacteria metabolism. However, further development of genetic engineering tools is required before more complicated strategies such as those proposed in our work [34] can be tested experimentally or in the higher-producing *P. acidipropionici* strains.

Though modest in scale, genetic engineering of propionibacteria have covered gene knockouts, gene overexpression as well as the expression of heterologous genes (Table 9).

4.3. Gene Knockouts

The first report on *Propionibacterium* genetic engineering aimed to construct a strain unable to produce acetate by knockout of the acetate kinase gene (*ack*) in *P. acidipropionici*. [64]. The *ack* gene interrupted by a tetracycline resistance cassette was electroporated into the wild-type strain of *P. acidipropionici* and ACK-Tet mutant strain was obtained by homologous recombination. While this work set the basis for molecular engineering, the subsequent release of the genome sequence [18] showed that the *ack* gene was not present in the genome. After careful review of the publication, it is evident that no validation of the knockout was performed, results were not statistically significant and acetate production was only reduced ~14% in the mutant strain compared to the wild-type strain.

Subsequently, homologous recombination was performed for gene inactivation in *P. acidipropionici*, targeting trehalose 6-phosphate synthase (*otsA*) and maltooligosyl trehalose synthase (*treY*) [66]. Single knockout strains ΔotsA and ΔtreY, and a double knockout strain ΔotsAΔtreY were constructed. The authors concluded that the OtsA-OtsB pathway is the major route for trehalose synthesis under acid stress in *P. acidipropionici* and overexpressed this pathway to improve propionate production; however, no significant change in the production of propionate or other fermentation end products was observed. Again, no appropriate validation of these knockouts was performed and results should be considered dubious with apparent yields of intracellular trehalose reported with no consideration for extracellular trehalose titre, which we estimate accounts for ~90% of the total trehalose (unpublished data). It has been claimed that Wild type *P. acidipropionici* can produce up to 27 g/L of extracellular trehalose [80] leading one to believe that the relatively small differences in intracellular yield observed by [66] could be insignificant. Furthermore, when trying to use the same strategy in our laboratory we failed to obtain mutant strains and learnt that introduction of a suicide vector to genetically engineer *P. acidipropionici* by homologous recombination, either by electroporation or conjugation with *E. coli*, is not a feasible approach. Our attempts resulted in a great number of false positives that could have led to erroneous results if not correctly validated. In our hands, the introduction of a suicide vector by electroporation or conjugation with *E.coli* has repeatedly failed to inactivate any gene in *P. acidipropionici*.

Acetate and lactate by-products were targeted in a gene knockout study in *P. jensenii* [65]. In this work, knock-out of the lactate dehydrogenase improved the final propionate titre slightly in fed-batch cultures but improvements in the productivity could not be observed. Deletion of the pyruvate oxidase severely reduced acetate production along with biomass generation; overall propionate production was negatively impacted with a much lower titre and productivity. The authors additionally overexpressed the phosphoenolpyruvate carboxylase from *Klebsiella pneumoniae* alone or in combination with the lactate dehydrogenase deletion. While a clear benefit in terms of final propionate titre and productivity was observed, there was no obvious benefit from the lactate dehydrogenase deletion.

A study showing clearly beneficial results from a gene knock-out in propionibacteria is yet to be achieved. It is difficult to extrapolate the potential implications of the failed acetate knock-out on glycerol [65] to the ability to use this strategy to overcome the yield challenges facing sugar fermentations. The over-expression of pathways that provide reduced cofactors, such as the pentose phosphate pathway, may be a more promising alternative as shown in our recent work [79], particularly since it may be impossible to eliminate acetate production by gene knock-out due to the promiscuity of the propionyl-CoA: succinate CoA-transferase (CoAT).

4.4. Gene Overexpression

Overexpression of enzymes as a way to increase propionate production has been tested in *P. shermanii* and *P. jensenii*. For example, a recent study upregulated the last step in propionic acid biosynthesis, the CoAT [68]. The native *coAT* gene was cloned into a replicative vector [60] under the control of a strong native promoter and used to transform *P. shermanii* by electroporation [15]. Even though the reaction catalysed by CoAT has been proposed to be the rate limiting step in propionic acid synthesis [73], the overexpression of this enzyme only slightly improved propionate production. The mutant strain showed improved propionic acid yield and productivity compared to the wild type strain, but the effects varied with fermentation conditions. Maximum yield and productivity were obtained during co-fermentation with glycerol and glucose as carbon sources (4:1 ratio). The strain also showed a marked decrease in the by-products succinate and acetate under this condition.

Another approach explored was the overexpression of three carboxylases, namely, pyruvate carboxylase (PYC), methylmalonyl-CoA carboxyltransferase (MMC) and methylmalonyl-CoA decarboxylase (MMD), from the dicarboxylic acid pathway controlling the carbon flux in the Wood-Werkman cycle in *P. shermanii* [69]. The effects of the overexpression on propionic acid fermentation were studied in serum bottles with glucose, glycerol or co-fermentation of both as substrates. Overall, only mutants overexpressing MMC and MMD showed increased propionic acid production relative to the wild-type strain. Maximal improvements were obtained with the MMC overexpression with a 14% improvement in propionate yield, obtained on glucose, and 17% increase in productivity, obtained on a 2:1 glycerol/glucose mix.

A novel approach for improving propionate production was explored by Guan et al. by over-expressing acid resistance mechanisms in *P. jensenii* [67]. Compared to the strain harbouring an empty plasmid, over-expression of components of the arginine deiminase and glutamate decarboxylase systems typically improved the yield of propionate and the absolute titre reached in a glycerol media in shake-flasks. This study demonstrates the potential importance of incorporating acid tolerance mechanisms in a rational strain design approach.

The introduction of heterologous enzymes has also been used to improve propionate biosynthesis [70,71]. Phosphoenolpyruvate carboxylase (PPC) from *E. coli* catalyses the conversion of phosphoenolpyruvate to oxaloacetate (OAA) by fixing CO_2. Not present in any known propionibacteria, expression of PPC might increase carbon flux towards OAA and thus propionate in *P. freudenrenchii*. Although the mutant showed better cell growth and propionate productivity than the wild type strain, propionate yield was not markedly improved under the conditions tested [70]. On glycerol, glycerol dehydrogenase (GDH) and malate dehydrogenase (MDH) are proposed to be rate-limiting steps in propionic acid synthesis from glycerol [71]. While the co-expression of GDH and MDH from *Klebsiella pneumoniae* in *P. jensenii* showed an increase in propionate synthesis, the mutant suffered from slow growth probably due to the extra burden of plasmid replication, transcription and expression [71]. While the GDH/MDH overexpression pair performed the best, the same study also experimented with fumarate hydratase overexpression in combination with GDH or MDH.

4.5. Propionic Acid Biosynthesis by Non-Native Producers

Genetic engineering limitations in propionibacteria have driven researchers to study *E. coli* as a platform for the heterologous production of propionic acid.

The so-called succinate pathway [81], encoded by the *sbm* operon in *E. coli*'s genome, is the most thoroughly studied approach for propionate production in *E. coli*. The succinate dissimilation pathway is similar to the methylmalonyl-CoA decarboxylase configuration previously described (Figure 1(AIV)) (see Section 3.1.3). However, this operon lacks a methylmalonyl-CoA epimerase [73] and should not be functional. Nevertheless, several groups have reported propionate production from overexpression of the operon in complex media [82–84]. We suspect that the reported propionate production is a result a ghost-peak interfering with propionate quantification [45].

The acrylate pathway from *Clostridium propionicum* has also been expressed in *E. coli* [45]. This system only produced up to 3.7 mM propionic acid under anaerobic conditions in complex media supplemented with glucose (20 g/L). The authors attributed their results to low enzymatic activity and a possible down-regulation of the *pfl* gene caused by the intermediates of the exogenous pathway. However, because *E. coli* contains the threonine degradation pathway (Figure 1(BI)), the low levels of propionate produced may result from the use of complex media.

Propionate yields and titres for heterologous expression are too low to meaningfully compare with propionibacteria. Further studies are needed to establish if substantial production is feasible and if *E. coli* can be engineered to tolerate propionate.

5. Concluding Remarks and Future Directions

Recent techno-economic analyses have set yield and productivity goals for the microbial production of propionate. Several challenges still need to be addressed for an economically feasible fermentation process. *Propionibacterium* remains the best candidate organism for the biological production of propionate at an industrial-scale. This is of little surprise given this species contains novel enzymes encoding the Wood-Werkman cycle that allow the production of propionate at the maximum theoretical yield with a high energy efficiency. While we identified many pathways that can achieve a similar maximum theoretical yield, none of them can compete with the energy yield obtained by the Wood-Werkman cycle.

Numerous studies have focussed on optimizing the fermentation process in order to improve propionate production, but often these focus on fermenters that have not yet been used on an industrial scale. Attention should be placed on identifying cheap, renewable feedstocks that provide high apparent propionate yields, such as enzymatically treated corn mash, or genetically modifying *P. acidipropionici*. While there have been many successful studies in native producers with lower propionate production such as *P. shermanii* and *P. jensenii*, these are still uncompetitive with wild-type *P. acidipropionici* strains. Rather, these organisms should be viewed as model propionibacteria and used to guide development of strategies ultimately to be implemented in *P. acidipropionici* when the tools become available. To this end, overcoming the transformation barrier by considering restriction modification systems may be a way forward. In the meantime, empirical strain design approaches such as genome-shuffling have proven to be the most successful approach to enhancing the performance of *P. acidipropionici* with new strains emerging that exceed the requirements of an economically viable fermentation.

While the heterologous production of propionate in model organisms remains an interesting possibility, yields remain low and discrepancies within the field, particularly with respect to propionate quantification, need to be resolved for further improvement. Nevertheless, the pathway analysis presented here revealed interesting alternative organisms with a potential for the production of propionate via the fixation of carbon dioxide through the 3HP and 4HB cycles, and via propanediol, a valuable commodity chemical, through the putative lactate pathway.

Acknowledgments: This work was supported by a grant from the Australian Research Council, partly funded by Dow (ARC LP120100517). Esteban Marcellin acknowledges funding from the Queensland Government for his accelerate fellowship. R. Axayacatl Gonzalez-Garcia acknowledges UQ and CONACYT (the Mexican council for science and technology for his PhD scholarship. Tim McCubbin was supported by an APA scholarship. We would like to acknowledge the contribution made by Clarence Jean at the early stages of the manuscript, particularly for his contribution to Table 7.

Author Contributions: Conceptualization, R. Axayacatl Gonzalez-Garcia, Tim McCubbin, Laura Navone; Chris Stowers, Lars K. Nielsen and Esteban Marcellin; Formal Analysis, R. Axayacatl Gonzalez-Garcia, Tim McCubbin, Laura Navone Investigation R. Axayacatl Gonzalez-Garcia, Tim McCubbin, Laura Navone, Lars K. Nielsen and Esteban Marcellin; Resources, Lars K. Nielsen and Esteban Marcellin; Writing. R. Axayacatl Gonzalez-Garcia,

Tim McCubbin, Laura Navone; Chris Stowers, Lars K. Nielsen and Esteban Marcellin; Supervision, Lars K. Nielsen and Esteban Marcellin; Project Administration, Lars K. Nielsen and Esteban Marcellin; Funding Acquisition, Chris Stowers, Lars K. Nielsen and Esteban Marcellin.

Conflicts of Interest: R. Axayacatl Gonzalez-Garcia, Tim McCubbin, Laura Navone, Esteban Marcellin, Lars K. Nielsen declare no conflict of interest. The Dow Chemical Company is a commercial propionic acid producer. Chris Stowers is an employee of Dow Agrosciences LLC.

References

1. Zidwick, M.J.; Chen, J.S.; Rogers, P. Organic acid and solvent production: Propionic and butyric acids and ethanol. In *The Prokaryotes*; Springer: Berlin/Heidelberg, Germany, 2013; pp. 135–167.
2. Wemmenhove, E.; van Valenberg, H.J.; Zwietering, M.H.; van Hooijdonk, T.C.; Wells-Bennik, M.H. Minimal inhibitory concentrations of undissociated lactic, acetic, citric and propionic acid for *Listeria monocytogenes* under conditions relevant to cheese. *Food Microbiol.* **2016**, *58*, 63–67. [CrossRef] [PubMed]
3. Álvarez-Chávez, C.R.; Edwards, S.; Moure-Eraso, R.; Geiser, K. Sustainability of bio-based plastics: General comparative analysis and recommendations for improvement. *J. Clean Prod.* **2012**, *23*, 47–56. [CrossRef]
4. Hebert, R.F.; Hebert Sam-E, L. Stable Indole-3-Propionate Salts of S-adenosyl-L-Methionine. U.S. Patent 9534010, 3 January 2017. Available online: https://www.google.com/patents/US9534010 (accessed on 25 February 2017).
5. Market Research Store. Propionic Acid Market for Animal Feed & Grain Preservatives, Calcium & Sodium Propionates, Cellulose Acetate Propionate and Other Applications: Global Industry Perspective, Comprehensive Analysis, Size, Share, Growth, Segment, Trends and Forecast, 2014–2020. Available online: http://www.marketresearchstore.com/report/propionic-acid-market-for-animal-feed-grain-z39993 (accessed on 25 February 2017).
6. Stowers, C.C.; Cox, B.M.; Rodriguez, B.A. Development of an industrializable fermentation process for propionic acid production. *J. Ind. Microbiol. Biotechnol.* **2014**, *41*, 837–852. [CrossRef] [PubMed]
7. Fitz, A. über spaltpilzgärungen, IV Bericht der Deutsch. *Chem. Ges.* **1878**, *11*, 1890. [CrossRef]
8. Swick, R.W.; Wood, H.G. The role of transcarboxylation in propionic acid fermentation. *Proc. Natl. Acad. Sci. USA* **1960**, *46*, 28–41. [CrossRef] [PubMed]
9. Rodriguez, B.A.; Stowers, C.C.; Pham, V.; Cox, B.M. The production of propionic acid, propanol and propylene via sugar fermentation: An industrial perspective on the progress, technical challenges and future outlook. *Green Chem.* **2014**, *16*, 1066–1076. [CrossRef]
10. Luna-Flores, C.H.; Cox, B.M.; Stowers, C.C.; Nielsen, L.K.; Marcellin, E. Isolated Propionibacterium strain used for producing propionic acid comprises a modified gene, relative to strain Propionibacterium acidipropionici selected from the group consisting of the ABC polar amino acid transporter gene. Patent WO2017055932-A2. Filed on September 30, 2015 (US234900P). Available online: http://apps.webofknowledge.com.ezproxy.library.uq.edu.au/full_record.do?product=DIIDW&search_mode=GeneralSearch&qid=1&SID=Q2PwF9TQkv7q9hDEN4a&page=1&doc=1&colname=DIIDW (accessed on 10 May 2017).
11. Liu, L.; Zhu, Y.; Li, J.; Wang, M.; Lee, P.; Du, G.; Chen, J. Microbial production of propionic acid from propionibacteria: Current state, challenges and perspectives. *Crit. Rev. Biotechnol.* **2012**, *32*, 374–381. [CrossRef] [PubMed]
12. Yang, S.T.; Zhu, H.; Li, Y.; Hong, G. Continuous propionate production from whey permeate using a novel fibrous bed bioreactor. *Biotechnol. Bioeng.* **1994**, *43*, 1124–1130. [CrossRef] [PubMed]
13. Barbirato, F.; Chedaille, D.; Bories, A. Propionic acid fermentation from glycerol: Comparison with conventional substrates. *Appl. Microbiol. Biotechnol.* **1997**, *47*, 441–446. [CrossRef]
14. Coral, J.; Karp, S.G.; de Souza Vandenberghe, L.P.; Parada, J.L.; Pandey, A.; Soccol, C.R. Batch fermentation model of propionic acid production by Propionibacterium acidipropionici in different carbon sources. *Appl. Biochem. Biotechnol.* **2008**, *151*, 333–341. [CrossRef] [PubMed]
15. Wang, Z.; Jin, Y.; Yang, S.T. High cell density propionic acid fermentation with an acid tolerant strain of Propionibacterium acidipropionici. *Biotechnol. Bioeng.* **2015**, *112*, 502–511. [CrossRef] [PubMed]
16. Eş, I.; Khaneghah, A.M.; Hashemi, S.M.; Koubaa, M. Current advances in biological production of propionic acid. *Biotechnol. Lett.* **2017**. [CrossRef] [PubMed]

17. Falentin, H.; Deutsch, S.M.; Jan, G.; Loux, V.; Thierry, A.; Parayre, S.; Maillard, M.B.; Dherbécourt, J.; Cousin, F.J.; Jardin, J.; et al. The complete genome of *Propionibacterium freudenreichii* CIRM-BIA1 T, a hardy Actinobacterium with food and probiotic applications. *PLoS ONE* **2010**, *235*, e11748.

18. Parizzi, L.P.; Grassi, M.C.; Llerena, L.A.; Carazzolle, M.F.; Queiroz, V.L.; Lunardi, I.; Zeidler, A.F.; Teixeira, P.J.; Mieczkowski, P.; Rincones, J.; et al. The genome sequence of *Propionibacterium acidipropionici* provides insights into its biotechnological and industrial potential. *BMC Genom.* **2012**, *13*, 562. [CrossRef] [PubMed]

19. Tufvesson, P.; Ekman, A.; Sardari, R.R.; Engdahl, K.; Tufvesson, L. Economic and environmental assessment of propionic acid production by fermentation using different renewable raw materials. *Bioresour. Technol.* **2013**, *149*, 556–564. [CrossRef] [PubMed]

20. Wang, Z.; Sun, J.; Zhang, A.; Yang, S.T. Propionic acid fermentation. In *Bioprocessing Technologies in Biorefinery for Sustainable Production of Fuels, Chemicals, and Polymers*; Wiley & Sons: New York, 2013; pp. 331–350.

21. Scott, K.P.; Martin, J.C.; Campbell, G.; Mayer, C.D.; Flint, H.J. Whole-genome transcription profiling reveals genes up-regulated by growth on fucose in the human gut bacterium "*Roseburia. inulinivorans*". *J. Bacteriol.* **2006**, *188*, 4340–4349. [CrossRef] [PubMed]

22. Zhuge, X.; Liu, L.; Shin, H.D.; Chen, R.R.; Li, J.; Du, G.; Chen, J. Development of a *Propionibacterium-Escherichia. coli* shuttle vector for metabolic engineering of *Propionibacterium jensenii*, an efficient producer of propionic acid. *Appl. Environ. Microbiol.* **2013**, *79*, 4595–4602. [CrossRef] [PubMed]

23. Feng, X.-H.; Chen, F.; Xu, H.; Wu, B.; Yao, J.; Ying, H.-J.; Ouyang, P.-K. Propionic acid fermentation by *Propionibacterium freudenreichii* CCTCC M207015 in a multi-point fibrous-bed bioreactor. *Bioprocess. Biosyst. Eng.* **2010**, *33*, 1077–1085. [CrossRef] [PubMed]

24. Liu, Y.; Zhang, Y.-G.; Zhang, R.-B.; Zhang, F.; Zhu, J. Glycerol/glucose co-fermentation: One more proficient process to produce propionic acid by *Propionibacterium acidipropionici*. *Curr. Microbiol.* **2011**, *62*, 152–158. [CrossRef] [PubMed]

25. Goswami, V.; Srivastava, A. Propionic acid production in an in situ cell retention bioreactor. *Appl. Micriobiol. Biotechnol.* **2001**, *56*, 676–680. [CrossRef]

26. Gu, Z.; Rickert, D.A.; Glatz, B.A.; Glatz, C.E. Feasibility of propionic acid production by extractive fermentation. *Le Lait* **1999**, *79*, 137–148. [CrossRef]

27. Zhu, Y.; Li, J.; Tan, M.; Liu, L.; Jiang, L.; Sun, J.; Lee, P.; Du, G.; Chen, J. Optimization and scale-up of propionic acid production by propionic acid-tolerant *Propionibacterium acidipropionici* with glycerol as the carbon source. *Bioresour. Technol.* **2010**, *101*, 8902–8906. [CrossRef] [PubMed]

28. Dishisha, T.; Ståhl, Å.; Lundmark, S.; Hatti-Kaul, R. An economical biorefinery process for propionic acid production from glycerol and potato juice using high cell density fermentation. *Bioresour. Technol.* **2013**, *135*, 504–512. [CrossRef] [PubMed]

29. Liu, Z.; Ma, C.; Gao, C.; Xu, P. Efficient utilization of hemicellulose hydrolysate for propionic acid production using *Propionibacterium acidipropionici*. *Bioresour. Technol.* **2012**, *114*, 711–714. [CrossRef] [PubMed]

30. Dishisha, T.; Alvarez, M.T.; Hatti-Kaul, R. Batch- and continuous propionic acid production from glycerol using free and immobilized cells of *Propionibacterium acidipropionici*. *Bioresour. Technol.* **2012**, *118*, 553–562. [CrossRef] [PubMed]

31. Zhu, L.; Wei, P.; Cai, J.; Zhu, X.; Wang, Z.; Huang, L.; Xu, Z. Improving the productivity of propionic acid with FBB-immobilized cells of an adapted acid-tolerant *Propionibacterium acidipropionici*. *Bioresour. Technol.* **2012**, *112*, 248–253. [CrossRef] [PubMed]

32. Wang, Z.; Yang, S.-T. Propionic acid production in glycerol/glucose co-fermentation by *Propionibacterium freudenreichii subsp shermanii*. *Bioresour. Technol.* **2013**, *137*, 116–123. [CrossRef] [PubMed]

33. Feng, X.; Chen, F.; Xu, H.; Wu, B.; Li, H.; Li, S.; Ouyang, P. Green and economical production of propionic acid by *Propionibacterium freudenreichii* CCTCC M207015 in plant fibrous-bed bioreactor. *Bioresour. Technol.* **2011**, *102*, 6141–6146. [CrossRef] [PubMed]

34. Navone, L.; McCubbin, T.; Gonzalez-Garcia, R.A.; Nielsen, L.K.; Marcellin, E. Genome-scale model guided design of *Propionibacterium* for enhanced propionic acid production. *Microb. Cell Fact.* **2017**, Manuscript submitted to Metabolic Engineering.

35. Ibarra, R.U.; Edwards, J.S.; Palsson, B.O. Escherichia coli K-12 undergoes adaptive evolution to achieve in silico predicted optimal growth. *Nature* **2002**, *420*, 186–189. [CrossRef] [PubMed]

36. Staib, L.; Fuchs, T.M. Regulation of fucose and 1, 2-propanediol utilization by *Salmonella enterica serovar Typhimurium*. *Front. Microbial.* **2006**. [CrossRef] [PubMed]

37. Saxena, R.K.; Anand, P.; Saran, S.; Isar, J.; Agarwal, L. Microbial production and applications of 1, 2-propanediol. *Indian J. Microbiol.* **2010**, *50*, 2–11. [CrossRef] [PubMed]
38. Cameron, D.C.; Cooney, C.L. A Novel Fermentation: The Production of R (–)–1, 2–Propanediol and Acetol by *Clostridium thermosaccharolyticum*. *Nat. Biotechnol.* **1986**, *4*, 651–654. [CrossRef]
39. Zhu, L.; Guan, X.; Xie, N.; Wang, L.; Yu, B.; Ma, Y. Fermentative production of enantiomerically pure S-1, 2-propanediol from glucose by engineered *E. coli* strain. *Appl. Microbiol. Biotechnol.* **2016**, *100*, 1241–1251. [CrossRef] [PubMed]
40. Elferink, S.J.O.; Krooneman, J.; Gottschal, J.C.; Spoelstra, S.F.; Faber, F.; Driehuis, F. Anaerobic conversion of lactic acid to acetic acid and 1, 2-propanediol by *Lactobacillus buchneri*. *Appl. Environ. Microbiol.* **2001**, *67*, 125–132. [CrossRef] [PubMed]
41. Niu, W.; Guo, J. Stereospecific microbial conversion of lactic acid into 1, 2-propanediol. *ACS Synth. Biol.* **2014**, *4*, 378–382. [CrossRef] [PubMed]
42. Altaras, N.E.; Cameron, D.C. Metabolic engineering of a 1, 2-propanediol pathway in *Escherichia coli*. *Appl. Environ. Microbiol.* **1999**, *65*, 1180–1185. [PubMed]
43. Hino, T.; Shimada, K.; Maruyama, T. Substrate preference in a strain of *Megasphaera. elsdenii*, a ruminal bacterium, and its implications in propionate production and growth competition. *Appl. Environ. Microbiol.* **1994**, *60*, 1827–1831. [PubMed]
44. Ladd, J.N.; Walker, D.J. The fermentation of lactate and acrylate by the rumen micro-organism LC. *Biochem. J.* **1959**, *71*, 364. [CrossRef] [PubMed]
45. Kandasamy, V.; Vaidyanathan, H.; Djurdjevic, I.; Jayamani, E.; Ramachandran, K.B.; Buckel, W.; Jayaraman, G.; Ramalingam, S. Engineering *Escherichia coli* with acrylate pathway genes for propionic acid synthesis and its impact on mixed-acid fermentation. *Appl. Environ. Microbiol.* **2013**, *97*, 1191–1200. [CrossRef] [PubMed]
46. Hilpert, W.; Dimroth, P. On the mechanism of sodium ion translocation by methylmalonyl-CoA decarboxylase from *Veillonella. alcalescens*. *Eur. J. Biochem.* **1991**, *195*, 79–86. [CrossRef] [PubMed]
47. Dimroth, P.; Jockel, P.; Schmid, M. Coupling mechanism of the oxaloacetate decarboxylase Na$^+$ pump. *BBA-Bioenergetics* **2001**, *1505*, 1–14. [CrossRef]
48. Di Berardino, M.; Dimroth, P. Aspartate 203 of the oxaloacetate decarboxylase β–subunit catalyses both the chemical and vectorial reaction of the Na$^+$ pump. *EMBO J.* **1996**, *15*, 1842. [PubMed]
49. Shen, C.R.; Liao, J.C. Metabolic engineering of *Escherichia coli* for 1-butanol and 1-propanol production via the keto-acid pathways. *Metab. Eng.* **2008**, *10*, 312–320. [CrossRef] [PubMed]
50. Choi, Y.J.; Park, J.H.; Kim, T.Y.; Lee, S.Y. Metabolic engineering of *Escherichia coli* for the production of 1-propanol. *Metab. Eng.* **2012**, *14*, 477–486. [CrossRef] [PubMed]
51. Jiang, M.; Pfeifer, B.A. Metabolic and pathway engineering to influence native and altered erythromycin production through *E. coli*. *Metab. Eng.* **2013**, *19*, 42–49. [CrossRef] [PubMed]
52. Tseng, H.C.; Harwell, C.L.; Martin, C.H.; Prather, K.L. Biosynthesis of chiral 3-hydroxyvalerate from single propionate-unrelated carbon sources in metabolically engineered *E. coli*. *Microb. Cell Fact.* **2010**, *9*, 1. [CrossRef] [PubMed]
53. Shen, C.R.; Liao, J.C. Synergy as design principle for metabolic engineering of 1-propanol production in *Escherichia coli*. *Metab. Eng.* **2013**, *17*, 12–22. [CrossRef] [PubMed]
54. Han, J.; Hou, J.; Zhang, F.; Ai, G.; Li, M.; Cai, S.; Liu, H.; Wang, L.; Wang, Z.; Zhang, S.; et al. Multiple propionyl coenzyme A-supplying pathways for production of the bioplastic poly (3-hydroxybutyrate-co-3-hydroxyvalerate) in *Haloferax. mediterranei*. *Appl. Environ. Microbiol.* **2013**, *79*, 2922–2931. [CrossRef] [PubMed]
55. Berg, I.A.; Kockelkorn, D.; Buckel, W.; Fuchs, G. A 3-hydroxypropionate/4-hydroxybutyrate autotrophic carbon dioxide assimilation pathway in Archaea. *Science* **2007**, *318*, 1782–1786. [CrossRef] [PubMed]
56. Mattozzi, M.D.; Ziesack, M.; Voges, M.J.; Silver, P.A.; Way, J.C. Expression of the sub-pathways of the *Chloroflexus. aurantiacus* 3-hydroxypropionate carbon fixation bicycle in *E. coli*: Toward horizontal transfer of autotrophic growth. *Metab. Eng.* **2013**, *16*, 130–139. [CrossRef] [PubMed]
57. Tang, K.H.; Barry, K.; Chertkov, O.; Dalin, E.; Han, C.S.; Hauser, L.J.; Honchak, B.M.; Karbach, L.E.; Land, M.L.; Lapidus, A.; et al. Complete genome sequence of the filamentous anoxygenic phototrophic bacterium *Chloroflexus. aurantiacus*. *BMC Genom.* **2011**, *12*, 1. [CrossRef] [PubMed]

58. McCubbin, T.; Palfreyman, R.W.; Stowers, C.; Nielsen, L.K.; Marcellin, E. A pan-genome guided *Propionibacterium* genome scale metabolic network reconstruction. *Microbiome* **2017**, Submitted.

59. Jore, J.P.M.; Van Luijk, N.; Luiten, R.G.M.; Van der Werf, M.J.; Pouwels, P.H. Efficient Transformation System for *Propionibacterium freudenreichii* Based on a Novel Vector. *Appl. Environ. Microbiol.* **2001**, *67*, 499–503. [CrossRef] [PubMed]

60. Kiatpapan, P.; Hashimoto, Y.; Nakamura, H.; Piao, Y.Z.; Ono, H.; Yamashita, M.; Murooka, Y. Characterization of pRGO1, a plasmid from *Propionibacterium acidipropionici*, and its use for development of a host-vector system in propionibacteria. *Appl. Environ. Microbiol.* **2000**, *66*, 4688–4695. [CrossRef] [PubMed]

61. Kiatpapan, P.; Murooka, Y. Genetic manipulation system in propionibacteria. *J. Biosci. Bioeng.* **2002**, *93*, 1–8. [CrossRef]

62. O'Connell Motherway, M.; O'Driscoll, J.; Fitzgerald, G.F.; Van Sinderen, D. Overcoming the restriction barrier to plasmid transformation and targeted mutagenesis in *Bifidobacterium. breve* UCC2003. *Microb. Biotechnol.* **2009**, *2*, 321–332. [CrossRef] [PubMed]

63. Zhang, G.; Wang, W.; Deng, A.; Sun, Z.; Zhang, Y.; Liang, Y.; Che, Y.; Wen, T. A mimicking-of-DNA-methylation-patterns pipeline for overcoming the restriction barrier of bacteria. *PLoS Genet.* **2012**, *8*, e1002987. [CrossRef] [PubMed]

64. Suwannakham, S.; Huang, Y.; Yang, S.T. Construction and characterization of *ack* knock-out mutants of *Propionibacterium acidipropionici* for enhanced propionic acid fermentation. *Biotechnol. Bioeng.* **2006**, *94*, 383–395. [CrossRef] [PubMed]

65. Liu, L.; Guan, N.; Zhu, G.; Li, J.; Shin, H.D.; Du, G.; Chen, J. Pathway engineering of Propionibacterium jensenii for improved production of propionic acid. *Sci. Rep.* **2016**, *6*, 19963. [CrossRef] [PubMed]

66. Jiang, L.; Cui, H.; Zhu, L.; Hu, Y.; Xu, X.; Li, S.; Huang, H. Enhanced propionic acid production from whey lactose with immobilized *Propionibacterium acidipropionici* and the role of trehalose synthesis in acid tolerance. *Green Chem.* **2015**, *17*, 250–259. [CrossRef]

67. Guan, N.; Li, J.; Shin, H.D.; Du, G.; Chen, J.; Liu, L. Metabolic engineering of acid resistance elements to improve acid resistance and propionic acid production of *Propionibacterium jensenii*. *Biotechnol. Bioeng.* **2015**, *113*, 1294–1304. [CrossRef] [PubMed]

68. Wang, Z.; Ammar, E.M.; Zhang, A.; Wang, L.; Lin, M.; Yang, S.-T. Engineering *Propionibacterium freudenreichii subsp. shermanii* for enhanced propionic acid fermentation: Effects of overexpressing propionyl-CoA: Succinate CoA transferase. *Metab. Eng.* **2015**, *27*, 46–56. [CrossRef] [PubMed]

69. Wang, Z.; Lin, M.; Wang, L.; Ammar, E.M.; Yang, S.-T. Metabolic engineering of *Propionibacterium freudenreichii subsp. shermanii* for enhanced propionic acid fermentation: Effects of overexpressing three biotin-dependent carboxylases. *Process. Biochem.* **2015**, *50*, 194–204. [CrossRef]

70. Ammar, E.M.; Jin, Y.; Wang, Z.; Yang, S.T. Metabolic engineering of *Propionibacterium freudenreichii*: Effect of expressing phosphoenolpyruvate carboxylase on propionic acid production. *Appl. Microbiol. Biotechnol.* **2014**, *98*, 7761–7772. [CrossRef] [PubMed]

71. Guan, N.; Liu, L.; Zhuge, X.; Xu, Q.; Li, J.; Du, C.; Chen, J. Genome-shuffling improves acid tolerance of *Propionibacterium acidipropionici*. In *Advances in Chemistry Research*; Taylor, J.C., Ed.; Nova: New York, NY, USA, 2012; Volume 15, pp. 143–152.

72. Guan, N.; Liu, L.; Shin, H.-D.; Chen, R.R.; Zhang, J.; Li, J.; Du, G.; Shi, Z.; Chen, J. Systems-level understanding of how *Propionibacterium acidipropionici* respond to propionic acid stress at the microenvironment levels: Mechanism and application. *J. Biotechnol.* **2013**, *167*, 56–63. [CrossRef] [PubMed]

73. Suwannakham, S.; Huang, Y.; Yang, S.T. Enhanced propionic acid fermentation by Propionibacterium acidipropionici mutant obtained by adaptation in a fibrous-bed bioreactor. *Biotechnol. Bioeng.* **2005**, *91*, 325–337. [CrossRef] [PubMed]

74. Zhang, A.; Yang, S.T. Propionic acid production from glycerol by metabolically engineered *Propionibacterium acidipropionici*. *Process. Biochem.* **2009**, *44*, 1346–1351. [CrossRef]

75. Luna-Flores, C.H.; Palfreyman, R.W.; Krömer, J.O.; Nielsen, L.K.; Marcellin, E. Improved production of propionic acid using genome shuffling. *Biotechnol. J.* **2016**, *12*, 1600120. [CrossRef] [PubMed]

76. Luna-Flores, C.H.; Stowers, C.C.; Cox, B.; Nielsen, L.K.; Marcellin, E. Scalable and economical process for propionic acid biosynthesis. *Biotechnol. Biofuels* **2017**, Submitted.

77. Biot-Pelletier, D.; Martin, V.J. Evolutionary engineering by genome shuffling. *Appl. Microbiol. Biotechnol.* **2014**. [CrossRef] [PubMed]

78. Zhang, Y.; Liu, J.-Z.; Huang, J.-S.; Mao, Z.W. Genome shuffling of *Propionibacterium shermanii* for improving vitamin B12 production and comparative proteome analysis. *J. Biotechnol.* **2010**, *148*, 139–143. [CrossRef] [PubMed]

79. Liu, L.; Zhuge, X.; Shin, H.D.; Chen, R.R.; Li, J.; Du, G.; Chen, J. Improved production of propionic acid via combinational overexpression of glycerol dehydrogenase and malate dehydrogenase from *Klebsiella. pneumoniae* in *Propionibacterium jensenii*. *Appl. Environ. Microbiol.* **2015**, *81*, 2256–2264. [CrossRef] [PubMed]

80. Stowers, C.; Dow AgroSciences LLC. Personal Comunication, Indianapolis, IN, USA, 2017.

81. Haller, T.; Buckel, T.; Rétey, J.; Gerlt, J.A. Discovering new enzymes and metabolic pathways: Conversion of succinate to propionate by *Escherichia coli*. *Biochemistry* **2000**, *39*, 4622–4629. [CrossRef] [PubMed]

82. Dellomonaco, C.; Rivera, C.; Campbell, P.; Gonzalez, R. Engineered respiro-fermentative metabolism for the production of biofuels and biochemicals from fatty acid-rich feedstocks. *Appl. Environ. Micriobiol.* **2010**, *76*, 5067–5078. [CrossRef] [PubMed]

83. Srirangan, K.; Akawi, L.; Liu, X.; Westbrook, A.; Blondeel, E.J.; Aucoin, M.G.; Moo-Young, M.; Chou, C.P. Manipulating the sleeping beauty mutase operon for the production of 1-propanol in engineered *Escherichia coli*. *Biotechnol. Biofuels* **2013**. [CrossRef] [PubMed]

84. Akawi, L.; Srirangan, K.; Liu, X.; Moo-Young, M.; Chou, C.P. Engineering *Escherichia coli* for high-level production of propionate. *J. Ind. Microbiol. Biotechnol.* **2015**, *42*, 1057–1072. [CrossRef] [PubMed]

fermentation

Review

Biochemical Production and Separation of Carboxylic Acids for Biorefinery Applications

Nanditha Murali [1,2], Keerthi Srinivas [1,2] and Birgitte K. Ahring [1,2,*]

1   Department of Biological Systems Engineering, Bioproducts, Sciences and Engineering Laboratory,
    Washington State University, Tri-Cities, 2710, Crimson Way, Richland, WA 99354, USA;
    nanditha.murali@tricity.wsu.edu (N.M.); keerthi.srinivas@tricity.wsu.edu (K.S.)
2   The Gene and Linda Voiland School of Chemical Engineering and Bioengineering,
    Washington State University, Pullman, WA 99163, USA
*   Correspondence: bka@tricity.wsu.edu; Tel.: +1-(509)-372-7682

Academic Editor: Gunnar Lidén
Received: 25 April 2017; Accepted: 16 May 2017; Published: 19 May 2017

Abstract: Carboxylic acids are traditionally produced from fossil fuels and have significant applications in the chemical, pharmaceutical, food, and fuel industries. Significant progress has been made in replacing such fossil fuel sources used for production of carboxylic acids with sustainable and renewable biomass resources. However, the merits and demerits of each carboxylic acid processing platform are dependent on the application of the final product in the industry. There are a number of studies that indicate that separation processes account for over 30% of the total processing costs in such processes. This review focuses on the sustainable processing of biomass resources to produce carboxylic acids. The primary focus of the review will be on a discussion of and comparison between existing biochemical processes for producing lower-chain fatty acids such as acetic-, propionic-, butyric-, and lactic acids. The significance of these acids stems from the recent progress in catalytic upgrading to produce biofuels apart from the current applications of the carboxylic acids in the food, pharmaceutical, and plastics sectors. A significant part of the review will discuss current state-of-art of techniques for separation and purification of these acids from fermentation broths for further downstream processing to produce high-value products.

Keywords: acetic acid; anaerobic fermentation; ion exchange resins; lactic acid; separations; volatile fatty acids

1. Introduction

There is significant interest among industrial and academic researchers around the world in replacing feedstocks for fuel and chemical production with sustainable biomass resources to supply the increasing population while using cutting-edge technologies to counteract environmental problems such as global warming. Biorefineries, in principle, work similarly to petrochemical refineries in that lignocellulosic biomass, which is a complex mixture containing sugars and aromatic components, are broken down into several high-value products [1]. Studies have also shown that certain functional groups added to naphtha in petrochemical refineries for the production of chemicals are usually naturally present in lignocellulosic biomass [2]. Also, the catalytic processing of petrochemical derivatives such as naphtha and syngas traditionally used for producing carboxylic acids involves high temperature and pressure conditions that result in high energy inputs [3].

The current North American acetic acid market was valued at $2.3 billion in 2014 and was expected to increase to $2.8 billion by 2019 [4]. Other reports have stated that the industry for biomass-based renewable chemicals was expected to increase at a calculated annual growth rate of 7.7%, resulting in a net market of $83.4 billion by 2018 [5]. Reports from the American Chemical Council have found that the global chemical production volume increased by approximately 10% between 2012 and 2016 [6]. This shows that the application of carboxylic acids, as chemicals or solvents, in industrial production has a large volume, as will be further discussed in Section 2. It should be noted that the current world production of lignocellulosic biomass is estimated at between 3 and 5 gigatons per year [1]. In fact, reports from a 2004 study indicated that existing forestry reserves around the world had the capacity to supply up to 9.2 billion tons of oil equivalents, which is enough to supply around 82% of the global energy demand [7]. Several routes have been developed for the conversion of lignocellulosic biomass to biofuels such as pyrolysis [8], gasification [9], liquefaction [10], and a combination of comparatively low-severity thermochemical pretreatment followed by concerted action of enzymes and microorganisms [11]. While each of these technologies has different merits and problems, as discussed in several reviews on their respective fields, the current state of the art in the latter platform is usually considered more specific and cost-efficient and results in high-value byproduct streams [2]. The biomass sugars obtained after thermochemical and enzymatic pretreatment have previously usually been converted to alcohols such as ethanol or butanol [12] and to a lesser degree anaerobically fermented to produce carboxylic acids for further catalytic upgrading to biofuels [13]. It, therefore, seems timely to update the current state of the art of the developments for biomass-based carboxylic acid production.

2. Carboxylic Acid: Formation and Applications

2.1. Current State of the Art

Carboxylic acids or short-chain fatty acids are, by definition, a group of aliphatic mono- and di-carboxylic acids [14] and include organic acids such as formic, acetic, propionic, iso-butytric, butyric, iso-valeric, valeric, iso-caproic, caproic, oxalic, lactic, succinic, malic, fumaric, itaconic, levulinic, citric, gluconic, ascorbic, etc. Another study describes carboxylic acids as "dissociated organic acids that are characterized by the presence of at least one carboxyl group" [15]. Each of the aforementioned organic acids has specific production pathways and applications in the market. For example, acetic acid is produced through carbonylation of methanol and its current prices are primarily controlled by natural gas markets [16]. Studies have indicated that the methanol carbonylation methods for producing acetic acid are accompanied by several drawbacks such as catalyst solubility limitations and the loss of expensive noble metal catalyst during separation steps [17]. Acetic acid has significant applications in the food industry as vinegar [18] and in several food preparations [19]. Acetic acid is also used in the production of vinyl acetate monomers for further polymerization to produce polyvinyl acetate or PVA, which is used in several plastics, and in the production of terephthalic acid and ethyl acetate, which replaces several industrial solvents as a "green" solvent [4]. Other applications for acetic acid include use as an etching agent [20], as a component in the manufacturing of hydrophobic and lipophobic papers in polymer industries [21], in production of cellulose acetate [22], etc.

Propionic acid is usually synthesized by hydrocarboxylation of ethylene in the presence of a catalyst such as nickel carbonyl or rhodium [23]. This study also indicated that this synthesis route has accounted for about 93,000 of the 102,000 ton capacity in the USA in 1991. However, it was found that the production capacity for propionic acid decreased to around 55,000 tons in the USA by 1998. Propionic acid has several uses but its primary application is in the preservation of food grains and animal fodder [24]. Its other applications include the manufacturing of esters [25], herbicides [26], and pharmaceuticals [27]. Moreover, it was found that propionic acid was a major byproduct in the oxidation of light distillate fuel to produce acetic acid in the petrochemical industry [23]. Since the shift in propionic acid production in the USA during the late 1990s, there has been a significant increase in the demand for propionic acid and its derivatives with the growth of the food industry. Market research has estimated the current worth of the propionic acid market to be $935.7 million in 2012, with an estimated increase of 7.8% until 2018 [28].

Similar to propionic acid, butyric acid has also been produced through an oxo-synthesis process, from the oxidation of butyraldehyde [29]. Butyric acid has significant applications in food and flavorings due to its butter-like taste and texture [30], in pharmaceuticals as a component in several anti-cancer drugs and other therapeutic treatments [31], and in perfumes in the form of esters due to its fruity fragrance [32]. Market research valued the butyric acid market at $124.6 million in 2014 with an estimated growth rate as high as 15.1% (significantly higher than any other bio-based chemical) by 2020 due to its varied applications [33].

Collectively, C1–C7 acids are referred to as "volatile fatty acids" (VFAs) due to their relatively high volatility and low vapor pressure compared to other carboxylic acids. However, the focus of the manuscript will be primarily on the short-chain carboxylic acids such as acetic, propionic, butyric (referred to as "VFAs" in this manuscript), and lactic acids, which are predominantly produced as reaction intermediates in the microbial conversion of sugars, as will be discussed in detail in the manuscript. As indicated previously, drawbacks related to poor efficiencies, expensive catalysts, and extreme reaction conditions involved in the current petro-catalysis methods for producing these organic acids are shifting the focus onto alternative biochemical methods.

An example of a carboxylic acid that is commercially produced from optimized biochemical methods, using corn sugars as a substrate, is lactic acid. Chemical synthesis of lactic acid was based on strong acid hydrolysis of lactonitrile, which is a minor byproduct of petrochemical processing [34, 35]. This was followed by base-catalyzed degradation of sugars, oxidation of propylene glycol, and high-temperature and high-pressure catalytic reactions between acetaldehyde, carbon monoxide and water [36]. Apart from the obvious drawbacks of these chemical synthesis methods such as high energy and cost requirements, these processes were also plagued by the production of a mixture of L- and D-lactic acid isomers, which affected the further processing of lactic acid for its varied applications. Some of the applications of lactic acid include the production of chemicals such as pyruvic acid, acrylic acid, 1,2-propane diol and ethyl lactate, which are currently replacing several toxic organic solvents in industries as "green" alternatives; and the polymerization to poly-lactic acid (PLA), which is used as a biodegradable plastic [37]. Currently, complete commercial production of lactic acid from pure sugar substrates is being done using biochemical methods, with an annual production capacity reported to be around 800,000 tons in 2013 [38].

2.2. Challenges and Considerations for the Production of Carboxylic Acids from Lignocellulosic Biomass

In order to better understand the steps required for developing cost-effective methods for producing organic acids from biomass instead of pure sugar substrates, it is important to understand the complex structure of the lignocellulosic biomass. The converted and unconverted components of the lignocellulosic biomass at different steps in a biorefinery will help engineers and scientists to develop optimal conversion techniques resulting in pure carboxylic acids that can be further converted to chemicals, polymers, or fuels.

Lignocellulosic biomass contains cellulose, hemicellulose, and lignin as its primary structural components. Studies have shown that the primary component of biomass is 35–50% cellulose along with 20–35% hemicellulose and 10–25% lignin [39]. Cellulose is a high molecular weight linear homo-polymer of repeated units of glucose held together by β-1, 4 glycosidic linkages and is the primary source of glucose or C–6 sugar for carboxylic acid production [40]. Hemicellulose is a linear and branched heterogeneous polymer containing different C5 and C6 sugars as well as other components such as ferulic, acetic, and glucuronic units, which are used for the linkage between sugar units or with lignin and cellulose [40]. While hemicellulose contains a mixture of C5 and C6 sugars and is amorphous in nature, it is easily hydrolysable compared to cellulose. While not the main carbon source for several microorganisms, studies have been done to engineer microbes to also take sugars developed from hemicellulose as the substrate for carboxylic acid production and, thereby, increase the biomass conversion efficiency to produce carboxylic acids. Unlike cellulose and hemicellulose, lignin is a complex structure constructed using phenyl–propane units and is bonded closely with cellulose and hemicellulose to provide rigidity and cohesion to the structure of the cell wall [41]. Lignin mainly consists of aromatic compounds and, due to its complex structure, is highly resistant to enzymatic and microbial action. However, recent studies have shown anaerobic digestibility of wet exploded lignin, resulting in increased methane production through conversion of around 44.4 wt % lignin in the pretreated material [42]. This study also found production of several phenolic compounds and fatty acids in the effluent after anaerobic digestion, which is an indication of the complexities of the fermentation broth, while designing separation processes fitted to produce carboxylic acid as the final product.

3. Biochemical Routes to Carboxylic Acid Production

The biochemical routes for producing carboxylic acids from sugar substrates are well known. However, these biochemical pathways are based on using glucose (or fructose) as the substrate and not on the use of a complex sugar mixture derived from biomass after pretreatment, which again will create many other components. The biochemical pathway for converting sugar substrates to different carboxylic acids is shown in Figure 1.

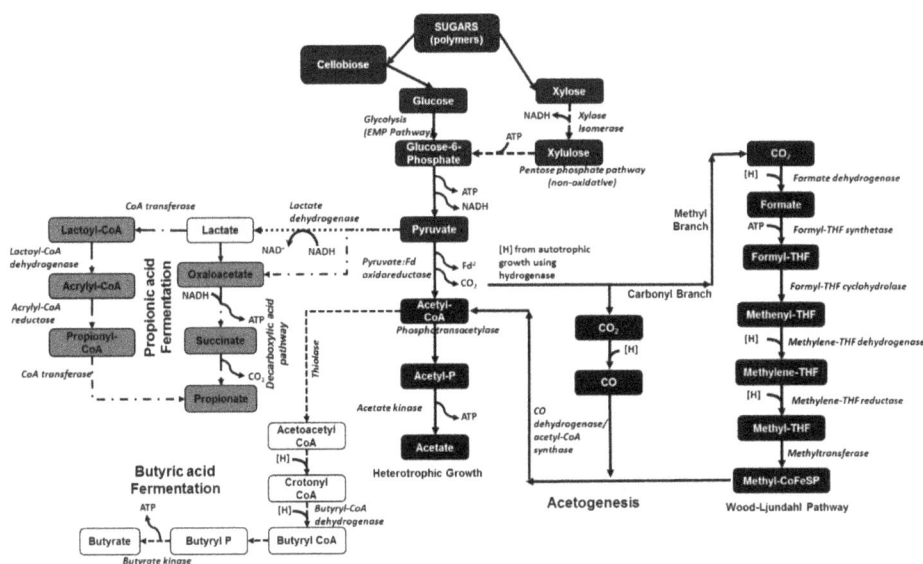

Figure 1. Different biochemical pathways in different microorganisms for the production of acetic, lactic, propionic, and butyric acid. The different pathways are represented using different line forms: (–) acetogenesis; (···) lactate pathway; (–·) propionate pathway; and (- -) butyrate pathway. [THF: tetrahydrofolate; P: phosphate; H: hydrogen; CoA: coenzyme A; ATP: adenosine triphosphate; CO_2: carbon dioxide; Fd^{2-}: ferredoxin ion; NADH: nicotinamide adenine dinucleotide].

3.1. Microbes Used for Acetic Acid Production

Homoacetogens are a group of bacteria that selectively produce acetic acid from organic matter. These homoacetogens or 'carbon dioxide reducing acetogens' are obligate anaerobes that use the Wood–Ljungdahl or reductive acetyl CoA pathway as their main mechanism for the synthesis of acetic acid [43]. One molecule of glucose is converted to two molecules of pyruvate using the Embden–Meyerhof–Parnas (EMP) pathway, which in turn is decarboxylated to two molecules of acetyl-CoA and finally into two molecules of acetate (see Figure 1) [44]. The acetogens convert the available H_2 and CO_2 to acetic acid through the Wood–Ljungdahl pathway [45], which has two branches, the carbonyl and the methyl branches, with CO and/or CO_2 substrates. The carbonyl branch reduces CO/CO_2 through cobalamine and tetrahydrofolate compounds into a methyl group, which binds to Coenzyme A to form acetyl CoA [43]. The net reaction is as follows:

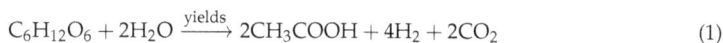

$$C_6H_{12}O_6 + 2H_2O \xrightarrow{\text{yields}} 2CH_3COOH + 4H_2 + 2CO_2 \tag{1}$$

Most of the known homoacetogens belong to the genera Clostridium and Acetobacetrium. Over 100 acetogens have been isolated and, out of this pool, genomic sequencing data for around 22 acetogens have been published [43]. Some of the most common and most studied acetogens include *Clostridium thermoaceticum* (or *Moorella thermoacetica*), *Acetobacterium woodii*, and *Clostridium ljungdahlii*. The biochemical pathways relating to these acetogenic bacteria, including their thermodynamics, have been intensively reviewed by Schuchmann and Muller [46]. *Moorella thermoacetica*, also known as *Clostridium thermoaceticum* ATCC 49707, is a gram-positive, endospore-forming, and strictly anaerobic thermophilic bacteria with an optimal growth temperature between 55 and 60 °C. There are several studies describing metabolic models of this strain's acetogenesis pathways [46,47]. The ability of *M. thermoacetica* to use a wide variety of substrates including sugars, gases such as CO_2 & H_2, other organic acids, methoxylated aromatic compounds, etc. has increased interest in studying and engineering the bacteria not only for research but also for industrial applications [46,48]. Previous studies have indicated that, with a complex mixture of sugars, *Moorella thermoacetica* (*M. thermoacetica*) initially consumes xylose followed by fructose and then glucose, making it a unique microbial strain [49]. Batch fermentation studies operated with 20 g/L xylose showed a maximum acetic acid yield using *M. thermoacetica* of 0.76 g acid/g xylose [50]. Other studies using pure glucose as a substrate have shown a conversion efficiency of only around 0.65 g acetic acid/g glucose. Very few studies have been done with lignocellulosic biomass substrates where acetic acid yield from an agricultural raw material such as sugarcane bagasse was found to be around 0.71 g acid/g biomass sugars while that from a forestry residue was found to not exceed 0.35 g acid/g biomass sugars [51]. These studies indicate that while the thermal degradation products of sugars and lignin obtained during thermochemical pretreatment (in this case, steam explosion) did not significantly inhibit *M. thermoacetica*, the primary source of inhibition was the glucose/xylose ratio and the inability of the strain to ferment other sugars such as arabinose, mannose, and galactose with high efficiency. The metabolic function of *M. thermoacetica* with a xylose substrate was studied using an in situ Nuclear Magnetic Resonance (NMR) bioreactor by Xue et al. [49], where around 12 metabolites were identified and quantified using in situ NMR capabilities and more than 40 metabolites were identified using ex situ HR-NMR (after sample preparation to remove spectral inhibition by cells and media). Real-time NMR analysis also showed production of formate, ethanol, and methanol by *M. thermoacetica* using xylose as a substrate [49].

C. ljundahlii (ATCC 49587) is similar in structure to *M. thermoacetica* but lacks the cytochromes and quinones that would normally result in salt imbalances in the microorganisms [46]. However, similar to *M. thermoacetica*, *C. ljundahlii* does not require Na^+ for growth and optimally grows between pH 5'and 7 under thermophilic conditions. Studies have indicated that this strain favors the production of ethanol over acetic acid at a pH below 4.5 and can produce a spectrum of products including acetic and butyric acids at higher pH [52]. However, most of the reported data on *C. ljundahlii* for the production of acetic acid focus on gaseous substrates such as synthesis gas (CO, CO_2 & H_2) and very few studies have been done with sugar substrates [53].

Unlike *M. thermoacetica* and *C. ljundhalii*, *A. woodii* is a non-spore-forming mesophilic bacterium that grows robustly on H_2 and CO_2 to produce predominantly acetate and has also been used in varying conditions to produce small amounts of other chemicals such as formate, methanol, ethanol, 1,2-propanediol, 2,3-butanediol, ethylene glycol, lactate, etc. [46]. Being mesophilic, *A. woodii* is significantly affected by contamination and the absence of cytochromes and quinones makes the strain dependent on Na^+ [54,55]. Metabolic modeling of the acetogenesis in *A. woodii* has shown that, during heterotrophic growth, 4 moles of hydrogen are produced and further used during autotrophic growth (producing 2 moles CO_2), resulting in the production of 1 mole of acetate and 0.3 moles of ATP [56].

Studies have also been done to understand the biochemical and energy profile of reductive acetogenesis in combination with the Wood–Ljundahl pathway for acetic acid production [57]. Theoretical models have shown that ATP is formed during acetate kinase reaction and during pyruvate formation but an equal amount of ATP is consumed for conversion of ATP to acetyl-CoA and during the formyl-THF synthase reaction (in the Wood–Ljundahl pathway). Hence, membrane-bound complexes are required to be coupled with the pathways to control the energetics resulting in active generation of ATP. Some metabolic modeling studies have reported that hexose fermentation in acetogens uses low-molecular-weight, iron–sulfur center-containing proteins known as ferredoxin for the conversion of pyruvate to acetyl-CoA [46,58] (shown in Figure 1). These metabolic studies hypothesized that this reaction through ferredoxin as an electron acceptor has a standard redox potential that is much more negative compared to the $NAD^+/NADH$ electron transport, resulting in an acceptable energy balance within the biochemical pathways of the acetogens. However, other studies [59] hypothesized another possible route through a reaction using formate-hydrogen lyase and methylene-tetrahydrofolate reductase where the reaction resulted in the transport of two protons out of the cell membrane, resulting in a proton gradient to facilitate electron transport in the acetogens. This model was investigated for *M. thermoacetica* iAI558 using CO_2 and H_2 as substrates to produce acetic acid and the metabolic studies indicated that both the routes (as described above) did not individually show a complete energy balance during the conversion of pyruvate to acetyl-coA in an acetogen. However, these studies concluded that the combination of both these routes could explain the increased efficiency of acetogens such as *M. thermoacetica* due to their ability to generate energy through thermodynamically infeasible cycles without a specific growth substrate [47]. Due to the absence of cytochromes in *C. ljundahlii*, a proton gradient facilitating electron transport and energy balance was hypothesized to be achieved through membrane-bound Rnf complexes [60]. Unlike thermophilic bacteria, energy balance analysis through metabolic modeling of mesophilic bacteria such as *A. woodii* indicated that the Na^+ gradient was used by the membrane-bound ATP synthase to produce ATP [56]. These studies also indicated that the acetogenesis began with H_2 being oxidized by an electron-bifurcating hydrogenase that coupled the reduction of ferredoxin with that of NAD to promote acetogenesis. Such metabolic information not only helps validate the biochemical pathway found by a pure microbe to consume a specific carbon substrate to produce a specific carboxylic acid, but also provides information regarding the energetics of specific parts of the pathway that will help us understand a mixed microbial consortium that contains one or more such microbes.

3.2. Microbes Used for Butyric Acid Production

Similar to acetogens, butyric acid production through fermentation is one of the oldest and most-studied techniques since the First World War. Several reviews have been published on different strains used for butyric acid production and seven different genera have been investigated for their potential industrial applications [56,61]. Some of the most commonly studied and industrially used strains includes *C. tyrobutyricum* [62,63], *C. acetobutylicum* [64,65], and *C. thermobutyricum* [66,67]. A major drawback of using clostridial strains for butyric acid production is the formation of other side products such as acetic acid, ethanol, butanol, acetaldehyde, and lactate, as well as gases such as CO_2 and H_2. Butyric acid production starts with the metabolism of glucose to pyruvate through the Embden–Mayerhof–Parnas (EMP) pathway followed by conversion to acetyl-CoA and then to acetoacetyl-CoA through a thiolase reaction (as shown in Figure 1). The butyryl CoA produced from acetoacetyl-CoA is then converted to butyryl phosphate by phosphotransbutyrylase and then converted to butyrate by butyrate kinase [62]. Metabolic studies showed that *C. tyrobutyricum*, *C. thermobutyricum*, and *C. acetobutylicum* followed different pathways with different efficiency, as shown in Equations (2)–(4):

$$C_6H_{12}O_6 \rightarrow 0.8C_4H_8O_2 + 0.4C_2H_4O_2 + 2CO_2 + 2.4H_2 \qquad (2)$$

$$C_6H_{12}O_6 \rightarrow 0.85C_4H_8O_2 + 0.1C_2H_4O_2 + 0.2C_3H_6O_3 + 2CO_2 + 2.4H_2 \qquad (3)$$

$$C_6H_{12}O_6 \rightarrow 0.7C_4H_8O_2 + 0.6C_2H_4O_2 + 2CO_2 + 2.7H_2 \qquad (4)$$

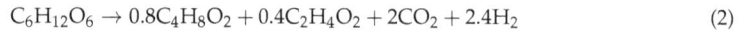

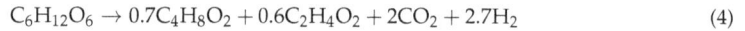

Similar to the acetogenic strains discussed in Section 3.1, the clostridial strains targeted for butyrate production usually do not possess cellulolytic enzymes and have a lesser ability to use cellulose as substrate; hence, they require extensive pretreatment and enzymatic hydrolysis to deconstruct biomass into simple sugars for carboxylic acid production. While some studies have indicated that certain clostridial species have cellulases encoding for genes allowing them to utilize polymers such as cellulose, most of these studies also indicated that these strains employ carbon catabolite repression (CCR), whereby preferred carbon substrates such as glucose will suppress the utilization of other sugars [56]. As can be seen from Equations (2) and (3), clostridial strains for butyrate production are also significantly pH-sensitive. Studies have indicated that at pH > 5 and with iron limitations, lactate can replace butyrate as the major product of fermentation in *C. butyricum* [68]. Other studies showed the highest butyrate production in *C. tyrobutyricum* between pH 6 and 6.7 [62]. These studies indicated that the shift in metabolic flux as a function of pH is related to the difference in activity of lactate dehydrogenase, phosphotransbutyrylase, and phosphotransacetylase.

Significant metabolic studies have been done on butyric acid production to understand the pathways leading to butanol production, especially after inhibiting hydrogen production by controlling the ferredoxin in the pathway (ferredoxin is produced during conversion of pyruvate to acetyl-CoA, as shown in Figure 1) [64]. These studies also showed that a proper energy (ATP) balance in the butyrate pathway (shown in Figure 1) could not be established for butyric-acid-producing strains such as *C. acetobutylicum* due to the absence of Rnf membranes or a similar membrane-bound electron transport system existing in its pathways. Other studies found that, in *C. butyricum*, the lower partial pressure of H_2 decreased the acetate/butyrate ratio, followed by an increased ATP yield, resulting in an energetically favorable pathway that would take up the produced acetate and convert it back to butyrate [69]. Such energetically favorable reversible reactions have been found to be of importance, especially in mixed microbial cultures compared to pure cultures.

3.3. Microbes Used for Propionic Acid Production

Paludibacter and *Propionibacterium* produce propionic acid as the major end product of fermentation. For example, *Paludibacter propionicigenes* produces propionate and acetate as major products when fermenting glucose [70]. *Paludibacter* belongs to the phylum Bacteroidetes and is one of the most abundant species in wastewater treatment plants [71]. *Paludibacter jiangxiensis* is another bacterium that utilizes sugars to produce propionate as the major product. However, compared with *P. propionicigenes*, it was found that yeast extract was not required for the growth of the bacterium [72]. Propionibacteria selectively produce propionic acid from organic matter and make up 1.4% of the ruminal microflora [73]. Propionate production from lactate occurs either via the 'direct reduction' pathway or the 'decarboxylic acid' pathway. In the decarboxylic acid pathway, succinate is initially formed (see Figure 1) from heterotrophic fixation of carbon dioxide, which is then decarboxylated to propionate [74].

Another bacterium that has the ability to produce propionic acid from sugar substrates following an alternate "acrylate pathway" is *Clostridium propionicum* (see Figure 1) [75,76]. Other studies have indicated that certain cellulolytic bacteria such as *C. saccharolyticum* and *C. thermocelluum* also follow a similar pathway during the production of propionic acid [56]. In this acrylate pathway, the sugars are initially converted to lactate, which is then converted to lactoyl-CoA through the action of CoA synthase, followed by dehydrogenase action leading towards acrylic acid and then propionic acid. Studies have indicated that, following the decarboxylic acid pathway, propionic acid production is hindered due to the production of lactic and succinic acid as byproducts [77]. Another study using glycerol as a substrate with *Propionibacterium acidipropionici* and *Propionibacterium freudenreichii* ssp. *shermanii* bacteria resulted in the production of 0.79 mol of propionic acid/mol substrate consumed at a productivity of around 0.42 g/L/h at optimized conditions, with acetic acid produced as the main byproduct along with lactic acid and succinic acid [78]. These studies also showed that when glycerol was replaced with glucose as the carbon substrate, a higher amount of acetic acid was produced as a byproduct, followed by a decrease in the propionate yield. Metabolic studies on the different biochemical pathways for propionic acid production showed that the optimized pathway had acrylic acid as an intermediate [79]. Metabolic studies using Biological Network Integrated Computational Explorer (BNICE) were one of the first successful attempts at using pure computational modeling to study around 16 different biochemical pathways towards propionate production in different propionate bacteria. These studies predicted that the rate-limiting step for propionate production was the reduction of acrylic acid to propionic acid through the "acrylate" pathway. Such computational methods, blended with actual metabolic data on the different biochemical pathways, allow for accurate prediction of the enzymatic activities of the different steps in a specific pathway.

3.4. Microbes Used for Lactic Acid Production

Lactic acid bacteria have been studied for several years, especially related to their use in the food industry. Taxonomically, lactic acid bacteria belong to two distinct phyla, *Firmicutes* (with *Lactobacillus* and *Lactococcus* as the most-studied genera) and *Actinobacteria* (with *Bifidobacterium* as the most-studied genus) [80]. *Lactobacillus* has been studied for biorefinery applications as they are facultative anaerobes and can also live under microaerophilic conditions in nature with the production of different fermentation end-products including lactic acid and sometimes further formic acid, acetic acid, ethanol, and carbon dioxide [81]. The lactate production follows glycolysis through the EMP pathway as shown in Figure 1, followed by the action of lactate dehydrogenase on pyruvate to produce lactate. The major advantage of the bacterium is its tolerance to low pH when compared to other carboxylic-acid-producing strains with optimal pH over 5 [82]. Another advantage of some *Lactobacillus* strains is their non-sporulating nature, which will be further discussed in Section 3.5. Based on the nature of the fermentation, lactic acid bacteria in general have been classified as homo-fermentative, resulting primarily in L-lactic acid as the only product, and hetero-fermentative, which produce small amounts of acetic acid and ethanol as byproducts. Major homo-fermentative products includes *Lactococcus lactic* [83], *Lactobacillus delbrueckii* [84]; *Lactobacillus helveticus* [85]; *Lactobacillus casei* [86], *Bacillus subtilis* [87], and *Bacillus coagulans* [88–90]. Similar to the VFA-producing bacteria, the *Lactobacillus* or *Bacillus* strains are rod-shaped bacteria that have a tendency to sporulate [80]. A high lactate-producing bacterial strain such as *B. coagulans* showed as much as 92% lactic acid yield from sucrose [88]. Continuous fermentation of lignocellulosic biomass using an isolated *B. coagulans* strain resulted in lactic acid yield of 0.95 g/g biomass sugars with a productivity as high as 3.69 g/L/h at the optimum pH of 6 with a residence time as low as 6 h [90]. Similar lactic acid yields were obtained with batch fermentation of pure glucose using the *B. coagulans* WCP10–4 strain with a productivity as high as 3.5 g/L/h [91]. These studies show that lactic acid bacteria are severely inhibited by product concentrations above a certain range and, hence, it would be beneficial to remove the product immediately to increase the efficiency of the fermentation.

3.5. Disadvantages of Pure Microbial Cultures in Biorefineries

The previous sections discussed several pure bacterial strains and pathways for the conversion of sugar substrates to different carboxylic acids. The advantages of using such pure strains include complete knowledge and control of the biochemical pathways towards carboxylic acid production from sugars, lesser contamination from other strains under sterile conditions, and easy genetic engineering of the strains to add other functionalities such as cellulose degradation (which allows the pure strains to digest cellulose to produce carboxylic acids instead of just simple sugars). However, there are several disadvantages related to using pure cultures, especially due to the risk of contamination from other strains that are more energetically favorable, resulting in the formation of other less-desired products instead of carboxylic acids. Also, in biorefineries, complex substrates such as lignocellulosic biomass have several components other than sugars such as aromatic and phenolic compounds coming from the lignin, which have antimicrobial properties and can inhibit some pure microbial strains more than others. Clostridial strains, as mentioned previously, have been indicated to suffer from low growth rates when lignocellulosic biomass is used as the substrate [56].

Another disadvantage with an acetogen such as *M. thermoacetica* is its significant dependence on pH. Studies have shown that the growth of *M. thermoacetica* (or *C. thermoaceticum*) strains stopped when the pH of the anaerobic glucose fermentation dropped under 5 [92]. Other studies indicated that *M. thermoacetica* strains survived well in a pH range of 6 to 7 with acetic acid yield between 0.8 and 0.95 g acid/g glucose but that the growth of *M. thermoacetica* decreased significantly with an increase in acetic acid concentration in the fermentation broth [93]. The pH sensitivity of butyrate producing strains has already been discussed in Section 3.3. Apart from pH sensitivity, these strains were also found to be sensitive to other factors such as the H_2 partial pressures, resulting in a significant effect on the product distribution. The sensitivity of such pure microbial strains to acid concentrations, system pH, gas partial pressures, and contamination from other microbes altering the end product of the fermentation point to the use of mixed microbial consortia specifically grown towards target acid product and the development and optimization of in situ separation techniques to remove acids from the fermentation broth as part of its production.

Another disadvantage of clostridial strains, in general, for biorefinery applications is the tendency of the strains to sporulate as a form of asymmetric cell division [94]. Studies have indicated that such spores tend to become metabolically inactive [95]. Studies on *C. acetobutylicum* have been done to change the sporulating pattern of this strain by inactivating sporulation-related sigma factors, resulting in mutants without the ability to sporulate at the stationary phase, leading to efficient acidogenesis and, hence, butyric acid production [96].

Other common engineered strains for biorefinery applications include *E. coli*, *S. cerevisiae*, and other yeasts. Several reviews and metabolic studies have been published on these mutants [97] but most of these studies indicated that such engineered strains were significantly limited by substrate utilization, especially the C6 and C5 sugars present in lignocellulosic biomass [98]. Mixed microbial consortia such as those present in the rumen, on the other hand, can combine both substrate utilization and targeted product formation through optimizing pathways and fermentation conditions, especially related to high carboxylic acid yields in biorefineries.

3.6. Mixed Bacterial Consortia for Carboxylic Acid Production

Ruminants are the most populous group of animals in the world, with more than 3.5 billion on each continent [99]. Ruminants (e.g. cows) are known to produce VFAs, especially acetic, propionic, and butyric acids, from organic matter [100]. They have the ability to digest cellulose and break it down into simple sugars, which can be attributed to the presence of a unique microflora in their rumen, including cellulolytic microorganisms [101] and consisting of about 10^{10} to 10^{11} bacterial cells per milliliter and 10^3 to 10^5 zoospores per milliliter, in addition to protozoans and bacteriophages. Many groups of microbes including bacteria, fungi, and archaea are involved in the production of carboxylic acids; of these the most important group of bacteria are the acetogens.

Ruminal fermentation of organic matter requires interaction between three groups of anaerobic microbes (Figure 2): fermentative bacteria, which degrade organic matter to hydrogen, carbon dioxide, and VFAs; acetogenic bacteria, which convert hydrogen, carbon dioxide, and acetate; and methanogenic bacteria, which reduce hydrogen, carbon dioxide, and formate to methane [102]. In the bovine rumen, methane production is primarily from the reduction of carbon dioxide with hydrogen as the electron donor [103]. Studies have indicated that although methanogenesis represents a 2–15% energy loss in cattle [104], a low concentration of hydrogen is required for complete fermentation of the organic matter, maximum yield of ATP, and optimal microbial growth in the rumen [105]. If methanogenesis in an artificial rumen reactor is inhibited, hydrogen accumulation increases through the inhibition of oxidation of the reduced NADH [106] and, in turn, VFA production including acetate and propionate production will further increase [107]. It has also been shown that high levels of propionic and butyric acids can inhibit methanogenesis [108]. In other words, it is ideal to prevent methanogenesis in the rumen microflora in order to increase the VFA production through anaerobic ruminal fermentation.

It is common knowledge from several in situ fermentation studies that the rumen microflora consists of several cellulose-degrading, acidogenic, acetogenic, and methanogenic bacteria that are necessary for the optimal function of the rumen in a ruminant animal [99–103]. Several studies have been done to identify and understand the microbial population of the ruminal microflora, especially related to biorefinery applications. One such study identified around 27,755 putative carbohydrate-active genes that accounted for as much as 57% of the cellulose-degrading enzymatic activity in the rumen microflora [109]. These metagenomic studies done with cellulose and switchgrass as substrates showed the presence of endoglucanases, β-glucosidases, and cellobiohydrolases in the ruminal microflora. Some of these cellulolytic strains present in the rumen include *Fibrobacter succinogenes*, *Ruminococcus flavefaciens*, and *Ruminococcus albus* [110]. Unlike pure microbial cultures used for carboxylic acid production, which would require genetic alterations using certain cellulolytic bacteria such as *C. thermocellum* (thermophilic) or *C. cellulolyticum* (mesophilic) [111], mixed microbial consortia have the advantage of the collaboration of several different cellulolytic bacteria that together produce a more diverse and optimal cellulolytic enzyme mixture for the conversion of cellulose in lignocellulosic materials. This internal enzyme capability is of major importance, since biorefineries are affected by the high enzyme costs required to completely convert the sugars present in pretreated biomass into C5 and C6 sugars for uptake by pure microbial cultures. Using a mixed rumen microflora, it is possible to significantly reduce the cost of the production process by harvesting the intrinsic capability for producing a complex enzyme mixture along with fermenting the resulting sugars into carboxylic acids.

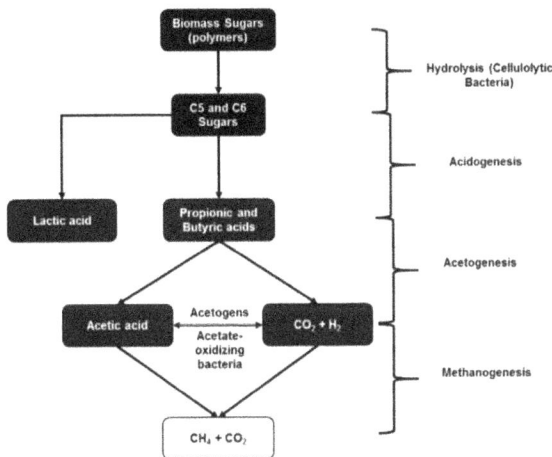

Figure 2. Different stages of biomass conversion through fermentation using a mixed microbial consortia such as rumen microflora.

Apart from cellulolytic bacteria and methanogenic archaea present in the rumen, a significant portion of the rumen microflora is composed of acidogenic bacteria. Studies have indicated that *Acetitomaculum ruminis* is the most predominant acetogen present in the ruminal microflora [112]. Other acetogens present in the rumen include *Eubacterium limosum*, *Ruminococcus productus* [113], and obligate hydrogen-producing acetogens such as *Syntrophomonas wolfeii* and *Syntrophobacter wolinii* [114]. *Ruminococcus schinkii*, which is phytogenically similar to the Clostridial species, has been isolated from the rumen in suckling lambs and has a greater propensity towards acetogenesis even when using O-methylated aromatic compounds [115]. The advantage of *A. ruminis* present in the rumen versus the other acetogens is its ability to produce acetate via heterotrophic growth (see Figure 1), while those acetogens that use autotrophic growth compete with methanogens for H_2 [116]. Reductive acetogenesis by acetogens following autotrophic growth in rumen microflora is usually similar to that shown in Equation (5), unlike the traditional acetogenic reaction shown in Equation (1):

$$2CO_2 + 4H_2 \rightarrow C_2H_4O_2 + 2H_2O \tag{5}$$

Davidson and Rehberger [117] showed that ruminal populations of *Propionibacteria* ranged from 10^3 to 10^4 cfu/mL and, out of 132 isolates studied, 126 were identified as *P. acidipropionici*, which is also capable of nitrate reduction. Initial studies have indicated that rumen microflora from a ruminant animal that has a high forage diet usually have *Streptococcus bovis* as the primary lactic-acid-producing bacteria [118]. However, other studies showed the metabolic shift in the lactate-producing ruminal population toward *Lactobacillus* when there was an increase in sugar substrates in the ruminant diet [119]. Hence, the *Lactobacillus* bacteria present in rumen microflora are essential to understand when using mixed microbial consortia for biorefinery applications. Studies have identified that *Lactobacillus ruminis* and *Lactobacillus vitulinus* are two prominent lactic acid bacteria in the rumen when feeding a high carbohydrate content in the diet [120]. Other studies isolated around 36 different lactate-producing bacteria from the rumen including *B. licheniformis*, *B. coagulans*, *B. circulans*, *B. laterosporous*, *B. pumilis*, etc. [121]. Even with the presence of such lactate-producing bacteria, one of the advantages (and, in certain cases, disadvantages) of using rumen mixed cultures for biorefinery applications is that the cellulolytic bacteria and the lactic acid bacteria work at different optimal pH ranges. It is commonly known that cellulolytic bacteria function optimally at a pH range between 6 and 7, while several studies found the optimal performance of lactic acid bacteria at a pH below 6. The advantage of this discrepancy is the ease of guiding mixed culture fermentation of biomass towards VFAs with negligible effect from the lactic acid bacteria present in the rumen microflora, simply by maintaining a pH of over 6. It is properties such as this that have led to the interpretations of ruminal mixed cultures as "habitat-simulated" mixed cultures [122], which can be selectively targeted towards specific carboxylic acid produced under specific controlled experimental conditions. Studies into isolating lactic acid bacteria from rumen microflora also resulted in isolating *Sphaerophorus prevot* species, which had the ability to produce butyric acid as a minor byproduct [123]. There is no specific literature on predominantly butyric-acid-producing bacteria in the rumen microflora, but butyric acid is produced as a minor byproduct by several acetogens and propionate-producing bacteria present in the rumen. These microbes include *Bacteriodes amylogenes* [124] and *Butyrivibrio fibisolvens* [125]. Also of interest to biorefinery applications is the effect of xylans as a substrate for rumen bacteria. Studies aimed at characterizing bovine ruminal bacteria isolated from the microflora after growth in a xylan-supplemented medium showed the presence of strains such as *Butyrivibrio fibrisolvens* and *Bacteriodes ruminicola*, with propionic acid as primary products [126]. The study found that there was a significant reduction in cellulolytic bacteria in the rumen microflora grown selectively on a xylan medium, and while the actual microbial population did not change significantly, the transition of the carbon substrate for growth from xylan to glucose resulted in a decrease in propionic acid production. Murali et al. [13] have successfully produced volatile fatty acids (acetic, propionic, and butyric acid) from pretreated corn stover with increased conversion of corn stover when solids loading was increased from 2.5% to 5%. This study showed good performance of the ruminal bacteria towards VFA production with a decrease in biomass carbohydrates (sugar polymers) due to the presence of cellulolytic bacteria in the rumen that can depolymerize cellulose and hemicellulose into sugars followed by acetogenesis. Such a microbial system can effectively reduce the costs of producing VFA by eliminating the need for external lignocellulolytic enzymes such as cellulases and hemicellulases.

In vitro studies were done in an anaerobic fermenter to study the dynamic response of the system to pulses of volatile fatty acids input into the system [127]. These studies found that the addition of butyrate and valerate to the anaerobic fermenter led to an increase in acetic acid production, with propionic acid as a minor byproduct. However, pulsed input of propionic acid to the anaerobic fermenter destabilized the system, resulting in a reduction in acetogenesis. Such effects of process imbalances have also been studied in batch cultures from anaerobic reactor systems [128–131]. Such studies are, again, sufficient evidence for the need for in situ product separation to control as well as prevent system instability during the anaerobic fermentation of lignocellulosic biomass to produce carboxylic acids.

4. Product Separation and Purification

As indicated earlier in this manuscript, separation processes are required in a biorefinery at different stages primarily to (a) separate and purify the product/intermediate for the next stage of processing and (b) remove biomass components that are inhibitory at a particular stage of processing. However, here we will primarily focus on the significance of in situ product recovery from biochemical fermentation in biorefineries that produce carboxylic acid as a product or an intermediate for fuels and/or chemicals. Mixed acid fermentation is usually inhibited by the product/products produced, i.e., in situ product recovery during fermentation will sufficiently remove the inhibitory effects of carboxylic acids produced during fermentation. For example, studies done by Garrett et al. [82] have shown that fermentation productivity using *B. coagulans* for the production of lactic acid was significantly inhibited above a lactate concentration of 20 g/L in the fermentation broth at 50 °C and pH 5.5. These studies using Amberlite IRA-67 for in situ extraction of lactic acid showed an increase in microbial productivity by at least 1.4-fold when compared to fed-batch fermentation with conventional salt precipitation. Several similar studies done on in situ carboxylic acid recovery from fermentation broths are shown in Table 1.

Table 1. Existing literature data on in situ carboxylic acid separation during fermentation.

Substrate	Fermentation	Product	Type of Separation	Optimized Conditions and Process Efficiency
In situ separation using electrodialysis				
Wheat Straw [132]	Continuous with *C. tyrobutyricum*	Butyric acid	Reverse electro-enhanced electrodialysis (REED)	19- and 53-fold higher sugar consumption in presence of REED resulting in butyric acid. Yield as high as 0.45 g/g sugars
Glucose [133]	Continuous with *L. plantarum*	Lactic acid	Bipolar electrodialysis	Lactate recovery of 69.5% (1.32 mol/L lactate) with current density of 40 mA/cm^2
Whey [134]	Batch with *P. shermanii*	Propionic and acetic acid	Electrodialysis	Increased acid yield by 1.4-fold and 1.31-fold for propionic and acetic acids, respectively, when compared to controls.
Sucrose and grass [135]	Fed-Batch with anaerobic sludge	Acetic and butyric acid	Electrodialysis	Up to 99% VFA removal from fermentation broth within 60 min containing 1.2 g/L initial VFAs
In situ separation using reactive extraction				
Sucrose [136]	Fed-Batch with *C. tyrobutyricum*	Butyric acid	Pertraction using 20% w/w Hostarex A327 in oleyl alcohol	0.30 g butyrate/g sugar with productivity of 0.21 g/L/h (pH 5.2 at 37 °C)
Glucose [137]	Batch with immobilized *L. delbrueckii*	Lactic acid	Alamine-336 in oleyl alcohol	Maximum yield of 25.5 g/L with Alamine-336 together with immobilized cells with 15% v/v sunflower oil ($V_{or}/V_{aq} = 0.5$ at 37 °C)
Lactose [138]	Hollow-fiber membrane extractor (Fed-batch) with *P. acidipropionici* ATCC 4875	Propionic acid	Adogen 283 (ditridecylamine) in oleyl alcohol	0.66 g propionate/g substrate with product concentration of 75 g/L and purity of ~90% (pH 5.3)
Switchgrass [139]	Hollow-fiber membrane extractor (Fed-batch) with *L. delbrueckii*	Lactic acid	Alamine 336 in oleyl alcohol with kerosene as diluent (20:40-40 wt%)	Lactate yield of 67% that of theoretical maximum (pH 5.0 at 43 °C)
Glucose [140]	Batch with *L. salivarius*	Lactic acid	Hoe F 2562, Cyanex 923 and Hostarex A327 with isodecanol and kerosene	Lactic acid yield as high as 87.5% with 10 wt % Hostarex A327 and 81% with 40 wt % Cyanex 927
Corn Stover [141]	Fed-Batch with *Megasphaera elsdenii*	Butyric and hexanoic acids	Pertraction with oleyl alcohol and 10% (v/v) trioctylamine	Carboxylic acid productivities were found to be increased by 3-fold for pertractive fermentation system when compared to batch and glucose conversion rates was also higher by ~3-fold
In situ separation using ion exchange resins				
Whey [142]	Batch with *L. casei*	Lactic acid	Amberlite IRA-400 (Cl$^-$)	Maximum concentration of 37.4 g/L with yield of 0.85 g lactate/g substrate and productivity of 0.984 g/L/h (pH 6.1 and 37 °C)
Corn Stover [82]	Fed-Batch with *B. coagulans*	Lactic acid	Amberlite IRA-67	0.94 g lactate/g biomass sugars obtained with productivity of 0.33 g/L/h (pH 5.5 at 50 °C)
Beet Molasses [143]	Continuous with *L. delbrueckii*	Lactic acid	Amberlite IRA-420 combined with Amberlite IR-120	Maximum lactate yield of 0.91 g/g sucrose at dilution rate of 0.1 h^{-1} (pH 6 at 49 °C)
Zizyhus oenoplia [144]	Batch with *L. amylophilus* GV6	Lactic acid	Amberlite IRA-96 combined with Amberlite IR-120	Maximum lactate recovery of 98.9% with optical purity of 99.17%. Maximum acid loading around 210.46 mg/g bead
Synthetic food waste [145]	Batch with mixed culture from food waste	Lactic, acetic and butyric acids	Amberlite IRA-67	Lactic, acetic and butyric acid loadings onto the resin of 84, 20.5, and 50.7 mg/g resin, respectively, with acid removal of around 75%

In situ or on-line product recovery of carboxylic acids from fermentation broths using dialysis, distillation, adsorption, and extraction have been attempted [146]. Adsorption and extraction are two fairly common techniques that have been used in the continuous acid recovery from anaerobic fermentation.

4.1. Separation Using Ion Exchange Resins

Adsorption with conventional adsorbents such as activated carbon usually results in contamination of the product, especially in the case of biomass fermentations due to the presence of unconverted biomass components that have an affinity towards activated carbon [146]. For example, polymeric adsorbent XAD-4 was used to adsorb furfural from acid-catalyzed biomass hydrolysate before ethanol fermentation for removal of inhibitory contaminant [147]. This study indicated that such polymeric adsorbents are a very good source of detoxification of the biomass hydrolysate before anaerobic fermentation. However, in the context of product separation and purification, adsorption by ion exchange seems to be far more efficient. Ion exchange resins are usually polymeric resins with a linked cation or anion exchange group [148]. For carboxylic acid separation, predominantly used resins are strong or weak base resins, which have tertiary or quaternary amines as the ion exchange group [149,150]. The carboxylic acids are usually recovered from the ion exchange resin through caustic elution and can be concentrated through evaporation and hydrolyzed to give a pure acid for further processing. As can be seen from Table 1, there are a number of studies on in situ separation of lactic acid from the fermentation broth. This is primarily because of the dependence of efficiency of ion exchange resin-mediated separation of carboxylic acid on the fermentation pH. The pK_a of acetic acid is 4.76, while that of lactic acid is 3.86. As discussed previously in the manuscript, anaerobic fermentation to produce acetic acid and other VFAs is usually optimal at a pH between 6 and 7, while that to produce lactic acid can function optimally at pH < 6. Hence, under these conditions, ion exchange resins will not function optimally for in situ VFA recovery from fermentation broth without the addition of an acidification column before the ion exchange resin. Studies have indicated, at its pK_a, that the maximum adsorption efficiency of Amberlite IRA-67 for acetic acid is around 61.36% at 25 °C, with acid loading on the resin as high as 33 g acetic acid/g resin [151]. Other studies showed maximum propionic acid loading onto Amberlite IRA-67 resin of around 36 g/g resin [152]. Currently, the only study on using ion exchange resin such as Amberlite IRA-67 for in situ product recovery was done by Garrett et al. [82] for anaerobic fermentation of corn stover using *B. coagulans* at a pH of 5.5 (well above the pK_a of 3.86) and a temperature of 50 °C. The resin loading for acetic acid and lactic acid were determined to be 3.81 mg/g resin and 170.2 mg/g resin, respectively. However, through the use of a continuous product recovery loop controlled by fermentation pH, the lactic acid produced during fermentation could be continuously separated by the resin and this process removed the necessity of a separate pH control for the fermentation, while also removing product inhibition due to lactate accumulation. However, if an acidification resin such as Amberlite IR-120 was connected to the product recovery loop before the ion exchange resin to decrease the pH of the effluent, the acid loading onto the weak/strong base resin could be sufficiently increased. Studies done by Bishai et al. [144] and Monteagudo & Aldavero [143], as shown in Table 1, have attempted this with strong and weak base resins, respectively, connected to lactate fermentation, resulting in increased product recovery and acid loading on the resin.

4.2. Separation Using Solvent Extraction

One of the primary disadvantages with using solvent extraction for in situ product recovery during fermentation is the difficulty of finding a common biocompatible solvent that has a high extraction coefficient for the product. Another problem is the ability of the solvents to function at the fermentation pH (ranging between 5.5 and 7), which is the interesting range for carboxylic acid production [153]. The primary problem is related to phase separation and, hence, different tertiary and quaternary amines need to be used for reactive extraction of the carboxylic acids from the fermentation broth. Unlike solvent extraction, in reactive extraction, the acid is extracted onto an organic phase and complexed with a carrier (additive) or complexing agent, resulting in higher extraction efficiencies. Several studies using amines with oleyl alcohol, iso-decanol, and kerosene as diluents or additives to the amines for effective in situ extraction of different carboxylic acids from fermentation broth are shown in Table 1. The most commonly used amines were Alamine-336 and Hostarex-A327, while other studies examined other amines and found their performance in carboxylic acid extraction to be comparatively low. Studies have further been done using tri-*n*-octylphosphine dissolved in methyl isobutyl ketone, which showed higher distribution coefficients for propionic and butyric acids when compared to these amines for reactive extraction from fermentation broth [154]. These studies also showed that the phase separation during reactive extraction could be increased by increasing the contact of the acids with the amines using membranes such as hollow fibers (see Table 1). While the reactive extraction using amines had similar efficiency when compared to ion exchange resins, there were problems related to the complete recovery of the amines during back-extraction [140] and separation of the diluent resulted in the need for additional separation steps during downstream processing. The loss of cells also presented a problem and can be controlled by immobilization, which can negatively affect the process operation costs.

Due to the volatile nature of the VFAs, reactive extraction of acids from fermentation broth has further been tested using pervaporation or supercritical fluid extraction. In pervaporation, the fermentation effluent containing two or more miscible components contacted a non-porous/ molecularly porous membrane with a vacuum applied on the other side to facilitate extraction [155]. In supercritical fluid extraction, supercritical fluids such as carbon dioxide were pressurized, heated, and allowed to diffuse through the fermentation effluent with the conditions controlled to selectively and efficiently remove VFAs [156]. Both these techniques are energy-intensive and, due to the harmful nature of pressure and temperature on the microbes, require an extra filtration step to prevent any contact with the microbes. However, these methods were found to have very high fluxes and were capable of achieving high yields. Pertraction was used with 20 wt % Hostarex in oleyl alcohol as the extractant, resulting in an increased butyric acid titer during fermentation from 7.3 g/L at control to as high as 20 g/L [136]. This study showed that in the absence of pertraction but with amines included, the extraction efficiency only increased by 1.4-fold, while with pertraction the increase was significantly higher. In another study with in situ separation using pertraction, butyric acid and hexanoic acid productivity was found to increase by at least 3-fold while using pertractive fermentation when compared to batch fermentation [141]. These studies indicated that oleyl alcohol with trioctylamine as the extractant was found to be less inhibitory to the microbial culture during pertractive fermentation when compared to octanol-trihexylamine extractant. There have been relatively fewer studies on the supercritical fluid extraction of carboxylic acids from the fermentation broths, which was found to work better at acidic pH [156]. Garrett et al. [157] used supercritical carbon dioxide to remove acetic acid from model fermentation broths, resulting in 93% acid recovery at optimized conditions of 2150 psi, 45 °C, and 5 h extraction time.

4.3. Effect of Product Concentration on Separation Processes

It should be noted that these separation procedures are significantly limited by concentration, i.e., the higher the concentration of the acid in the solution the higher the extraction efficiency at optimized conditions. Studies done by Garrett et al. [157] showed that the acetic acid extraction from fermentation broth decreased from 93% at an initial acid concentration of 92.4 g/L to around 81% at an initial concentration of 10 g/L. These studies showed that there was a significant effect of initial acid concentration on the amount of carbon dioxide used for extraction. The studies also showed that the carboxylic acid yield as a function of amount of CO_2 used for extraction decreased from around 6.5 mg acetic acid extracted per g CO_2 used when the initial acetic acid concentration was 92.4 g/L to as low as 0.5 mg acetic acid extracted per g CO_2 used at an initial acetic acid concentration of 10 g/L. The concentration effect is also applicable, in a slightly different manner, in the case of ion exchange resin-mediated separation. The concentration of acid ions in the fermentation broth is a function of the pH of the fermentation and with an increased fermentation pH above the pK_a of the carboxylic acid, there is a decreased concentration of the acids in the form of ions versus that in the form of inactive salts. For example, at pH 5.5, approximately 95% of the lactic acid produced through fermentation exists in the form of sodium lactate, which has a lower tendency to bind to the ion exchange resin, whereas only 5% exists as lactate ions that can be extracted using ion exchange resin. This usually results in the requirement of an extra acidification step in the product recovery loop to extract all the lactic acid from the fermentation broth. Hence, a concentration step is necessary before product separation (as shown in Figure 3), which can vary between acidification in the case of ion exchange-mediated separation and evaporation or membranes in the case of reactive extraction or precipitation (an example of a membrane is the hollow fibers used concurrently with pertraction for separation from fermentation broths).

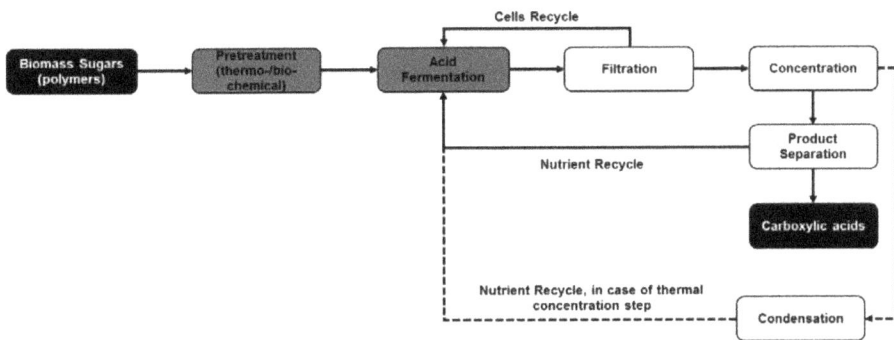

Figure 3. Efficient processing pathway outlining different stages in biomass conversion with fermentation and in situ product separation for biorefinery applications. (– shows that processing pathway that uses direct nutrient cycle due to non-thermal separation processes while - - shows the nutrient recycle loop in processing pathways that uses thermal separation processes such as evaporation or distillation requiring an extra processing step before nutrient recycle).

4.4. Separation Using Electrodialysis

Another important technology for in situ product recovery of carboxylic acids is electrodialysis using an electrical gradient to facilitate ionic transport towards electrodes for optimal acid removal. Several electrodialysis processes have been tested for acid recovery including conventional electrodialysis: electro-metathesis, electro-ion substitution, electro-deionization, bipolar membrane electrodialysis, etc. [158]. While industries currently use crystallization and distillation techniques for product recovery, which are inefficient and can be energy-intensive and in some cases result in product loss or transformation [159], electrodialysis can result in high acid purity, as shown in the studies elucidated in Table 1. However, electrodialysis is also concentration-limiting but not in a similar manner to the other techniques discussed above. Higher acid and cell loading in the fermenter effluent can result in concentration polarization during electrodialysis after a certain time (depending on the current density), resulting in inefficient separation performance [158]. Hence, the concentration step from Figure 3 would be ill-suited to using electrodialysis for acid separation. One of the primary disadvantages of electrodialysis is the electricity requirements, which can result in high processing costs. For example, studies have indicated that the power requirements for lactic acid separation using electrodialysis can vary from 0.21 to 0.71 kWh/kg acid produced under optimized conditions for maximum lactate recovery [160,161]. However, even with the membrane fouling problems and electricity requirements, electrodialysis has shown significant promise (through improvements in membrane quality) in in situ extraction of carboxylic acid from fermentation broths [133,159].

An extension of electrodialysis known as reverse electro-enhanced electrodialysis (REED) has been successfully used to minimize fouling problems during the in situ extraction of butyric acid [132] and lactic acid [162]. As indicated previously, both these carboxylic acids have a significant inhibitory effect on the microbial productivity during fermentation at high concentrations and require in situ removal. The REED process is equipped with a mechanism that allows for polarity reversal during times of imposed electrical potential gradient that will help clear concentration polarization during continuous operation, allowing for Donnan dialysis operation [163]. Such systems can significantly reduce the electricity costs during operation since they reduce the time of less-than optimal performance of the electrodialytic separation during continuous operation. However, studies are currently being done to further reduce processing costs by developing low-cost, robust materials to be used as electrodes and membranes during electrodialysis.

5. Conclusions and Future Perspectives

In this review we have discussed different microbes and in situ product recovery processes for carboxylic acid production from lignocellulosic biomass for biorefinery applications. Only limited focus was devoted to the different types of fermentation processes used since a number of studies have concluded that, for most biochemical processes, continuous fermentation (by controlling feeding rate and retention time) has significant advantages over fed-batch or batch fermentations [90,164]. The production of specific carboxylic acids usually necessitates specific pure cultures. Pure cultures are, however, easily affected by microbial contamination, resulting in a requirement for high sterility, which can add to the production costs. Using a mixed microbial consortium such as a rumen to develop a selective culture for the production of carboxylic acids works well without the need for strain engineering of specific pure microbes. The presence of cellulose- and hemicellulose-degrading strains in mixed cultures such as a mixed rumen culture also serves as an advantage, resulting in minimizing or even eliminating the need for external enzymes in biorefineries, which can further reduce processing costs [13]. Carbohydrate and protein metabolism in the rumen during cellulose fermentation was studied to understand the effect of fermentation conditions on the sugar-degrading bacteria; such studies indicated an increase in the concentration of cellulolytic bacteria when the system was left unperturbed [165]. However, further studies are required to fully understand the nature of the enzymes and microbial populations active during the anaerobic fermentation of biomass using mixed microbial cultures. Such studies can provide scientific input for increasing validation and data reproducibility and can pave the way for the application of mixed culture fermentations in biorefineries.

Apart from optimization of the microbial strains (or consortia) used during fermentation, further improvements are required to facilitate in situ product recovery during fermentation. We have discussed the advantages and disadvantages of different separation processes for the in situ separation of carboxylic acids from fermentation effluent. The ideal separation process will sufficiently and selectively separate the carboxylic acid while having little or no effect on the fermentation process itself. There is sufficient evidence that in situ product recovery has a positive effect on both microbial productivity (through the elimination or minimization of product inhibition) and product purity (through optimization for selective removal of the desired carboxylic acid product), thereby producing a carboxylic acid intermediate that can be directly upgraded to fuels and/or used as chemicals. However, it is important to gain further knowledge on how to integrate the product recovery with the fermentation processes, especially related to the effects of product concentrations and membrane fouling. The development and optimization of such in situ product separation processes can enhance the quality of the product produced from the biorefineries while reducing the overall cost to the biorefineries.

Acknowledgments: This work was done with support from the United States Department of Agriculture (USDA)-National Institute of Food and Agriculture (NIFA) under the Northwest Advanced Renewables Alliance (NARA) (Grant# 2011–68005–30416) and Denmark's Energinet program ForskEL-funded XEL2Gas project (#12437).

Author Contributions: All the authors contributed equally to the review article.

Conflicts of Interest: The authors declare no conflict of interest.

References

1. Lange, J.-P. Lignocellulose conversion: An introduction to chemistry, process and economics. *Biofuels Bioprod. Bioref* **2007**, *1*, 39–48. [CrossRef]
2. Baumann, I.; Westermann, P. Microbial production of short chain fatty acids from lignocellulosic biomass: Current processes and market. *BioMed Res. Intl.* **2016**, 1–15. [CrossRef] [PubMed]
3. Worrell, E.; Phylipsen, D.; Einstein, D.; Martin, N. Energy Use and Energy Intensity of the US Chemical Industry. US DOE Report LBNL-44314. 2000. Available online: http://ateam.lbl.gov/PUBS/doc/LBNL-44314.png (accessed on 8 November 2016).
4. Micromarket Monitor. North America Acetic Acid Market by Application (Vinyl Acetate Monomer (VAM), Purified Terephthalic Acid (PTA), Acetic Anhydride, Ester Solvents & Others) & by Country-Trends & Forecasts to 2019. Report Code AC 1086. 2015. Available online: http://www.micromarketmonitor.com/market/north-america-acetic-acid-9541396535.html?utm_source=NL-NAAAM&utm_medium=NL-NAAAM&utm_campaign=NL-NAAAM (accessed on 25 May 2015).
5. Renewable Chemicals Market–Alcohols (Ethanol, Methanol), Biopolymers (Starch Blends, Regenerated Cellulose, PBS, bio-PET, PLA, PHA, bio-PE, and Others), Platform Chemicals & Others-Global Trends & Forecast to 2020. Report Code CH 2063. 2015. Available online: http://www.marketsandmarkets.com/Market-Reports/renewable-chemical-274.html (accessed on 1 October 2015).
6. Brindle, F. American chemistry council: Global chemical production looks positive. *Energy Glob.* **2016**. Available online: https://www.energyglobal.com/downstream/petrochemicals/13042016/american-chemistry-council-positive-growth-cpri-global-chemical-production-3023/ (accessed on 13 April 2016).
7. Metzger, J.O.; Huttermann, A. Sustainable global energy supply based on lignocellulosic biomass from afforestation of degraded areas. *Naturwissenschaften* **2009**, *96*, 279–288. [CrossRef] [PubMed]
8. Kan, T.; Strezov, V.; Evans, T.J. Lignocellulosic biomass pyrolysis: a review of product properties and effects of pyrolysis parameters. *Ren. Sust. Energy Rev.* **2016**, *57*, 1126–1140. [CrossRef]
9. Kumar, A.; Jones, D.D.; Hanna, M.A. Thermochemical biomass gasification: A review of the current status of the technology. *Energies* **2009**, *2*, 556–581. [CrossRef]
10. Elliott, D.C.; Biller, P.; Ross, A.B.; Schmidt, A.J.; Jones, S.B. Hydrothermal liquefaction of biomass: Developments from batch to continuous process. *Bioresour. Technol.* **2015**, *178*, 147–156. [CrossRef] [PubMed]
11. Ghosh, D.; Dasgupta, D.; Agrawal, D.; Kaul, S.; Adhikari, D.K.; Kurmi, A.K.; Arya, P.K.; Bangwal, D.; Negi, M.S. Fuels and chemicals from lignocellulosic biomass: An integrated biorefinery approach. *Energy Fuels* **2015**, *29*, 3149–3157. [CrossRef]
12. Liao, J.C.; Mi, L.; Pontrelli, S.; Luo, S. Fuelling the future: Microbial engineering for the production of sustainable biofuels. *Nat. Rev. Microbiol.* **2016**, *14*, 288–304. [CrossRef] [PubMed]
13. Murali, N.; Fernandez, S.; Ahring, B.K. Fermentation of wet-exploded corn stover for the production of volatile fatty acids. *Bioresour. Technol.* **2017**, *227*, 197–204. [CrossRef] [PubMed]
14. Raposo, F.; Borja, R.; Cacho, J.A.; Mumme, J.; Orupold, K.; Esteves, S.; Nogruel-Arias, J.; Picard, S.; Nielfa, A.; Scherer, P.; et al. First international comparative study of volatile fatty acids in aqueous samples by chromatographic techniques: Evaluating sources of error. *Trends Anal. Chem.* **2013**, *51*, 127–143. [CrossRef]
15. Agler, M.T.; Wrenn, B.A.; Zinder, S.H.; Angenent, L.T. Waste to bioproducts conversion with undefined mixed cultures: the carboxylate platform. *Trends Biotechnol.* **2011**, *29*, 70–78. [CrossRef] [PubMed]
16. GBIR Research. Acetic Acid Global Market to 2020-Surge in VAM and PTA Sectors in Asia Pacific to Drive Global Demand. Report Code GBICH0082MR. 2013. Available online: http://www.gbiresearch.com/report-store/market-reports/archive/acetic-acid-global-market-to-2020-surge-in-vam-and-pta-sectors-in-asia-pacific-to-drive-global-demand (accessed on 1 February 2013).
17. Yoneda, N.; Kusano, S.; Yasui, M.; Pujado, P.; Wilcher, S. Recent advances in processes and catalysts for the production of acetic acid. *Appl. Catal. A* **2001**, *221*, 253–265. [CrossRef]
18. Wood, B.J.B. *Microbiology of Fermented Foods*, 2nd ed.; Springer: New York, NY, USA, 1998.

19. Sengun, I.Y.; Karabiyikli, S. Importance of acetic acid bacteria in food industry. *Food Control* **2011**, *22*, 647–656. [CrossRef]

20. Chavez, K.L.; Hess, D.W. A novel method of etching copper oxide using acetic acid. *J. Electrochem. Soc.* **2001**, *148*, G640–G643. [CrossRef]

21. Gandini, A. Polymers from renewable resources: a challenge for the future of macromolecular materials. *Macromolecules* **2008**, *41*, 9491–9504. [CrossRef]

22. Ganster, J.; Fink, H.-P. Cellulose and cellulose acetate. In *Bio-Based Plastics: Materials and Applications*, 1st ed.; Kabasci, S., Ed.; Wiley: West Sussex, UK, 2014; pp. 35–62.

23. Weissermel, K.; Arpe, H.-J. *Industrial Organic Chemistry*, 4th ed.; Wiley-VCH: Weinheim, Germany, 2003.

24. Luck, E.; Jager, M. Propionic acid. In *Antimicrobial Food Additives*, 2nd ed.; Luck, E., Jager, M., Eds.; Springer: Berlin, Germany, 1997; pp. 145–151.

25. Ali, S.H.; Tarakmah, A.; Merchant, S.Q.; Al-Sahhaf, T. Synthesis of esters: Development of the rate expression for the Dowex 50 Wx8-400 catalyzed esterification of propionic acid with 1-propanol. *Chem. Eng. Sci.* **2007**, *62*, 3197–3217. [CrossRef]

26. El-Shahawy, T.A.E.-G. Chemicals with a natural reference for controlling water hyacinth, *Eichhornia crassipes* (Mart.) *Solms. J. Plant Protec. Res.* **2015**, *55*, 294–300. [CrossRef]

27. Ihre, H.; Hult, A.; Soderlind, E. Synthesis, characterization and 1H NMR self-diffusion studies of dendritic aliphatic polyesters based on 2,2-bis (hydroxymethyl) propionic acid and 1,1,1-Tris(hydroxyphenyl)ethane. *J. Am. Chem. Soc.* **1996**, *118*, 6388–6395. [CrossRef]

28. Propionic Acid & Derivatives Market by Applications (Animal Feed & Grain Preservatives, Food Preservatives, Herbicides, Cellulose Acetate Propionate) & Geography–Global Trends & Forecasts to 2018. Report Code CH 1533. 2013. Available online: http://www.marketsandmarkets.com/PressReleases/propionic-acid-derivatives.asp (accessed on 1 July 2013).

29. Playne, M.J. Propionic and butyric acids. In *Comprehensive Biotechnology*; Moo-Young, M., Ed.; Pergamon Press: Oxford, UK, 1985.

30. Jha, A.K.; Li, J.; Yuan, Y.; Baral, N.; Ai, B. A review on bio-butyric acid production and its optimization. *Int. J. Agric. Biol.* **2014**, *16*, 1019–1024.

31. Entin-Meer, M.; Rephaeli, A.; Yang, X.; Nudelman, A.; VandenBerg, S.R.; Haas-Kogan, D.A. Butyric acid prodrugs are histone deacetylase inhibtiors that show antineoplastic activity and radiosensitizing capacity in the treatment of malignant gliomas. *Mol. Cancer Ther.* **2005**, *4*, 1952–1961. [CrossRef] [PubMed]

32. Armstrong, D.W.; Yamazaki, H. Natural flavours production: A biotechnological approach. *Trends Biotechnol.* **1986**, *4*, 264–268. [CrossRef]

33. Butyric Acid Market by Application (Animal Feed, Chemical Intermediate, Foods & Flavors, Pharmaceuticals, Perfumes, Others), by Type (Synthetic Butyric Acid, Renewable Butyric Acd) by Geography (APAC, North America, Europe, ROW)—Global Analysis and Forecast to 2020. Report Code CH 3662. 2015. Available online: http://www.marketsandmarkets.com/Market-Reports/butyric-acid-market-76962011.html (accessed on 8 January 2015).

34. Holten, C.H.; Muller, A.; Rehbinder, D. *Lactic Acid*; International Research Association, Verlag Chemie: Copenhagen, Denmark, 1971.

35. John, R.P.; Nampoothiri, K.M.; Pandey, A. Fermentative production of lactic acid from biomass: An overview on process developments and future perspectives. *Appl. Microb. Biotechnol.* **2007**, *74*, 524–534. [CrossRef] [PubMed]

36. Datta, R.; Tsai, S.-P.; Bonsignore, P.; Moon, S.-H.; Frank, J.R. Technological and economical of poly (lactic acid) and lactic acid derivatives. *FEMS Mircob. Rev.* **1995**, *16*, 221–231. [CrossRef]

37. Datta, R.; Henry, M. Lactic acid: Recent advances in products, processes and technologies—A review. *J. Chem. Technol. Biotechnol.* **2006**, *81*, 1119–1129. [CrossRef]

38. Lactic Acid Market by Application (Biodegradable Polymer, Food & Beverage, Personal Care & Pharmaceutical) & Polylactic Acid Market by Application (Packaging, Agriculture, Automobile, Electronics, Textile), & by Geography-Global Trends & Forecasts to 2020. 2015. Available online: http://www.rnrmarketresearch.com/lactic-acid-market-by-application-biodegradable-polymer-food-beverage-personal-care-pharmaceutical-polylactic-acid-market-by-application-packaging-agriculture-automobile-electronics-tex-market-report.html (accessed on 12 March 2015).

39. Dence, C.W.; Lin, S.Y. General structural features of lignin. In *Methods in Lignin Chemisry*; Lin, S.Y., Dence, C.W., Eds.; Springer: Berlin/Heidelberg, German, 1992.

40. Cao, S.; Pu, Y.; Studer, M.; Wyman, C.; Ragauskas, A.J. Chemical transformations of *Populus trichocarpa* during dilute acid pretreatment. *RSC Adv.* **2012**, *2*, 10925–10936. [CrossRef]

41. Tejado, A.; Pena, C.; Labidi, J.; Echeverria, J.M.; Mondragon, I. Physico-chemical characterization of lignins from different sources for use in phenol-formaldehyde resin synthesis. *Bioresour. Technol.* **2007**, *98*, 1655–1663. [CrossRef] [PubMed]

42. Ahring, B.K.; Biswas, R.; Ahamed, A.; Teller, P.J.; Uellendahl, H. Making lignin accessible for anaerobic digestion by wet-explosion pretreatment. *Bioresour. Technol.* **2015**, *175*, 182–188. [CrossRef] [PubMed]

43. Schiel-Bengelsdorf, B.; Durre, P. Pathway engineering and synthetic biology using acetogens. *FEBS Lett.* **2012**, *586*, 2191–2198. [CrossRef] [PubMed]

44. Ragsdale, S.W.; Pierce, E. Acetogenesis and the Wood-Ljundahl pathway of CO_2 fixation. *Biochim. Biophys. Acta.* **2008**, *1784*, 1873–1898. [CrossRef] [PubMed]

45. Ni, B.J.; Liu, H.; Nie, Y.Q.; Zeng, R.J.; Du, G.C.; Chen, J.; Yu, H.Q. Coupling glucose fermentation and homoacetogenesis for elevated acetate production: experimental and mathematical approaches. *Biotechnol. Bioeng.* **2010**, *108*, 345–353. [CrossRef] [PubMed]

46. Schuchmann, K.; Muller, V. Autotrophy at the thermodynamic limit of life a model for energy conservation in acetogenic bacteria. *Natur. Rev. Microbiol.* **2014**, *12*, 809–821. [CrossRef] [PubMed]

47. Islam, M.A.; Zengler, K.; Edwards, E.A.; Mahadevan, R.; Stephanopoulos, G. Investigating *Moorella thermoacetica* metabolism with a genome-scale constraint-based metabolic model. *Integr. Biol.* **2015**, *7*, 869–882. [CrossRef] [PubMed]

48. Drake, H.L.; Daniel, S.L. Physiology of the thermophilic acetogen *Moorella thermoacetica*. *Res. Microbiol.* **2004**, *155*, 869–883. [CrossRef] [PubMed]

49. Xue, J.; Isern, N.G.; Ewing, R.J.; Liyu, A.V.; Sears, J.A.; Knapp, H.; Iversen, J.; Sisk, D.R.; Ahring, B.K.; Majors, P.D. New generation NMR bioreactor coupled with high-resolution NMR spectroscopy leads to novel discoveries in *Moorella thermoacetica* metabolic profiles. *Envrion. Biotechnol.* **2014**, *98*, 8367–8375. [CrossRef] [PubMed]

50. Balasubramaniam, N.; Kim, J.S.; Lee, Y.Y. Fermentation of xylose into acetic acid by *Clostridium thermoaceticum*. *J. Appl. Biochem. Biotechnol.* **2001**, 367–376. [CrossRef]

51. Ehsanipour, M.; Suko, A.V.; Bura, R. Fermentation of lignocellulosic sugars to acetic acid by *Moorella thermoacetica*. *J. Ind. Microbiol. Biotechnol.* **2016**, *43*, 807–816. [CrossRef] [PubMed]

52. Gaddy, J.L.; Clausen, E.C. *Clostridium ljungdahlii*, an anaerobic ethanol and acetate producing microorganism. U.S. Patent 5,173,429, 1992.

53. Phillips, J.R.; Clausen, E.C.; Gaddy, J.L. Synthesis gas as substrate for the biological production of fuels and chemicals. *Appl. Biochem. Biotechnol.* **1994**, *45*, 145–157. [CrossRef]

54. Heise, R.; Muller, V.; Gottschalk, G. Sodium dependence of acetate formation by the acetogenic bacterium *Acetobacterium wooddii*. *J. Bacteriol.* **1989**, *171*, 5473–5478. [CrossRef] [PubMed]

55. Muller, V.; Bowien, S. Differential effects of sodium ions on motility in the homoacetogeniic bacteria *Acetobacterium wooddii* and *Sporomusa sphaeroides*. *Arch. Microbiol.* **1995**, *164*, 363–369. [CrossRef]

56. Tracy, B.P.; Jones, S.W.; Fast, A.G.; Indurthi, D.C.; Papoutsakis, E.T. Clostridia: The importance of their exceptional substrate and metabolite diversity for biofuel and biorefinery applications. *Curr. Opin. Biotechnol.* **2012**, *23*, 364–381. [CrossRef] [PubMed]

57. Das, A.; Ljungdahl, L.G. Electron-transport system in acetogens. In *Biochemistry and Physiology of Anaerobic Bacteria*; Ljungdahl, L.G., Adams, M.W., Barton, L., Ferry, J.G., Johnson, M.K., Eds.; Springer: New York, NY, USA, 2003.

58. Huang, H.; Wang, S.; Moll, J.; Thauer, R.K. Electron bifurcation involved in the energy metabolism of the acetoggenic bacterium *Moorella thermoacetica* growing on glucose or H_2 plus CO_2. *J. Bacteriol.* **2012**, *194*, 3689–3699. [CrossRef] [PubMed]

59. Mock, J.; Wang, S.; Huang, H.; Kahnt, J.; Thauer, R.K. Evidence for a hexaheterometric methylenetetrahydrofolate reductase in *Moorella thermoacetica*. *J. Bacteriol.* **2014**, *196*, 3303–3314. [CrossRef] [PubMed]

60. Kopke, M.; Held, C.; Hujer, S.; Liesegang, H.; Wiezer, A.; Wollherr, A.; Ehrenreich, A.; Liebl, W.; Gottschalk, G.; Durre, P. *Clostridium ljungdahlii* represents a microbial production platform based on syngas. *Proc. Natl. Acad. Sci. USA* **2010**, *107*, 13087–13092. [CrossRef] [PubMed]

61. Zhang, C.; Yang, H.; Yang, F.; Ma, Y. Current progress on butyric acid production by fermentation. *Curr. Microbiol.* **2009**, *59*, 656–663. [CrossRef] [PubMed]

62. Zhu, Y.; Yang, S.T. Effects of pH on metabolic pathway shift in fermentation of xylose by *Clostridium tyrobutyricum*. *J. Biotechnol.* **2004**, *110*, 143–157. [CrossRef] [PubMed]

63. Liu, X.; Zhu, Y.; Yang, S.T. Construction and characterization of ack deleted mutant of *Clostridium tyrobutyricum* for enhanced butyric acid and hydrogen production. *Biotechnol. Prog.* **2006**, *22*, 1265–1275. [CrossRef] [PubMed]

64. Lutke-Eversloh, T.; Bahl, H. Metabolic engineering of *Clostridium acetobutylicum*: Recent advances to improve butanol production. *Curr. Opin. Biotechnol.* **2011**, *22*, 634–647. [CrossRef] [PubMed]

65. Lee, J.; Jang, Y.S.; Choi, S.J.; Im, J.A.; Song, H.; Cho, J.H.; Seung, D.Y.; Papoutsakis, T.; Bennett, G.N.; Lee, S.Y. Metabolic engineering of *Clostridium acetobutylicum* ATCC 824 for isopropanol-butanol-ethanol fermentation. *App. Environ. Microbiol.* **2012**, *78*, 1416–1423. [CrossRef] [PubMed]

66. Wiegel, J.; Kuk, S.U.; Kohring, G.W. *Clostridium thermobutyricum* sp. nov., a moderate thermophile isolated from a cellulolytic culture, that produces butyrate as the major product. *Int. J. Sys. Bacteriol.* **1989**, *39*, 199–204. [CrossRef]

67. Zhang, C.H.; Ma, Y.J.; Yang, F.X.; Liu, W.; Zhang, Y.-D. Optimization of medium composition for butyric acid production by *Clostridium thermobutyricum* using response surface methodology. *Bioresour. Technol.* **2009**, *100*, 4284–4288. [CrossRef] [PubMed]

68. Bahl, H.; Gottschalk, G. Parameters affecting solvent production by *Clostridium acetobutylicum* in continuous culture. *Biotechnol. Bioeng. Symp.* **1984**, *14*, 215–233.

69. Vandak, D.; Tomaska, M.; Zigova, J.; Sturdik, E. Effect of growth supplements and whey pretreatment on butyric-acid production by *Clostridium butyricum*. *World J. Microbiol. Biotechnol.* **1995**, *11*, 363. [CrossRef] [PubMed]

70. Ueki, A.; Akasaka, H.; Suzuki, D.; Ueki, K. *Paludibacter propionicigenes* gen. nov., sp. nov., a novel strictly anaerobic, gram-negative, propionate-producing bacterium isolated from plant residue in irrigated rice-field soil in Japan. *Int. J. Sys. Evol. Microbiol.* **2006**, *56*, 39–44. [CrossRef] [PubMed]

71. Narihiro, T.; Kim, N.K.; Mei, R.; Nobu, M.K.; Liu, W.-T. Microbial community analysis of anaerobic reactors treating soft drink wastewater. *PLoS ONE* **2015**, *10*, e0119131. [CrossRef] [PubMed]

72. Qiu, Y.L.; Kuang, X.Z.; Shi, X.S.; Yuan, X.Z.; Guo, R.B. *Paludibacter jiangxiensis* sp. nov., a strictly anaerobic, propionate-producing bacterium isolated from rice paddy field. *Arch. Microbiol.* **2014**, *196*, 149–155. [CrossRef] [PubMed]

73. Oshio, S.; Tahata, I.; Minato, H. Effect of diets differing in ratios of roughage to concentrate on microflora in the rumen of heifers. *J. Gen. Appl. Microbiol.* **1987**, *33*, 99–111. [CrossRef]

74. Paynter, M.J.; Elsdent, S.R. Mechanism of propionate formation by *Selenomonas ruminatium*, a rumen microorganism. *Microbiology* **1970**, *61*, 1–7.

75. Akedo, M.; Cooney, C.L.; Sinskey, A.J. Direct demonstration of lactate-acrylate interconversion in *Clostridium propionicum*. *Natur. Biotechnol.* **1983**, *1*, 791–794. [CrossRef]

76. Johns, A.T. The mechanism of propionic acid formation by *Clostridium propionicum*. *J. Gen. Microbiol.* **1952**, *6*, 123–127. [CrossRef] [PubMed]

77. Rodriguez, B.A.; Stowers, C.C.; Pham, V.; Cox, B.M. The production of propionic acid, propanol and propylene via sugar fermentation: an industrial perspective on the progress, technical challenges and future outlook. *Green Chem.* **2014**, *16*, 1066–1076. [CrossRef]

78. Himmi, E.H.; Bories, A.; Boussaid, A.; Hassani, L. Propionic acid fermentation of glycerol and glucose by *Propionibacterium acidipropionici* and *Propionibacterium freudenreichii* ssp. *shermanii*. *Appl. Microbiol. Biotechnol.* **2000**, *53*, 435–440. [CrossRef] [PubMed]

79. Stine, A.; Zhang, M.; Ro, S.; Clendennen, S.; Shelton, M.C.; Tyo, K.E.J.; Broadbelt, L.J. Exploring de novo metabolic pathways from pyruvate to propionic acid. *Biotechnol. Prog.* **2016**, *32*, 303–311. [CrossRef] [PubMed]

80. Holzapfel, W.H.; Wood, B.J.B. *Lactic Acid Bacteria—Biodiversity and Taxonomy*; John Wiley & Sons, Wiley Blackwell Publishing: Oxford, UK, 2014.

81. Claesson, M.J.; van Sinderen, D.; O'Toole, P.W. The genus *Lactobacillus*–a genomic basis for understanding its diversity. *FEMS Microbiol. Lett.* **2007**, *269*, 22–28. [CrossRef] [PubMed]

82. Garrett, B.G.; Srinivas, K.; Ahring, B.K. Performance and stability of Amberlite IRA-67 ion exchange resin for product extraction and pH control during homolactic fermentation of corn stover sugars. *Biochem. Eng. J.* **2015**, *94*, 1–8. [CrossRef]

83. Nolasco-Hipolito, C.; Matsunaka, T.; Kobayashi, G.; Sonomoto, K.; Ishizaki, A. Synchronized fresh cell bioreactor system for continuous L-(+)-lactic acid production using *Lactococcus lactis* IO-1 in hydrolysed sago starch. *J. Biosci. Bioeng.* **2002**, *93*, 281–287. [CrossRef]

84. Kadam, S.R.; Patil, S.S.; Bastawde, K.B.; Khire, J.M.; Gokhale, D.V. Strain improvement of *Lactobacillus delbruecki* NCIM 2365 for lactic acid production. *Proc. Biochem.* **2006**, *41*, 120–126. [CrossRef]

85. Tango, M.; Ghaly, A. A continuous lactic acid production system using an immobilized packed bed of *Lactobacillus helveticus*. *Appl. Microbiol. Biotechnol.* **2002**, *58*, 712–720. [PubMed]

86. Hujanen, M.; Linko, S.; Linko, Y.-Y.; Leisola, M. Optimization of media and cultivation conditions for L (+)(S)-lactic acid production by *Lactobacillus casei* NRRL B-441. *Appl. Microbiol. Biotechnol.* **2001**, *56*, 126–130. [CrossRef] [PubMed]

87. Romero-Garcia, S.; Hernandez-Bustos, C.; Merino, E.; Gosset, G.; Martinez, A. Homolactic fermentation from glucose and cellobiose using *Bacillus subtilis*. *Microb. Cell Fac.* **2009**, *8*, 23. [CrossRef] [PubMed]

88. Payot, T.; Chemaly, Z.; Fick, M. Lactic acid production by *Bacillus coagulans*–kinetic studies and optimization of culture medium for batch and continuous fermentations. *Enzym. Microb. Technol.* **1999**, *24*, 191–199. [CrossRef]

89. Pol, E.C.; Eggink, G.; Weusthuis, R.A. Production of L(+)-lactic acid from acid pretreated sugarcane bagasse using *Bacillus coagulans* DSM2314 in a simultaneous saccharification and fermentation strategy. *Biotechnol. Biofuels* **2016**, *9*, 248. [CrossRef] [PubMed]

90. Ahring, B.K.; Traverso, J.J.; Murali, N.; Srinivas, K. Continuous fermentation of clarified corn stover hydrolysate for the production of lactic acid at high yield and productivity. *Biochem. Engg. J.* **2016**, *109*, 162–169. [CrossRef]

91. Zhou, X.; Ye, L.; Wu, J.C. Efficient production of L-lactic acid by newly isolated thermophilic *Bacillus coagulans* WCP 10-4 with high glucose tolerance. *Appl. Microbiol. Biotechnol.* **2013**, *97*, 4309–4314. [CrossRef] [PubMed]

92. Baranofsky, J.J.; Schreurs, W.J.; Kashket, E.R. Uncoupling by acetic acid limits growth of and acetogenesis by *Clostridium thermoaceticum*. *Appl. Environ. Microbiol.* **1984**, *48*, 1134–1139.

93. Schwartz, R.D.; Keller, F.A. Isolation of a strain of *Clostridium thermoaceticum* capable of growth and acetic acid production at pH 4.5. *Appl. Environ. Microbiol.* **1982**, *43*, 117–123. [PubMed]

94. Paredes, C.J.; Alsaker, K.V.; Papoutsakis, E.T. A comparative genomic view of clostridial sporulation and physiology. *Nat. Rev. Microbiol.* **2005**, *3*, 969–978. [CrossRef] [PubMed]

95. Lee, K.S.; Bumbaca, D.; Kosman, J.; Setlow, P.; Jedrzejas, M.J. Structure of a protein-DNA complex essential for DNS protection in spores of *Bacillus* spores. *Proc. Nat. Acad. Sci. USA* **2008**, *105*, 2806–2811. [CrossRef] [PubMed]

96. Hu, S.; Zheng, H.; Gu, Y.; Zhao, J.; Zhang, W.; Yang, Y.; Wang, S.; Zhao, G.; Yang, S.; Jiang, W. Comparative genomic and transcriptomic analysis revealed genetic characteristics related to solvent formation and xylose utilization in *Clostridium acetobutylicum* EA 2018. *BMC Genom.* **2011**, *12*, 93. [CrossRef] [PubMed]

97. Liu, P.; Jarboe, L.R. Metabolic engineering of biocatalysts for carboxylic acids production. *Comput. Struct. Biotechnol. J.* **2012**, *3*, e201210011. [CrossRef] [PubMed]

98. Kim, T.W.; Chokhawala, H.A.; Nadler, D.C.; Blanch, H.W.; Clark, D.S. Binding modules alter the activity of chimeric cellulases: Effects of biomass pretreatment and enzyme source. *Biotechnol. Bioeng.* **2010**, *107*, 601–611. [CrossRef] [PubMed]

99. Hackmann, T.J.; Spain, J.N. Invited review: Ruminant ecology and evolution: Perspectives useful to ruminant livestock research and production. *J. Dairy Sci.* **2010**, *93*, 1320–1324. [CrossRef] [PubMed]

100. Dijkstra, J.; Forbes, J.M.; France, J. *Quantitative Aspects of Ruminant Digestion and Metabolism*; CABI Publishing: Cambridge, MA, USA, 1994.

101. Saleem, F.; Bouatra, S.; Guo, A.C.; Psychogios, N.; Mandal, R.; Dunn, S.M.; Ametaj, B.N.; Wishart, D.S. The bovine ruminal fluid metabolome. *Metabolomics* **2013**, *9*, 360–378. [CrossRef]

102. Ahring, B.K. Perspectives for anaerobic digestion. *Adv. Biochem. Eng. Biotechnol.* **2003**, *81*, 1–30. [PubMed]

103. Mah, R.A.; Ward, D.M.; Baresi, L.; Glass, T.L. Biogenesis of methane. *Annu. Rev. Microbiol.* **1977**, *31*, 309–341. [CrossRef] [PubMed]

104. Johnson, K.A.; Johnson, D.E. Methane emissions from cattle. *J. Anim. Sci.* **1995**, *73*, 2483–2492. [CrossRef] [PubMed]

105. Schulman, M.D.; Valentino, D. Factors influencing rumen fermentation: Effect of hydrogen on formation of propionate. *J. Dairy Sci.* **1976**, *59*, 1444–1451. [CrossRef]

106. Miller, T.L.; Wolin, M.J. Inhibition of growth of methane-producing bacteria of the ruminant forestomach by hydroxymethylglutaryl-SCoA reductase inhibitors. *J. Dairy Sci.* **2001**, *84*, 1445–1448. [CrossRef]

107. Van Kessel, J.A.S.; Russell, J.B. The effect of pH on ruminal methanogenesis. *FEMS Microbiol. Ecol.* **1996**, *20*, 205–210. [CrossRef]

108. Wang, Y.; Zhang, Y.; Wang, J.; Meng, L. Effects of volatile fatty acid concentrations on methane yield and methanogenic bacteria. *Biomass Bioenergy* **2009**, *33*, 848–853. [CrossRef]

109. Hess, M.; Sczyrba, A.; Egan, R.; Kim, T.W.; Chokhawala, H.; Schroth, G.; Luo, S.; Clark, D.S.; Chen, F.; Zhang, T.; et al. Metagenomic discovery of biomass-degrading genes and genomes from cow rumen. *Science* **2011**, *331*, 463–467. [CrossRef] [PubMed]

110. Weimer, P.J. Manipulating ruminal fermentation: a microbial ecological perspective. *J. Anim. Sci.* **1998**, *76*, 3114–3122. [CrossRef] [PubMed]

111. Perret, S.; Casalot, L.; Fierobe, H.-P.; Tardif, C.; Sabathe, F.; Belaich, J.-P.; Belaich, A. Production of heterologous and chimeric scaffoldins by *Clostridium acetobutylicum* ATCC 824. *J. Bacteriol.* **2004**, *186*, 253–257. [CrossRef] [PubMed]

112. Le Van, T.D.; Robinson, J.A.; Ralph, J.; Greening, R.C.; Smolenski, W.J.; Leedle, J.A.; Schaefer, D.M. Assessment of reductive acetogenesis with indigenous ruminal bacterium populations and *Acetitomaculum ruminis*. *Appl. Environ. Microbiol.* **1998**, *64*, 3429–3436. [PubMed]

113. Lopez, S.; Valdes, C.; Newbold, C.J.; Wallace, R.J. Influence of sodium fumarate addition on rumen fermentation in vitro. *Br. J. Nutr.* **1999**, *81*, 59–64. [PubMed]

114. Christy, P.M.; Gopinath, L.R.; Divya, D. A review on anaerobic decomposition and enhancement of biogas production through enzymes and microorganisms. *Ren. Sus. Energ. Rev.* **2014**, *34*, 167–173. [CrossRef]

115. Rieu-Lesme, F.; Morva, B.; Collins, M.D.; Fonty, G.; Willems, A. A new H_2/CO_2-using acetogenic bacterium from the rumen: Description of *Ruminococcus schinkii* sp. nov. *FEMS Microbiol. Lett.* **2006**, *140*, 281–286.

116. Leedle, J.A.; Greening, R.C. Postprandial changes in methanogenic and acidogenic bacteria in the rumens of steers fed high- or low-forage diets once daily. *Appl. Environ. Microbiol.* **1988**, *54*, 502–506. [PubMed]

117. Davidson, C.A.; Rehberger, T.G. Characterization of *Propionibacterium* isolated from rumen of lactating dairy cows. *Proc. Am. Soc. Microbiol.* **1995**, *1*, 6.

118. Russell, J.R.; Hino, T. Regulation of lactate production in *Streptococcus bovis*: A spiraling effect that contributes to rumen acidosis. *J. Dairy Sci.* **1985**, *68*, 1712–1721. [CrossRef]

119. Wells, J.E.; Krause, D.O.; Callaway, T.R.; Russell, J.B. A bacteriocin-mediated antagonism by ruminal lactobacilli against *Streptococcus bovis*. *FEMS Microbiol. Ecol.* **1997**, *22*, 237–243. [CrossRef]

120. Sharpe, M.E.; Latham, M.J.; Garvie, E.I.; Zirngibl, J.; Kandler, O. Two new species of *Lactobacillus* isolated from the bovine rumen, *Lactobacillus ruminis* sp. nov. and *Lactobacillus vitulinus* sp. nov. *J. Gen. Microbiol.* **1973**, *77*, 37–49. [CrossRef] [PubMed]

121. Williams, A.G.; Withers, S.E. Changes in the rumen microbial population and its activities during the refaunation period after the reintroduction of ciliate protozoa into the rumen of defaunated sheep. *Can. J. Microbiol.* **1993**, *39*, 61–69. [CrossRef] [PubMed]

122. Hobson, P.N.; Stewart, C.S. *The rumen microbial ecosystem*; Blackie academic & professional, Chapman & Hall: London, UK, 1997.

123. Bryant, M.P. Bacterial species of the rumen. *Bacteriol. Rev.* **1959**, *23*, 125–153. [PubMed]

124. Doetsch, R.N.; Howard, B.H.; Mann, S.O. Physiological factors in the production of an iodophilic polysaccharide from pentose by a sheep rumen bacterium. *J. Gen. Microbiol.* **1957**, *16*, 156–168. [CrossRef] [PubMed]

125. Bryant, M.P.; Small, N. Characteristics of two new genera of anaerobic curved rods isolated from the rumen of cattle. *J. Bacteriol.* **1956**, *72*, 22–26. [PubMed]

126. Dehority, B.A. Characterization of several bovine rumen bacteria isolated with a xylan medium. *J. Bacteriol.* **1966**, *91*, 1724–1729. [PubMed]

127. Pind, P.F.; Angelidaki, I.; Ahring, B.K. Dynamics of the anaerobic process: effects of volatile fatty acids. *Biotechnol. Bioeng.* **2003**, *82*, 791–801. [CrossRef] [PubMed]
128. Ahring, B.K.; Sandberg, M.; Angelidaki, I. Volatile fatty acids as indicators of process imbalance in anaerobic digestors. *Appl. Microbiol. Biotechnol.* **1995**, *43*, 559–565. [CrossRef]
129. Ahring, B.K. Status on science and application of thermophilic anaerobic digestion. *Water Sci. Technol.* **1994**, *30*, 241–249.
130. Ahring, B.K.; Wstermann, P. Kinetics of butyrate, acetate, and hydrogen metabolism in a thermophilic, anaerobic, butyrate-degrading triculture. *Appl. Environ. Microbiol.* **1987**, *53*, 434–439. [PubMed]
131. Angelidaki, I.; Ahring, B.K. Establishment and characterization of an anaerobic thermophilic 55 (deg) C enrichment culture degrading long-chain fatty acids. *Appl. Environ. Microbiol.* **1995**, *61*, 2442–2445. [PubMed]
132. Baroi, G.N.; Skiadas, I.V.; Westermann, P.; Gavala, H.N. Continuous fermentation of wheat straw hydrolysate by *Clostridium tyrobutyircum* with in-situ acids removal. *Wast. Biomass Valor.* **2015**, *6*, 317–320. [CrossRef] [PubMed]
133. Wang, X.; Wang, Y.; Zhang, X.; Feng, H.; Xu, T. In-situ combination of fermentation and electrodialysis with bipolar membranes for the production of lactic acid: continuous operation. *Bioresour. Technol.* **2013**, *147*, 442–448. [CrossRef] [PubMed]
134. Zhang, S.T.; Matsuoka, H.; Toda, K. Production and recovery of propionic and acetic acids in electrodialysis culture of *Propionibacterium shermanii*. *J. Ferm. Bioeng.* **1993**, *75*, 276–282. [CrossRef]
135. Jones, R.J.; Massanet-Nicolau, J.; Guwy, A.; Premier, G.C.; Dinsdale, R.M.; Reilly, M. Removal and recovery of inhibitory volatile fatty acids from mixed acid fermentations by conventional electrodialysis. *Bioresour. Technol.* **2015**, *189*, 279–284. [CrossRef] [PubMed]
136. Zigova, J.; Sturdik, E.; Vandak, D.; Schlosser, S. Butyric acid production by *Clostridium butyricum* with integrated extraction and pertraction. *Proc. Biochem.* **1999**, *34*, 835–843. [CrossRef]
137. Tik, N.; Bayraktar, E.; Mehmetoglu, U. In situ reactive extraction of lactic acid from fermentation media. *J. Chem. Technol. Biotechnol.* **2001**, *76*, 764–768. [CrossRef]
138. Jin, Z.; Yang, S.-T. Extractive fermentation for enhanced propionic acid production from lactose by *Propionibacterium acidipropionici*. *Biotechnol. Prog.* **1998**, *14*, 457–465. [CrossRef] [PubMed]
139. Chen, R.; Lee, Y.Y. Membrane-mediated extractive fermentation for lactic acid production from cellulosic biomass. *Appl. Biochem. Biotechnol.* **1997**, *63*, 435. [CrossRef] [PubMed]
140. Von Frieling, P.; Schugerl, K. Recovery of lactic acid from aqueous model solutions and fermentation broths. *Proc. Biochem.* **1999**, *34*, 685–696. [CrossRef]
141. Nelson, R.S.; Peterson, D.J.; Karp, E.M.; Beckham, G.T.; Salvachua, D. Mixed carboxylic acid production by *Megasphaera elsdenii* from glucose and lignocellulosic hydrolysate. *Fermentation* **2017**, *3*, 10. [CrossRef]
142. Ataei, S.A.; Vasheghani-Farahani, E.J. In situ separation of lactic acid from fermentation broth using ion exchange resins. *J. Ind. Microbiol. Biotechnol.* **2008**, *35*, 1229–1233. [CrossRef] [PubMed]
143. Monteagudo, J.M.; Aldavero, M. Production of L-lactic acid by *Lactobacillus delbrueckii* in chemostat culture using an ion exchange resins system. *J. Chem. Technol. Biotechnol.* **1999**, *74*, 627–634. [CrossRef]
144. Bishai, M.; de, S.; Adhikari, B.; Banerjee, R. A platform technology of recovery of lactic acid from a fermentation broth of novel substrate *Zizyphus oenophlia*. *Biotech* **2015**, *5*, 455–463. [CrossRef] [PubMed]
145. Yousuf, A.; Bonk, F.; Bastidas-Oyandel, J.-R.; Schmidt, J.E. Recovery of carboxylic acids produced during dark fermentation of food waste by adsorption on Amberlite IRA-67 and activated carbon. *Bioresour. Technol.* **2016**, *217*, 137–140. [CrossRef] [PubMed]
146. Freeman, A.; Woodley, J.M.; Lilly, M.D. In-situ product removal as a tool for bioprocessing. *Biotechnol.* **1993**, *11*, 1007–1012. [CrossRef]
147. Weil, J.R.; Dien, B.; Bothast, R.; Hendrickson, R.; Mosier, N.S.; Ladisch, M.R. Removal of fermentation inhibitors formed during pretreatment of biomass by polymeric adsorbents. *Ind. Eng. Chem. Res.* **2002**, *41*, 6132–6138. [CrossRef]
148. Xu, T. Ion exchange membranes: State of their development and perspective. *J. Membr. Sci.* **2005**, *263*, 1–29. [CrossRef]
149. Gonzalez, M.I.; Alvarez, S.; Riera, F.A.; Alvarez, R. Purification of lactic acid from fermentation broths by ion-exchange resins. *Ind. Eng. Chem. Res.* **2006**, *45*, 3242–3247. [CrossRef]
150. Dethe, M.J.; Marathe, K.V.; Gaikar, V.G. Adsorption of lactic acid on weak base polymeric resins. *J. Separ. Sci. Technol.* **2006**, *41*, 2947–2971. [CrossRef]

151. Uslu, H.; Inci, I.; Bayazit, S.S. Adsorption equilibrium data for acetic acid and glycolic acid onto Amberlite IRA-67. *J. Chem. Eng. Data* **2010**, *55*, 1295–1299. [CrossRef]

152. Uslu, H.; Inci, I.; Bayazit, S.S.; Demir, G. Comparison of solid-liquid equilibrium data for the adsorption of propionic acid and tartaric acid from aqueous solution onto Amberlite IRA-67. *Ind. Eng. Chem. Res.* **2009**, *48*, 7767–7772. [CrossRef]

153. Yang, S.T.; White, S.A.; Hsu, S.T. Extraction of carboxylic acids with tertiary and quarternary amines: effect of pH. *Ind. Eng. Chem. Res.* **1991**, *30*, 1335–1342. [CrossRef]

154. Keshav, A.; Wasewar, K.L.; Chand, S. Reactive extraction of propionic acid using tri-*n*-octylamine, tri-*n*-butyl phosphate and aliquat 336 in sunflower oil as diluent. *J. Chem. Technol. Biotechnol.* **2009**, *84*, 484–489. [CrossRef]

155. Vane, L.M. A review of pervaporation of product recovery from biomass fermentation processes. *J. Chem. Technol. Biotechnol.* **2005**, *80*, 603–629. [CrossRef]

156. Cockrem, M.C.M. Process for Recovering Organic Acids From Aqueous Salt Solutions. U.S. Patent 5,522,995, 1996.

157. Garrett, B.G.; Srinivas, K.; Ahring, B.K. Design and optimization of a semi-continuous high pressure carbon dioxide extraction system for acetic acid. *J. Supercrit. Fluids* **2014**, *95*, 243–251. [CrossRef]

158. Huang, C.; Xu, T.; Zhang, Y.; Xue, Y.; Chen, G. Application of electrodialysis to the production of organic acids: state-of-the-art and recent developments. *J. Membr. Sci.* **2007**, *288*, 1–12. [CrossRef]

159. Akerberg, C.; Zacchi, G. An economic evaluation of the fermentative production of lactic acid from wheat flour. *Bioresour. Technol.* **2000**, *75*, 119–126. [CrossRef]

160. Habova, V.; Melzoch, K.; Rychtera, M.; Sekavova, B. Electrodialysis as a useful technique for lactic acid separation from a model solution and a fermentation broth. *Desal.* **2004**, *162*, 361–372. [CrossRef]

161. Lee, E.G.; Moon, S.H.; Chang, Y.K.; Yoo, I.-K.; Chang, H.N. Lactic acid recovery using two-stage electrodialysis and its modelling. *J. Membr. Sci.* **1998**, *145*, 53–66.

162. Prado-Rubio, O.A.; Jorgensen, S.B.; Jonsson, G. Reverse electro-enhanced dialysis for lactate recovery from a fermentation broth. *J. Membr. Sci.* **2011**, *374*, 20–32. [CrossRef]

163. Garde, A. Production of Lactic Acid From Renewable Resources Using Electrodialysis for Product Recovery. Ph.D. Thesis, Technical University of Denmark, Copenhagen, Denmark, 2002.

164. Dwidar, M.; Lee, S.; Mitchell, R.J. The production of biofuels from carbonated beverages. *Appl. Energy* **2012**, *100*, 47–51. [CrossRef]

165. Boaro, A.A.; Kim, Y.M.; Konopka, A.E.; Callister, S.J.; Ahring, B.K. Integrated omics analysis for studying the microbial community response to a pH perturbation of a cellulose-degrading bioreactor culture. *FEMS Microbiol. Ecol.* **2014**, *90*, 802–815. [CrossRef] [PubMed]

fermentation

MDPI

Review

Succinic Acid: Technology Development and Commercialization

Nhuan P. Nghiem [1,*], Susanne Kleff [2] and Stefan Schwegmann [2]

[1]   Eastern Regional Research Center, Agricultural Research Service, U.S. Department of Agriculture, Wyndmoor, PA 19446, USA
[2]   Bioeconomy Institute, Michigan State University, Lansing, MI 48910, USA; kleff@msu.edu (S.K.); sschwegm@msu.edu (S.S.)
*   Correspondence: john.nghiem@ars.usda.gov; Tel.: +1-215-233-6753

Academic Editor: Gunnar Lidén
Received: 31 March 2017; Accepted: 5 June 2017; Published: 9 June 2017

Abstract: Succinic acid is a precursor of many important, large-volume industrial chemicals and consumer products. It was once common knowledge that many ruminant microorganisms accumulated succinic acid under anaerobic conditions. However, it was not until the discovery of *Anaerobiospirillum succiniciproducens* at the Michigan Biotechnology Institute (MBI), which was capable of producing succinic acid up to about 50 g/L under optimum conditions, that the commercial feasibility of producing the compound by biological processes was realized. Other microbial strains capable of producing succinic acid to high final concentrations subsequently were isolated and engineered, followed by development of fermentation processes for their uses. Processes for recovery and purification of succinic acid from fermentation broths were simultaneously established along with new applications of succinic acid, e.g., production of biodegradable deicing compounds and solvents. Several technologies for the fermentation-based production of succinic acid and the subsequent conversion to useful products are currently commercialized. This review gives a summary of the development of microbial strains, their fermentation, and the importance of the down-stream recovery and purification efforts to suit various applications in the context of their current commercialization status for biologically derived succinic acid.

Keywords: succinic acid; fermentation; platform chemicals; bio-based products; industrial solvents; biodegradable plastics

1. Introduction

Succinic acid, which is also known as butanedioic acid, 1,2-ethanedicarboxylic acid and amber acid, occurs in nature as such or in various forms of its esters. Before the development of fermentation processes for its production, succinic acid was manufactured by catalytic hydrogenation of maleic anhydride, which is a fossil-based chemical. Traditional applications of succinic acid include food additives, detergents, cosmetics, pigments, toners, cement additives, soldering fluxes, and pharmaceutical intermediates. In the past, succinic acid had a relatively small market size. It was estimated that the annual total world production in 1990 was between 16,000 and 18,000 metric tons (MT) [1]. Production of succinic acid from renewable feedstocks started to gain attention due to the increasing oil prices, dwindling oil supplies, and most importantly, the potential of converting succinic acid to a range of industrial chemicals with very large markets, such as 1,4-butanediol and other organic solvents. As an intermediate of several biochemical pathways, succinic acid is produced by many microorganisms. In fact, it is one of the metabolic co-products of ethanol fermentation by the yeast *Saccharomyces cerevisiae* in addition to glycerol, lactic acid, and acetic acid. The challenge of producing succinic acid biologically at commercial scale and potentially for the commodity chemical market called for development of microorganisms that could accumulate the product at high concentrations to justify economically feasible recovery. The first isolated microorganism with commercial potential was *Anaerobiospirillum succiniciproducens*, which was discovered by Michigan Biotechnology Institute [2,3]. At about the same time, the United States Department of Energy (US DOE) established the Alternative Feedstocks Program (AFP), which directed a team of national laboratories to develop alternative commodity chemical platforms based on renewable feedstocks. After several industrial workshops and reviews, and initial economic analysis performed by the National Renewable Energy Laboratory (NREL), succinic acid was selected as the first target platform chemical. Interests in biologically derived succinic acid steadily increased and resulted in the development of additional microbial strains, recovery and purification processes, and methods for conversion of succinic acid to important industrial chemicals and consumer products. Currently, there are four manufacturers of bio-based succinic acid, which have formed joined ventures with several other commercial entities. It has been estimated that the global succinic acid market would steadily grow at a compound annual growth rate (CAGR) of around 27.4% to reach $1.8 billion (768 million MT at $2.3/kg) in 2025 [4]. This report reviews the microbial strains for production of succinic acid, the processes for its recovery from the fermentation broths and conversion to important industrial chemicals, and commercialization of the fermentative succinic acid technology.

2. Succinic Acid-Producing Microorganisms

Succinic acid lends itself to biological production, because it is part of every organism's central metabolism. This was recognized decades ago, when the US DOE ranked succinic acid, along with other dicarboxylic acids, on its top value-added chemicals list [5], and sparked the isolation and development of multiple microorganisms for succinic acid production. The various routes to succinic acid are summarized in Figure 1.

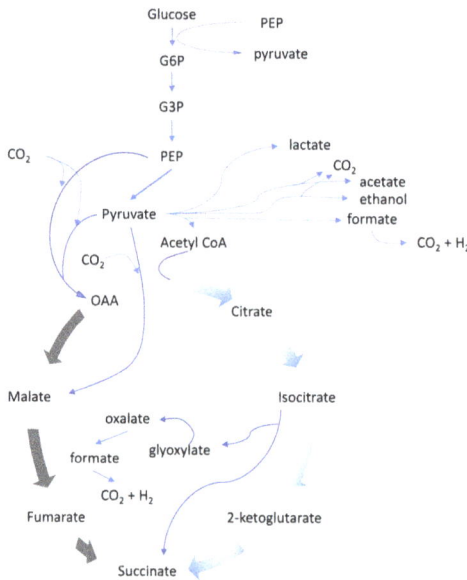

Figure 1. Metabolic pathways to succinic acid. Thick grey arrows showing the reductive (anaerobic) and thick blue arrows indicating the oxidative (aerobic) tricarboxylic acid (TCA) cycle. Thin blue arrows indicate contributing reactions and alternative pathways to succinic acid. Co-factors and electron balances were omitted for clarity. Notes: G6P: glucose-6-phosphate; G3P: glyceraldehyde-3-phosphate; OAA: oxaloacetate; PEP: phosphoenolpyruvate.

Succinic acid is a metabolite of the tricarboxylic acid (TCA) cycle. The use of the reductive direction of the TCA cycle for succinic acid production is appealing, and can theoretically produce two molecules of succinic from each molecule of a 6-carbon sugar with incorporation of 2 carbon dioxide molecules in the conversion of phosphoenolpyruvate (PEP) to oxaloacetate (OAA), offering the potential for an excellent fermentation yield. The required carbon dioxide can be provided by using an alkali or alkali-earth carbonate solution for pH control or sparging carbon dioxide directly into the fermentor while controlling the pH with other bases such as NaOH, KOH and NH_4OH or alkali-earth hydroxides [6]. Succinic acid is a more reduced molecule than its sugar fermentation feedstocks. This implies that in conversion to succinic acid, some of the sugar substrate must also be used for the production of reducing power. The theoretical yield of succinic acid from 6-carbon sugars such as glucose ($C_6H_{12}O_6$) and CO_2 via the reductive TCA route is 1.71 mole/mole sugar, or 1.12 g/g sugar, which is summarized in the following reaction [7]:

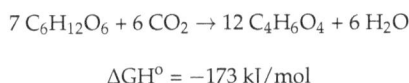

$$7\,C_6H_{12}O_6 + 6\,CO_2 \rightarrow 12\,C_4H_6O_4 + 6\,H_2O$$

$$\Delta GH^\circ = -173\ kJ/mol$$

Many independent and parallel efforts have been made to fermentatively produce succinic acid, with many choices for which organism was most suitable as a host. Regardless of which organism was chosen, each needed corrections that were often similar among the various organisms. The reduction or elimination of fermentation by-products other than succinic acid was the first step for improvement of most host organisms. For many of them the by-products were other organic acids such as lactic, acetic, or formic acid, and ethanol. The control and regulation of carbon flow surrounding pyruvate as the central node was also a common target for genetic manipulation.

A second challenge in the enhancement of succinic acid-producing hosts was the alleviation of an electron imbalance in the host organism. Succinic acid is a more reduced, or electron rich molecule, than its commonly used carbon feedstock, i.e., sugar, and many different approaches were taken to harness reducing equivalents or to restore co-factors. This is the underlying reason that it is not possible to produce two molecules of succinic from one C6-sugar molecule.

An important difference among the various microorganisms involves the optimum pH of the succinic acid fermentation. The accumulation of succinic acid in the fermentation broth inevitably leads to lowering of the pH in the fermentation medium. The pKa values for succinic acid, $pKa_1 = 4.61$ and $pKa_2 = 5.61$, lie below the optimal growth and production pH of many organisms. Maintaining the pH of the fermentation in a range suitable for the microorganism, and the choice of base to neutralize the acid, has a significant impact on the overall succinic acid production costs. Furthermore, the choice of base determines what succinate salt is produced, which can influence the target market for succinic acid. For example, the use of ammonium hydroxide as base, leading to the production of di-ammonium succinate, lends itself to post-fermentative production of pyrrolidinones. Several inventions surrounding succinic and other organic acids targeted the recovery and recycling of the neutralizing agent. The recognition that neutralization and base recycling contributed significantly to succinic acid production costs prompted the engineering of host organisms that can tolerate a lower pH and have reduced base requirements.

Natural succinic acid producers are organisms which produce succinic acid as the main fermentation product, and many of them are capable of fumarate respiration. All natural succinic acid-producing microorganisms were isolated from rumen fluids, an environment rich in microorganisms that produce volatile fatty acids (VFAs), which are directly and readily absorbed into the bloodstream of the ruminant host animal and used for energy production. These microorganisms are anaerobes, thrive at a near neutral pH, are capnophilic, and naturally use the reductive TCA cycle for succinic acid production. All natural succinic acid-producing organisms isolated from the rumen require pH control for a fast and efficient fermentation.

Anaerobiospirillum succiniciproducens (ANS) was one of the first bacteria isolated and shown to produce succinic acid [3,8]. Fermentations with this strict anaerobe reached titers of 50 g/L succinic acid. A classical selection for reduction of by-products, which included formic and acetic acid, was successful. However, the organism experienced instability and deterioration, was sensitive to high glucose concentrations and exposure to air, which made it unsuitable for industrial application.

The same isolation scheme used for ANS was also used to identify *Actinobacillus succinogenes*, a more robust organism, and a facultative anaerobe [9]. The pathway to succinic via the reductive TCA cycle in this gram-negative bacterium was biochemically characterized [10], and the fermentation by-products, acetic and formic acid, were nearly eliminated using a classical selection method [3]. Variants capable of reaching titers above 100 g/L succinate were reported. However, the organism did not thrive at pH values below pH 6.0, necessitating the neutralization of the produced acid, and the best performances were reported when using magnesium Mg^{++} as a counterion. Consistent with its rumen origin, the organism was versatile in its use of carbon substrates. Most notable was its simultaneous utilization of 5- and 6-carbon sugars, as well as polyols such as sorbitol and glycerol. Fermentations using cellulosic biomass sugars, specifically corn stover hydrolysates, reached titers approaching 50 g/L in simultaneous saccharification fermentations (SSF) [11]. In recent years, metabolic flux analyses and genome analysis confounded the understanding of the organism [12] and metabolic engineering to achieve better electron balance enhanced the yield by channeling carbon through the pentose phosphate pathway [13]. Fermentation performance reaching a titer of 109 g/L with productivity of 2.0 g/L.h and yield of 0.96 g succinic acid/g glucose was achieved in medium containing small amounts of complex nutrient sources, i.e., less than 0.5 g/L yeast extract or corn steep liquor (Guettler and Kleff, personal communications). Growth in defined minimal medium was possible but the low titer of 9 g/L did not meet industrial performance standards under these conditions [14].

Genome analysis of *Actinobacillus* revealed many shared features with other natural succinic acid producers, especially its closest relative *Mannheimia*, and for these organisms phosphoenolpyruvate (PEP) is a key branch point. PEP carboxykinase, encoded by the *pckA* gene, is responsible for fixing CO_2 to PEP and generating oxaloacetate, while harnessing the energy in form of GTP. *A. succinogenes* lacks a complete TCA cycle, missing key enzymes such as isocitrate dehydrogenase and α-ketoglutarate dehydrogenase, indicating a metabolism geared towards succinic acid production.

Mannheimia succiniciproducens was isolated from the rumen of a Korean cow, and initially referenced as a mixed acid producer. The organism uses the same reductive TCA production pathway to succinic acid as *A. succinogenes*. By-products such as formate and acetate were eliminated through gene disruption of pyruvate formate lyase (*pflB*), acetate kinase (*ackA*), and phosphotransacetylase (*pta*) [15]. Additional paths to acetic acid are absent in the *M. succiniciproducens* genome. The original *Mannheimia* isolate also produced lactate, when grown in CO_2-poor conditions, which was precluded through disruption of the *ldhA* gene, thereby generating a homosuccinate producing daughter strain, LPK7. A thorough genome analysis of *Mannheimia* and comparison to engineered organisms elucidated predilection of *Mannheimia* for succinic acid production, e.g., the reliance on PEP-ckA as a CO_2-fixing enzyme. This is possible in *Mannheimia*, because glucose import is accomplished by a glucokinase activity that transfers phosphate from ATP to glucose [16]. The organism lacks an NAD^+-dependent malic enzyme, pyruvate kinase, and pyruvate oxidase, thereby reducing the need for a large pyruvate pool. *Mannheimia* has a complete TCA cycle, which may indicate that its metabolism may have different and more complex controls for channeling carbon towards succinic acid. *Mannheimia* has been shown to demonstrate high rates of succinate production, although often in medium containing complex nitrogen sources such as yeast extract [17]. A chemically defined medium has been developed [18].

Isolation of *Basfia succiniciproducens* followed the knowledge gained from the isolations of *Mannheimia* and *Actinobacillus* and was geared to isolate a strain belonging to the family of the *Pasteurellaceae*. The genome of the original *Basfia* isolate (DD1) was compared to the genome of *Mannheimia*, and showed high genetic identity [19], yet it was classified as a distinct organism.

Similar to *Mannheimia*, *Basfia* also produces lactic acid as a major by-product under fermentative conditions. Lactic and formic acid production were eliminated through targeted deletions [20]. *Basfia* can utilize a variety of carbon feedstocks, and continuous fermentations using crude glycerol from biodiesel production were successful [21].

With regard to genetically engineered succinic acid producers, many research teams developed *Escherichia coli* for succinic acid production, utilizing the genetic tools and knowledge available for extensive metabolic engineering approaches for this organism. The efforts at the Argonne National Laboratory (ANL) resulted in development of several promising *Escherichia coli* strains for succinic acid production. Two of the most efficient strains were designated AFP111 and AFP184. Subsequently, a two-stage fermentation process was developed at the Oak Ridge National Laboratory (ORNL) for production of succinic acid using these *E. coli* strains [22,23]. The developed technology was licensed to Applied Carbochemicals, Inc. (ACC), a start-up company in the United States. ACC became the first company that attempted to develop a commercially feasible fermentation process for the manufacture of succinic acid. Earlier efforts built on *E. coli*'s mixed acid fermentation pathways to strengthen the reductive route to succinic acid, and built on the above mentioned starting strains AFP111 and AFP184 [24]. The first improvement steps eliminated the production of lactic acid and formic acid, the latter through deletions in pyruvate formate lyase. *E. coli* can use its PEP-carboxylase to enter the reductive TCA cycle, but the energy from PEP is lost as inorganic phosphate in the carboxylation step. This energy loss is undesirable under anaerobic, ATP limited conditions. Furthermore, the import of glucose through the phosphoenolpyruvate:carbohydrate phosphotransferase system (PTS) system, where PEP is the source of phosphate for the generation of glucose-6-phosphate, leads to a large pool of pyruvate, making it a key biological branch point in the organism, where multiple enzymatic activities compete for the pyruvate substrate [25]. The control of the pyruvate node in *E. coli* included changes in the glucose import mechanism, the introduction of heterologous genes with more favorable enzymatic characteristics, which either supported or replaced the corresponding endogenous genes, including the overexpression of malic enzyme [26]. In contrast to natural anaerobic succinic producers, *E. coli* was amenable to the separation of growth and succinic production, by combining fast aerobic growth, followed by a switch to anaerobic conditions for succinic production [27], and the use of microaerobic conditions.

In an alternative approach, *E. coli* was engineered for succinic production under fully aerobic conditions, thereby harnessing the higher production of cell-biomass, and faster carbon throughput to succinic production to favor a high productivity fermentation [28]. However, this reduced the theoretical yield to one mole succinic per mole glucose. These innovative strain improvements channeled the carbon flow in *E. coli* through either the glyoxylate shunt or a two-pronged route using the glyoxylate shunt and the oxidative TCA cycle [29].

In contrast to many natural succinic producers *E. coli* can grow to high cell densities in defined mineral media, while maintaining high succinic production capabilities [30]. This broadened the technology deployment options, eliminated the potential need for costly complex nutrients and facilitated product recovery. *E. coli* is capable of using a variety of sugars, and several engineered strains were designed to use sucrose or molasses as feedstock. However, the organism naturally shows a preference for glucose, which is consumed first or preferentially, when multiple sugars are present. Even among the 5-carbon sugars *E. coli* demonstrates a preference for arabinose over xylose [31]. The organism is also sensitive to high acetate concentrations, a commonly found inhibitor in cellulosic sugar streams. *E. coli* may not be the best choice as host for the use of cellulosic mixed sugar streams, but this is a lesser concern currently, when other carbon sources are available for the bio-based production of succinic acid.

Corynebacterium glutamicum is an established industrial organism for the production of amino acids, and many tools have been developed for its genetic manipulation. *C. glutamicum* can grow aerobically and anaerobically, and high succinic acid titers have been reached with an engineered strain of this bacterium under fed-batch conditions [32]. Interestingly, the fed-batch process used glucose as carbon source and formic acid as a source of reducing equivalents to support succinic acid production under anaerobic conditions. The reported production yields consider only the succinic production phase, and do not account for the aerobic production of cell biomass. The small amounts of by-products seen in this two-phase system included α-ketoglutarate, pyruvate, and acetic acid. The process was run at a near neutral pH, and the succinic acid produced was neutralized with potassium hydroxide.

All bacterial succinic acid producers described above require the neutralization of the fermentation product. The final production process may include a base recycle, or recover the base-salt as a fermentation by-product. It is worth noting that the salt by-product in pH neutral succinic acid fermentations is generated at an equivalent scale to the amount of succinic acid produced. Such a high volume market is available for ammonium hydroxide neutralized fermentations, in which the diammonium succinate salt (DAS) is acidified with sulfuric acid to form free succinic acid and ammonium sulfate. The salt by-product, ammonium sulfate, can serve the fertilizer market.

The cost and effort associated with the use of a neutralizing base fostered the development of succinic acid production at a low pH by using a low-pH-tolerant host. *Saccharomyces cerevisiae*, or other yeasts such as *Yarrowia*, can thrive under slightly acidic conditions and methods for their metabolic engineering are well established [33]. *S. cerevisiae* is an industrial organism capable of fermentative production and tolerance of high sugar concentrations. Similar to *E. coli*, the organism has the potential to produce succinic acid both aerobically and anaerobically [34]. However, fermentative conditions favor ethanol as well as glycerol production in yeast, and the elimination of these natural fermentation products was more complicated than in prokaryotic systems [35]. This is due to the presence of multiple genes encoding alcohol dehydrogenases, and the deletion of genes leading to natural products had unfavorable side effects, such as low osmotolerance. Furthermore, the genes involved in anaerobic succinate production, using the reductive TCA cycle starting from oxaloacetate, underlie glucose repression, and the route is thermodynamically unfavorable.

Succinic acid production under aerobic conditions via the glyoxylate shunt is possible in yeast. As described for *E. coli*, this route has the benefit of faster growth, faster carbon metabolism, and better ATP balance, but also leads to a lower theoretical yield. Dual production pathways can be implemented and recent publications show improvements, but the performance of all three fermentation parameters, titer, productivity and yield, remain unclear. It is also worth noting that the fermentative production of succinic acid at a pH of 4.6, the pKa_1 of succinic acid, will reduce the amount of base needed for neutralization, but will not eliminate the need for base entirely. At this pH 25% of the acid produced will remain in the mono-salt form, which needs to be considered in the recovery. Nevertheless, low pH succinic production enjoys the benefits of reduced base and reduced salt formation.

Many succinic acid-producing microorganisms have been developed and much effort was put into characterizing and improving the organisms. However, a direct performance comparison between the strains is difficult because the culture conditions were different and in many cases were insufficiently described. In addition, the fermentation process used for several organisms included a growth phase, during which there was none or negligible succinic acid synthesis, followed by a production phase, whereas the growth phase was not separated from the production phase for others. Calculations of succinic acid yield and productivity, therefore, could not be performed on the same basis. Despite these problems, one publication has provided a thorough summary of many succinic acid-producing bacterial strains [36].

3. Succinic Acid Recovery

The recovery and purification of succinic acid from the fermentation broth can be a complex, multi-step, and expensive process. Knowledge of purification procedures, and the required purity of the final product, also guided the selection of inputs for the production of succinic acid. Due to the expense that the recovery and purification process can impart on the final product and its uses, each of the current succinic acid producers developed and patented their own recovery and purification processes. These processes vary depending on the organism used for production, feedstock used, nutrients supplied, acids and bases used to maintain pH, solubility of the intermediates, final product titer achievable, and intended applications of the recovered succinic acid products. The first step, which is common in all the described and patented processes, is removal of cells and insoluble solids through the use of standard equipment such as filtration or centrifugation. Subsequent recovery and purification steps described in the recent patents and patent applications are highlighted.

BASF describes a recovery in patents and publications surrounding claims for its proprietary microorganism. The first step describes the concentration of clarified broth through multistage evaporation to reduce processing volumes, followed by a cation exchange chromatography, in which the succinic acid salt is reacted with a strong acid cation exchange resin at a temperature of 46 to 60 °C. This achieves the conversion of the succinate salt to succinic acid, which can be crystallized to form the final product. Alternatively, if the fermentation is neutralized by a calcium base, the calcium succinate product has low solubility and can be separated from the broth by filtration. The precipitate is treated with sulfuric acid to form soluble succinic acid and calcium sulfate (gypsum). The latter has very low solubility and can be separated by filtration. The succinic acid solution may be further purified by the same cation exchange chromatography as described previously [37,38].

For BioAmber, the starting material for the recovery is diammonium succinate (DAS), which is produced by neutralizing the succinic acid in the fermentation broth with NH_3. The filtered broth undergoes a reactive evaporation, in which it is heated to 135 °C and 50 psig, converting the DAS to monoammonium succinate (MAS) and resulting in roughly a two-fold concentration. The MAS solution is sent to a three stage evaporative crystallization system, which cools the solution to the point when MAS is no longer soluble enabling up to 95 wt % recovery. The crystalline MAS can be dissolved in reverse osmosis water, and the solution may be converted to succinic acid by either bipolar membrane electrodialysis, or through the use of ion exchange chromatography. The succinic acid solution undergoes another evaporative crystallization step in which 95 wt % is recovered as a solid, with the remaining succinic acid in the mother liquor recycled back to the reactive evaporation step [39,40].

The DSM/Roquette process begins with an evaporation phase carried out between 65 and 80 °C followed by a crystallization phase in which the concentrated solution is cooled to between 1 and 25 °C to produce intermediate crystals and a mother liquor. The intermediate crystals are separated from the mother liquor by centrifugation and the mother liquor is subjected to microfiltration and nanofiltration. The small pore (100–300 Da) membrane allows succinic acid to pass through, while higher molecular weight soluble materials are retained. The filtered mother liquor is recycled to the evaporation step in the beginning of the process to produce additional intermediate crystals. The intermediate crystals are dissolved into a minimum volume of 40–90 °C water. The solution is passed through a series of purification steps including activated carbon column, cation exchange, and anion exchange. The purified solution undergoes a final crystallization to produce a succinic acid product of high purity [41].

Mitsubishi's recovery method employs crystallization, which is applicable to solutions of succinic acid of either petrochemical or biological origin. The solution containing succinic acid at near saturated concentrations is fed to a crystallization tank where the overhead pressure is reduced to below atmospheric to cause a drop in temperature, which subsequently causes precipitation of succinic acid. Stirring is carried out at pre-determined rates to ensure uniform size of the crystals. The crystals are removed from the tank and pulverized to form a product having particle size between 100 and 300 mm [42].

For Myriant, the purification process begins with a fermentation broth that has been neutralized with NaOH to produce diammonium succinate. The clarified broth is concentrated by evaporation under vacuum and subsequently acidified with H_2SO_4 to produce succinic acid and ammonium sulfate. The temperature of the acidified broth is further lowered to cause the succinic acid to crystallize and precipitate, while the ammonium sulfate remains soluble. The succinic acid crystals are harvested by centrifugation. The crystals stream may either be dissolved again and recrystallized to improve purity, or be sent to an esterification process to produce a stream that is suitable for the further production of 1,4-butanediol, γ-butyrolactone, or tetrahydrofuran. The mother liquor, which contains primarily ammonium sulfate and the remaining succinic acid, is separated into two streams by simulated moving bed chromatography. The ammonium sulfate stream can be sold as fertilizer in liquid or crystallized solid form, and the succinic acid stream can be polished by nanofiltration before being crystallized and dried to form the final product [43].

The process patented by Purac focuses on recovery of magnesium or calcium succinate. Following broth clarification, a monovalent base is added to convert the divalent succinate to the monovalent succinate salt. The best conversion is obtained using sodium hydroxide in the case of magnesium succinate, and sodium carbonate for calcium succinate, which yielded a 99.8% and 99.3% conversion, respectively. While these are the preferred monovalent bases, other monovalent bases may be used for economic reasons or to produce a different end product. The monovalent succinate salt and the magnesium or calcium hydroxide/carbonate can be separated by filtration. The divalent base may be washed, to reduce product losses, before being recycled back into the fermentation process. The succinate salt is further purified via ion exchange to reduce the calcium/magnesium ion content to levels that would allow the use of bipolar electrodialysis. This would yield a very high-purity succinic acid stream ready for crystallization, or for conversion to succinate esters and alternate product synthesis [44].

4. Succinic Acid Applications

Traditionally, succinic acid was produced from petroleum sources to meet a relatively small market for pharmaceutical and food applications. The potential use of succinic acid as starting feedstock for production of industrial chemicals and consumer products with large markets was the main driver that launched the development of technology for bio-based succinic acid production. The potential succinic acid value chain is shown in Figure 2 [45]. The projected market volume and market share of various applications of succinic acid, which include both traditional and potential new applications, are given in Table 1 [46]. Among the potential applications of bio-based succinic acid, 1,4-butanediol (BDO) is the largest market. BDO is an industrial chemical with many uses. It can be used as a solvent starting material for several other important industrial chemicals such as γ-butyrolactone (GBL), tetrahydrofuran (THF), 2-pyrrolidone (2-P), and *N*-methyl-2-pyrrolidone (NMP), and can be combined with succinic acid to make polybutylene succinate (PBS) and poly(butylene succinate-co-butylene terephthalate), which are two biodegradable plastics with many potential applications. Development of catalysts for conversion of succinic acid to BDO, GBL and THF has been an area of strong interest and active research, especially for catalysis in aqueous media. Catalysts for conversion of succinic acid to pyrrolidones have also been extensively studied [47,48]. High-value rare metal-containing catalysts normally are used in BDO synthesis. This was recognized by Genomatica, a California-based company, which engineered *E. coli* directly for BDO production [49]. Compared to succinic acid, BDO is less acidic, has reduced base requirement during the fermentation, and requires a different recovery method. It may have a lesser purity requirement, since it eliminates the need for expensive and impurity-sensitive hydrogenation catalysts. The biological production of BDO has an even greater demand for biological reducing equivalents than succinic acid, thereby refuting the good theoretical yields seen for anaerobic production of succinic acid. Technologies for production of other potential products from succinic acid have also been actively pursued by the current bio-based succinic acid manufacturers [50–53].

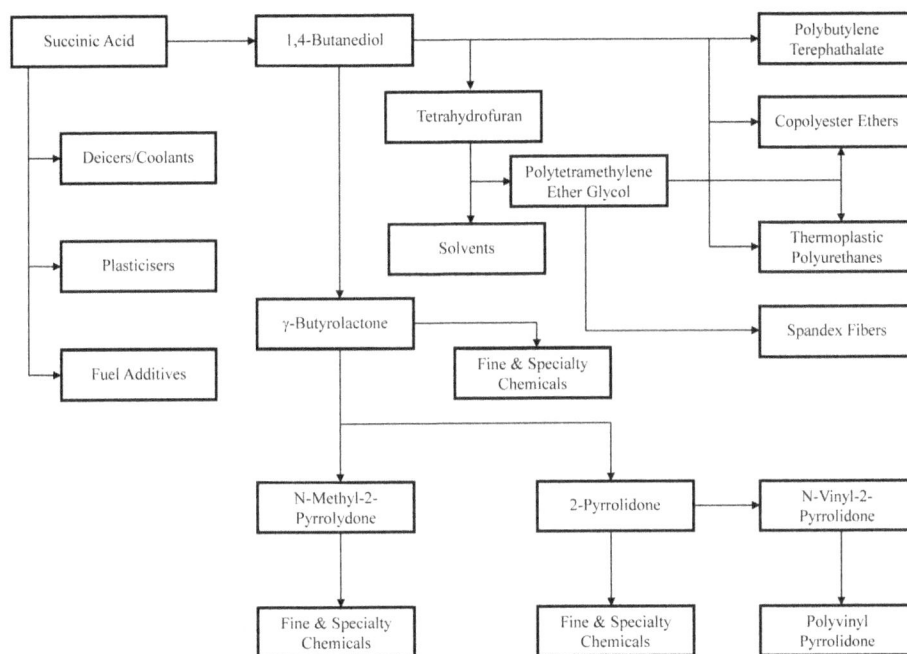

Figure 2. The potential succinic acid value chain [45].

Table 1. Projected market volume and market share for succinic acid applications for 2020 [46].

Product Applications	Market Volume (1000 MT)	Market Share (%)
BDO [1]	316	52.7
PBS, PBST [2]	82	13.7
Polyester polyols	51	8.5
Plasticizers	37	6.2
Food	26	4.3
Pharmaceutical	21	3.5
Alkyd resins	21	3.5
Resins, coatings	12	2.0
Cosmetics	12	2.0
Solvents & lubricants	6	1.0
De-icer solutions	3	0.5
Others	13	2.2

Notes: [1] BDO: 1,4-butanediol; [2] PBS: polybutylene succinate; PBST: poly(butylene succinate-co-butylene terephthalate). Data used to set up this table were extracted from the reference cited [46].

5. Commercialization of Bio-Based Succinic Acid

Bio-based succinic acid currently is produced commercially by four companies, which include BioAmber, Myriant, Reverdia and Succinity. Each of these manufacturers of bio-based succinic acid has collaborations with a host of other companies on development of new technologies for production of succinic acid and its derivatives.

As discussed previously, ACC was the first company that was formed to commercialize bio-based succinic acid using the *E. coli* strains and fermentation process developed by the US DOE. ACC became Diversified Natural Products (DNP) Green Technology, which subsequently formed a collaboration with the France-based Agro-Industrie Recherches et Développements (ARD) to develop and commercialize bio-based succinic acid. In 2010, DNP Green Technology acquired 100 percent of the joint venture from ARD and changed its corporate name to BioAmber, Inc. (Plymouth, MN, USA). BioAmber (and its predecessors) developed succinic acid technology based on the US DOE's *E. coli* strains. The company recently collaborated with Cargill to develop yeast strains for succinic acid production [54]. BioAmber also has collaborated with Mitsui to build a 30,000 MT/year succinic acid production plant in Sarnia, ON, Canada, which is now in operation. In addition to the Sarnia plant, BioAmber is planning to build a second plant in North America that will produce 1,4-butanediol (BDO), tetrahydrofuran (THF) and succinic acid. The nameplate capacity is expected to be 91,000 MT of BDO/THF and 63,500 MT of succinic acid per year. A third succinic acid plant is being planned for in Thailand in a partnership with PTT-MCC Biochem, which is a joint venture between PTT PLC and Mitsubishi Chemical. The succinic acid produced at this plant is intended for use exclusively for production of PBS by PTT-MCC Biochem [55].

Myriant is another bio-based succinic acid producer in the United States. The company initially licensed the *E. coli* strains developed at the University of Florida [56,57] then continued their development to obtain *E. coli* strains capable of utilizing sugars derived from lignocellulosic feedstocks for succinic acid production [58]. Myriant currently is operating a 13.6 MT succinic acid production plant in Lake Providence, Louisiana. The intended feedstocks for use in this plant include glucose derived from grain sorghum and sugars derived from lignocellulosic biomass [59]. A second succinic acid production plant with initial annual output of 500 MT and plan for expansion to 5000 MT is located in Leuna, Germany and is operated by Myriant's partner ThyssenKrupp Uhde [46,59]. There have been reports on potential partnership with China National BlueStar to build a third succinic acid production plant in Nanjing, China with annual output of 100,000 MT. The succinic acid produced is intended for use as feedstock for BDO production [46,60]. In addition to succinic acid, Myriant also attempts to develop technologies for production of other organic acids, which include lactic, acrylic, muconic and fumaric acid [59].

Reverdia is a joint venture between Dutch chemical company DSM and French starch derivatives producer Roquette. Reverdia successfully operated a demonstration plant with annual capacity of 500 MT in Lestrem, France from 2010 to 2012. In December 2012, commercial succinic acid production began at the 10,000 MT/year plant in Cassano Spinola, AL, Italy [61]. The fermentation process is based on a recombinant *S. cerevisiae* strain developed by DSM [62]. DSM also developed a recombinant *S. cerevisiae* strain for co-production of ethanol and succinic acid. In this strain, the energy generated by ethanol production in the form of ATP is sufficient to support succinic acid synthesis [63]. There has been no indication that this strain would be used in a commercial process.

Succinity is a joint venture between Germany-based BASF and Dutch company Corbion Purac, which was formed to produce succinic acid using a proprietary strain of *Basfia succiniciproducens*. Succinity successfully operated a 500 MT/year demonstration plant in Barcelona, Spain [64]. Commercial production of succinic acid started in 2014 at a 10,000 MT/year plant in Montmeló, Spain. In addition to this plant, planning for a second production plant was reported [46,65].

There have been only one LCA study and one sustainability study performed on the commercial processes for succinic acid production. In the LCA study of the Myriant's process, real data from the plant in Louisiana were used [66]. In the base case in this study, sorghum grains were used as feedstock. Comparing to the base case, the use of glucose as feedstock would increase the global warming potentials (GWP) and the non-renewable fossil cumulative energy demand (non-ren CED) by 1.72 times and 1.86 times, respectively. The GWP and non-ren CED in the manufacture of succinic acid by the petrochemical route using maleic anhydride as feedstock were found to be 3.85 and 10.44 times higher than in the base case. A sustainability study was performed on the Myriant's process using sorghum grains and sugar beet as feedstocks and the Reverdia's process for co-production of ethanol and succinic acid [67]. The material efficiency ratios for bio-based succinic acid produced from sorghum and sugar beet were calculated to be 13% in both cases, compared to 76% calculated for petrochemical-based succinic acid. The corresponding energy efficiency ratios were calculated to be

30%, 31%, and 23%, respectively. The calculated energy efficiency ratios for the Reverdia's process using sorghum grains and sugar beet were 51% and 54%, respectively. The calculated production cost in all cases of succinic acid manufactured by either bio-based process was significantly lower than the calculated cost for petrochemical-based succinic acid. Even in the worst-case scenario, the production cost of bio-based succinic acid was only 41% of the production cost of petrochemical-based succinic acid ($1.17/kg vs. $2.86/kg). The results of the two aforementioned studies clearly indicated the advantages of bio-based succinic acid.

Conflicts of Interest: The authors declare no conflict of interest.

References

1. Fumagalli, C. Succinic acid and succinic anhydride. In *Kirk-Othmer Encyclopedia of Chemical Technology*, 4th ed.; Kroschwitz, J.I., Howe-Grant, M., Eds.; Wiley: New York, NY, USA, 1997; Volume 22, pp. 1074–1102.
2. Glassner, D.A.; Datta, R. Process for the Production and Purification of Succinic Acid. U.S. Patent 5143834, 1992.
3. Guettler, M.V.; Jain, M.K. Method for Making Succinic Acid, *Anaerobiospirillum Succiniciproducens* Variants for Use in Process and Methods for Obtaining Variants. U.S. Patent 5521075, 1996.
4. Global Succinic Acid Market Analysis & Trends 2013–2017 - Industry Forecast to 2025: $1.76 Billion Growth Opportunities/Investment Opportunities—Research and Markets. Available online: https://ceo.ca/@newswire/global-succinic-acid-market-analysis-trends-2013-2017 (accessed on 6 March 2017).
5. Werpy, T.; Petersen, G. *Top Value Added Chemicals from Biomass–Volume 1: Results of Screening for Potential Candidates from Sugars and Synthesis Gas*; Office of Biomass Program, United States Department of Energy: Washington, DC, USA, 2004.
6. Nghiem, N.P.; Hicks, K.B.; Johnston, D.B. Integration of succinic acid and ethanol production with potential application in a corn or barley biorefinery. *Appl. Biochem. Biotechnol.* **2010**, *162*, 1915–1928. [CrossRef] [PubMed]
7. McKinlay, J.B.; Vieille, C.; Zeikus, J.G. Prospects for a bio-based succinate industry. *Appl. Microbiol. Biotechnol.* **2007**, *76*, 727–740. [CrossRef] [PubMed]
8. Samuelov, N.S.; Lamed, R.; Lowe, S.; Zeikus, J.G. Influence of CO_2-HCO_3^-levels and pH on growth, succinate production, and enzyme activities of *Anaerobiospirillum succiniciproducens*. *Appl. Environ. Microbiol.* **1991**, *57*, 3013–3019. [PubMed]
9. Guettler, M.V.; Rumler, D.; Jain, M.K. *Actinobacillus succinogenes* sp. nov., a novel succinic acid-producing strain from the bovine rumen. *Int. J. Syst. Evol. Microbiol.* **1999**, *49*, 207–216. [CrossRef] [PubMed]
10. Der Werf, M.J.V.; Guettler, M.V.; Jain, M.K.; Zeikus, J.G. Environmental and physiological factors affecting the succinate product ratio during carbohydrate fermentation by *Actinobacillus* sp. 130Z. *Arch. Microbiol.* **1997**, *167*, 332–342. [CrossRef]
11. Zheng, P.; Fang, L.; Xu, Y.; Dong, J.J.; Ni, Y.; Sun, Z.H. Succinic acid production from corn stover by simultaneous saccharification and fermentation using *Actinobacillus succinogenes*. *Bioresour. Technol.* **2010**, *101*, 7889–7894. [CrossRef] [PubMed]
12. McKinlay, J.B.; Laivenieks, M.; Schindler, B.D.; McKinlay, A.A.; Siddaramappa, S.; Challacombe, J.F.; Lowry, S.R.; Clum, A.; Lapidus, A.L.; Burkhart, K.B.; et al. A genomic perspective on the potential of *Actinobacillus succinogenes* for industrial succinate production. *BMC Genom.* **2010**, *1*, 680. [CrossRef] [PubMed]
13. Jian, Y.; Kleff, S.; Guettler, M.V. Recombinant Microorganisms for Increased Production of Organic Acids. U.S. Patent 8431373, 2013.
14. McKinlay, J.B.; Zeikus, J.G.; Vieille, C. Insights into *Actinobacillus succinogenes* fermentative metabolism in a chemically defined growth medium. *Appl. Environ. Microbiol.* **2005**, *71*, 6651–6656. [CrossRef] [PubMed]
15. Lee, S.J.; Song, H.; Lee, S.Y. Genome-based Metabolic Engineering of *Mannheimia succiniciproducens* for Succinic Acid Production. *Appl. Environ. Microbiol.* **2006**, *72*, 1939–1948. [CrossRef] [PubMed]
16. Hong, S.H.; Kim, J.S.; Lee, S.Y.; In, Y.H.; Choi, S.S.; Rih, J.-K.; Kim, C.H.; Jeong, H.; Hur, C.G.; Kim, J.J. The genome sequence of the capnophilic rumen bacterium *Mannheimia succiniciproducens*. *Nat. Biotechnol.* **2004**, *22*, 1275–1281. [CrossRef] [PubMed]

17. Lee, S.Y.; Kim, J.M.; Song, H.; Lee, J.W.; Kim, T.Y.; Jang, Y.-S. From genome sequence to integrated bioprocess for succinic acid production by *Mannheimia succiniciproducens*. *Appl. Microbiol. Biotechnol.* **2008**, *79*, 11–22. [CrossRef] [PubMed]

18. Song, H.; Kim, T.Y.; Choi, B.-K.; Choi, S.J.; Nielsen, L.K.; Chang, H.N.; Lee, S.Y. Development of a chemically defined medium for *Mannheimia succiniciproducens* based on its genome sequence. *Appl. Microbiol. Biotechnol.* **2008**, *79*, 263–272. [CrossRef] [PubMed]

19. Kuhnert, P.; Scholten, E.; Haefner, S.; Mayor, D.; Frey, J. *Basfia succiniciproducens* gen. nov., sp. nov., a new member of the family *Pasteurellaceae* isolated from bovine rumen. *Int. J. Syst. Evol. Microbiol.* **2010**, *60*, 44–50. [CrossRef] [PubMed]

20. Becker, J.; Reinefeld, J.; Stellmacher, R.; Schaefer, R.; Lange, A.; Meyer, H.; Lalk, M.; Zelder, O.; Abendroth, G.; Schroeder, H.; et al. System-wide analysis and engineering of metabolic pathway fluxes in bio-succinate producing *Basfia succiniciproducens*. *Biotechnol. Bioeng.* **2013**, *110*, 3013–3023. [CrossRef] [PubMed]

21. Scholten, E.; Renz, T.; Thomas, J. Continuous cultivation approach for fermentative succinic acid production from crude glycerol using *Basfia succiniciproducens*. *Biotechnol. Lett.* **2009**, *31*, 1947–1951. [CrossRef] [PubMed]

22. Nghiem, N.P.; Donnelly, M.; Millard, C.S.; Stols, L. Method for the Production of Dicarboxylic Acids. U.S. Patent 5869301, 1999.

23. Donnelly, M.I.; Sanville-Millard, C.Y.; Nghiem, N.P. Method to Produce Succinic Acid from Raw Hydrolysates. U.S. Patent 6743610, 2004.

24. Stols, L.; Donnelly, M.I. Production of succinic acid through overexpression of NAD^+-dependent malic enzyme in an *Escherichia coli* mutant. *Appl. Environ. Microbiol.* **1997**, *63*, 2695–2701. [PubMed]

25. Gokarn, R.R.; Eiteman, M.A.; Altman, E. Expression of pyruvate carboxylase enhances succinate production in *Escherichia coli* without affecting glucose uptake. *Biotechnol. Lett.* **1998**, *20*, 795–798. [CrossRef]

26. Vemuri, G.N.; Eiteman, M.A.; Altman, E. Effects of growth mode and pyruvate carboxylase on succinic acid production by metabolically engineered strains of *Escherichia coli*. *Appl. Environ. Microbiol.* **2002**, *68*, 1715–1727. [CrossRef] [PubMed]

27. Vemuri, G.N.; Eiteman, M.A.; Altman, E. Succinate production in dual-phase *Escherichia coli* fermentations depends on the time of transition from aerobic to anaerobic conditions. *J. Ind. Microbiol. Biotechnol.* **2002**, *28*, 325–332. [CrossRef] [PubMed]

28. Lin, H.; Bennett, G.N.; San, K.-Y. Metabolic engineering of aerobic succinate production systems in *Escherichia coli* to improve productivity and achieve the maximum theoretical succinate yield. *Metab. Eng.* **2005**, *7*, 116–127. [CrossRef] [PubMed]

29. San, K.-Y.; Bennett, G.N.; Lin, H.; Sanchez, A. High Succinate Producing Bacteria. WO Patent application 2006034156 A2, 2006.

30. Zhang, X.; Jantama, K.; Moore, J.C.; Jarboe, L.R.; Shanmugam, K.T.; Ingram, L.O. Metabolic evolution of energy-conserving pathways for succinate production in *Escherichia coli*. *Proc. Nat. Acad. Sci. USA* **2009**, *106*, 20180–20185. [CrossRef] [PubMed]

31. Desai, T.A.; Rao, C.V. Regulation of Arabinose and Xylose metabolism in *E. coli*. *Appl. Environ. Microbiol.* **2010**, *76*, 1524–1532. [CrossRef] [PubMed]

32. Okino, S.; Noburyu, R.; Suda, M.; Jojima, T.; Inui, M.; Yukawa, H. An efficient succinic acid production process in a metabolically engineered *Corynebacterium glutamicum* strain. *Appl. Microbiol. Biotechnol.* **2008**, *81*, 459–464. [CrossRef] [PubMed]

33. Yuzbachev, T.V.; Yuzbasheva, E.Y.; Laptev, I.A.; Sobolevskaya, T.I.; Vybornaya, T.V.; Larina, A.S.; Gvilava, I.T.; Antonova, S.V.; Sineoky, S.P. Is it possible to produce succinic acid at a low pH? *Bioeng. Bugs* **2011**, *2*, 115–119. [CrossRef] [PubMed]

34. Camarasa, C.; Grivet, J.-P.; Dequin, S. Investigation by ^{13}C-NMR and tricarboxylic acid (TCA) deletion mutant analysis of pathways for succinate formation in *Saccharomyces cerevisiae* during anaerobic fermentation. *Microbiology* **2003**, *149*, 2669–2678. [CrossRef] [PubMed]

35. Van Maris, A.J.; Geertman, J.-M.; Vermeulen, A.; Groothuizen, M.K.; Winkler, A.A.; Piper, M.D.W.; Van Dijken, J.P.; Pronk, J.T. Directed evolution of pyruvate decarboxylase-negative *Saccharomyces cerevisiae* yielding a C_2-independent, glucose-tolerant, and pyruvate-hyperproducing yeast. *Appl. Environ. Microbiol.* **2004**, *70*, 159–166. [CrossRef] [PubMed]

36. Beauprez, J.J.; De Mey, M.; Soetaert, W.S. Microbial succinic acid production: Natural versus metabolic engineered producers. *Process Biochem.* **2010**, *45*, 1103–1114. [CrossRef]

37. Schröder, H.; Haefner, S.; Von Abendroth, G.; Hollmann, R.; Raddatz, A.; Ernst, H.; Gurski, H. Microbial Succinic Acid Producers and Purification of Succinic Acid. U.S. Patent 9023632 B2, 2015.
38. Krawczyk, J.M.; Haefner, S.; Schröder, H.; Costa, E.D.; Zelder, O.; Von Abendroth, G.; Wittmann, C.; Stellmacher, R.; Lange, A.; Lyons, B. J.; Lyons, T.J.; Crecy, E.; Hughes, E. Microorganisms for Succinic Acid Production. U.S. Patent application 20160348082, 2016.
39. Dunuwila, D.; Cockrem, M. Methods and Systems of Producing Dicarboxylic Acids. U.S. Patent application 20120259138, 2012.
40. Fruchey, O.S.; Manzer, L.E.; Dunuwila, D.; Keen, B.T.; Albin, B.A.; Clinton, N.A.; Dombek, B.D. Processes for the Production of Hydrogenated Products. U.S. Patent 8193375, 2012.
41. Boit, B.; Fiey, G.; van, D.G.M.J. Process for Manufacturing Succinic Acid from A Fermentation broth Using Nano Filtration to Purify Recycled Mother Liquor. WO Patent application 2016087408, 2016.
42. Mori, Y.; Takahashi, G.; Suda, H.; Yoshida, S. Processes for Producing Succinic Acid. U.S. Patent 9035095, 2015.
43. Tosukhowong, T. A Process for Preparing Succinic Acid and Succinate Ester. WO Patent application 2015085198, 2015.
44. Krieken, J.V.; Breugel, J.V. Process for the Preparation of A Monovalent Succinate Salt. U.S. Patent application 20110244534, 2011.
45. De Jong, E.; Higson, A.; Walsh, P.; Wellisch, M. Bio-based Chemicals–Value Added Products from Biorefineries. IEA Bioenergy–Task 42 Biorefinery. Available online: http://www.ieabioenergy.com/wp-content/uploads/2013/10/Task-42-Biobased-Chemicals-value-added-products-from-biorefineries.png (accessed on 17 March 2017).
46. Bio-conversion and Separation Technology (2013). WP 8.1. Determination of Market Potential for Selected Platform Chemicals–Itaconic acid, Succinic acid, 2,5-Furandicarboxylic acid. Available online: http://www.bioconsept.eu/wp-content/uploads/BioConSepT_Market-potential-for-selected-platform-chemicals_report1.png (accessed on 17 March 2017).
47. Varadarajan, S.; Miller, D.J. Catalytic upgrading of fermentation-derived organic acids. *Biotechnol. Prog.* **1999**, *15*, 845–854. [CrossRef] [PubMed]
48. Delhomme, C.; Weuster-Boltz, D.; Kühn, F.E. Succinic acid from renewable resources as a C_4 building-block chemical—A review of the catalytic possibilities in aqueous media. *Green Chem.* **2008**, *11*, 13–26. [CrossRef]
49. Burgard, A.P.; van Dien, S.J.; Burk, M. Methods and Organisms for the Growth-Coupled Production of 1,4-butanediol. U.S. Patent 7947483, 2011.
50. Berglund, K.A.; Alizadeh, H.; Dunuwila, D.D. Deicing Compositions and Methods of Use. U.S. Patent 6287480, 2001.
51. Facklam, T. Mixed Alkyl Benzyl Esters of Succinic Acid Used as Plasticizers. US Patent 9080032, 2015.
52. Broz, J.; Seon, A.; Simoes-Nunes, C. Use of Succinic Acid. U.S. Patent application 20110189347, 2011.
53. Tosukhowong, T. A Process for Preparing Succinic Acid and Succinate Ester. U.S. Patent application 20160304431, 2016.
54. Rush, B.J.; Watts, K.T.; McIntosh, V.L., Jr.; Fosmer, A.M.; Poynter, G.M.; McMullin, T.W. Yeast Cells Having Reductive TCA Pathway from Pyruvate to Succinate and Overexpressing An Exogenous $NAD(P)^+$ Transhydrogenase Enzyme. U.S. Patent application 20150203877, 2015.
55. BioAmber website. Available online: https://www.bio-amber.com/bioamber/en/company (accessed on 20 March 2017).
56. Jantama, K.; Haupt, M.J.; Svoronos, S.A.; Zhang, X.; Moore, J.C.; Shanmugham, K.T.; Ingram, L.O. Combining metabolic engineeringand metabolic evolutions to develop nonrecombinant strains of *Escherichia coli* C that produce succinate and malate. *Biotechnol. Bioeng.* **2008**, *99*, 1140–1153. [CrossRef] [PubMed]
57. Jantama, K.; Zhang, X.; Moore, J.C.; Shanmugham, K.T.; Svoronos, S.A.; Ingram, L.O. Eliminating side products and increasing succinate yields in engineered strains of *Escherichia coli* C. *Biotechnol. Bioeng.* **2008**, *101*, 881–893. [CrossRef] [PubMed]
58. Grabar, T.; Gong, W.; Yocum, R. Metabolic Evolution of *Escherichia coli* Strains That Produce Organic Acids. U.S. Patent 8871489, 2014.
59. Myriant Website. Available online: http://myriant.com/our-company/index.cfm (accessed on 20 March 2017).
60. Myriant in Talks to Build Bio-BDO Plant in Asia. Available online: https://www.icis.com/resources/news/2013/07/02/9684067/myriant-in-talks-to-build-bio-bdo-plant-in-asia-exec/ (accessed on 20 March 2017).

61. Reverdia Website. Available online: http://www.reverdia.com/ (accessed on 20 March 2017).

62. Verwaal, R.; Wu, L.; Damveld, R.A.; Sagt, C.M.J. Dicarboxylic Acid Production in Eukaryotes. U.S. Patent 9340804, 2016.

63. Jansen, M.L.A.; Van De Graaf, M.J.; Verwaal, R. Dicarboxylic acid production process. U.S. Patent application US 20120040422, 2012.

64. Succinity Website. Available online: http://succinity.com/ (accessed 20 March 2017).

65. Succinity Produces First Commercial Quantities of Biobased Succinic Acid. Available online: https://www.basf.com/en/company/news-and-media/news-releases/2014/03/p-14-0303-ci.html (accessed 20 March 2017).

66. Moussa, H.I.; Elkamel, A.; Young, S.B. Assessing energy performance of bio-based succinic acid production using LCA. *J. Clean. Prod.* **2016**, *139*, 761–769. [CrossRef]

67. Pinazo, J.M.; Domine, M.E.; Parvulescu, V.; Petru, F. Sustainability metrics for succinic acid production: A comparison between biomass-based and petrochemical routes. *Catal. Today* **2015**, *239*, 17–24. [CrossRef]

fermentation

MDPI

Article

Direct Succinic Acid Production from Minimally Pretreated Biomass Using Sequential Solid-State and Slurry Fermentation with Mixed Fungal Cultures

Jerico Alcantara [1,2], Andro Mondala [1,*], Logan Hughey [1] and Shaun Shields [1]

[1] Department of Chemical and Paper Engineering, Western Michigan University, Kalamazoo, MI 49008, USA; jerico.alcantara@wmich.edu (J.A.); logan.r.hughey@wmich.edu (L.H.); shaun.p.shields@wmich.edu (S.S.)

[2] Department of Chemical Engineering, University of the Philippines Los Baños, College, Laguna 4031, Philippines

[*] Correspondence: andro.mondala@wmich.edu; Tel.: +1-269-276-3508

Received: 25 May 2017; Accepted: 26 June 2017; Published: 30 June 2017

Abstract: Conventional bio-based succinic acid production involves anaerobic bacterial fermentation of pure sugars. This study explored a new route for directly producing succinic acid from minimally-pretreated lignocellulosic biomass via a consolidated bioprocessing technology employing a mixed lignocellulolytic and acidogenic fungal co-culture. The process involved a solid-state pre-fermentation stage followed by a two-phase slurry fermentation stage. During the solid-state pre-fermentation stage, *Aspergillus niger* and *Trichoderma reesei* were co-cultured in a nitrogen-rich substrate (e.g., soybean hull) to induce cellulolytic enzyme activity. The ligninolytic fungus *Phanerochaete chrysosporium* was grown separately on carbon-rich birch wood chips to induce ligninolytic enzymes, rendering the biomass more susceptible to cellulase attack. The solid-state pre-cultures were then combined in a slurry fermentation culture to achieve simultaneous enzymatic cellulolysis and succinic acid production. This approach generated succinic acid at maximum titers of 32.43 g/L after 72 h of batch slurry fermentation (~10 g/L production), and 61.12 g/L after 36 h of addition of fresh birch wood chips at the onset of the slurry fermentation stage (~26 g/L production). Based on this result, this approach is a promising alternative to current bacterial succinic acid production due to its minimal substrate pretreatment requirements, which could reduce production costs.

Keywords: consolidated bioprocessing; fungi; solid-phase fermentation; bio-based chemicals; lignocellulose biomass; mixed cultures

---

1. Introduction

Succinic acid is a four-carbon 1,4-dicarboxylic acid listed in the US Department of Energy's top 12 bio-based molecules [1]. It is a "platform" chemical with wide applications in food and pharmaceuticals, surfactants, detergents, green solvents, and biodegradable plastics [2]. Succinic acid is traditionally commercially synthesized from petroleum-derived precursors; specifically, through hydrogenation of maleic acid, oxidation of 1,4-butanediol, or carbonylation of ethylene glycol [3]. An alternative pathway for bio-based succinic acid production involves anaerobic bacterial fermentation of pure sugars under CO_2-rich conditions [4,5]. Lignocellulosic biomass can be used but requires substantial pretreatment and hydrolysis to liberate fermentable sugars that can be directly utilized by succinate-producing bacteria, which can significantly add to the product cost.

This study presents a proof-of-concept for a new direct bioconversion technology for succinic acid production from minimally-pretreated non-hydrolyzed biomass. This method utilizes a mixed culture of lignocellulolytic and acidogenic fungi in a sequential solid-state and slurry fermentation process. As shown in Figure 1, the process begins with solid-state pre-cultivation of the fungal strains on moist substrates. The goal is to initially induce and maximize cellulase, hemicellulase, and ligninase activities under static conditions, which also serves to biologically pretreat the biomass substrates in preparation for simultaneous saccharification and fermentation in the slurry fermentation stage. In the solid-state cultivation stage, nitrogen-rich (e.g., soybean hulls) and carbon-rich substrates (e.g., birch wood chips) were pre-fermented separately. The N-rich substrate was pre-fermented using a *Trichoderma reesei* and *Aspergillus niger* co-culture to induce cellulase and hemicellulase production. Prior studies have shown that *Trichoderma reesei* and *Aspergillus* spp. co-cultures are more effective for maximizing cellulolytic enzyme activities than single cultures by establishing synergistic levels of exocellulases (produced primarily by *T. reesei*) and endocellulases and β-glucosidases (produced mainly by *A. niger*) for sustained enzymatic hydrolysis and minimization of product inhibition effects [6–9]. On the other hand, the C-rich substrate was pre-fermented using *Phanerochaete chrysosporium*, which has been shown to be effective in delignifying agricultural biomass residues [10,11]. Separate solid-state pre-fermentation was done to prevent negative competitive interactions between the cellulolytic *T. reesei* and *A. niger* co-culture and the ligninolytic *P. chrysosporium*. After a sufficient cultivation time wherein optimal enzyme activities are established, the separate solid-state pre-cultures containing the fermented C-rich and N-rich substrates, fungal mycelia, and generated enzymes and other metabolites are combined (to a certain C:N ratio), submerged in a buffered medium, and incubated under continuous agitation during the slurry fermentation stage. Under the submerged two-phase fermentation condition, simultaneous biomass saccharification and fermentation of liberated biomass sugars into dicarboxylic acids such as succinic acid will occur. While few *A. niger* strains are known to produce succinic acid in low concentrations [12], typically as a minor co-product of citric acid production [13], the *T. reesei*-*A. niger*-*P. chrysosporium* triple cultures tested in this study unexpectedly favored the overproduction and accumulation of succinic acid.

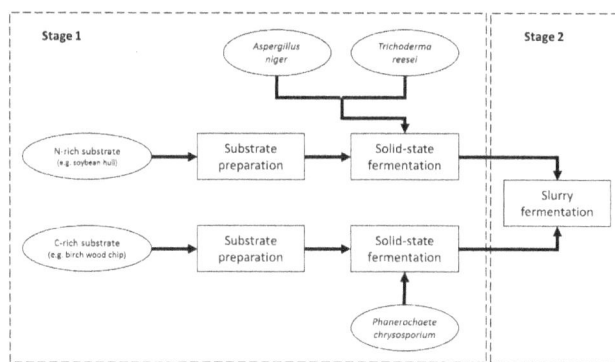

Figure 1. Sequential solid-state fermentation and slurry fermentation for direct succinic acid production from biomass using mixed fungal culture.

In this paper, we present our preliminary findings demonstrating the feasibility of this new method for direct succinic acid production from biomass substrates. Batch and fed-batch kinetic data of cellulase activity, residual sugars, and succinic acid production using *T. reesei*–*A. niger*–*P. chrysosporium* triple cultures in a sequential solid-state and slurry fermentation process are discussed.

2. Materials and Methods

2.1. Substrates

Soybean hulls and birch wood chips were used as fermentation substrates in this study. Both materials were not subjected to any physicochemical pretreatment other than size reduction and autoclaving for sterilization purposes. Soybean hulls were obtained from Archer Daniels Midland, MI, USA. Soybean hulls typically consist of 38.4% cellulose, 10.2% hemicellulose, 2.8% lignin, and 10.7% protein [14]. Birch wood chips were provided by the Fiber and Pulp Bleaching Laboratory at Western Michigan University. Birch wood chips are typically composed of approximately 40% cellulose, 33% hemicellulose, and 20% lignin [15]. Birch wood chips were first milled to 2000-μm sizes using a Wiley laboratory mill (ED-5, Arthur H. Thomas Company, Philadelphia, PA, USA). Both soybean hulls and pre-milled birch wood chips were then further milled using a kitchen blender. The ground biomass was then screened on a stainless-steel sieve to collect substrates of approximately 500–1000 μm nominal particle diameters.

2.2. Fungi

A. niger (ATCC 15475), *T. reesei* (ATCC 56765), and *P. chrysosporium* (ATCC 48746) were obtained from the American Type Culture Collection (ATCC), Manassas, VA, USA. Fungal inocula were prepared as follows: Spores from seven-day old potato dextrose agar (PDA) slant stock cultures were suspended in sterile deionized water, poured and spread on PDA plates, and incubated at 30 °C for seven days. The spores formed on the plates were washed with sterile modified Mandel's medium containing 2 g glucose, 1.4 g $(NH_4)_2SO_4$, 2.0 g KH_2PO_4, 0.3 g $MgSO_4 \cdot 7H_2O$, 0.4 g $CaCl_2 \cdot 2H_2O$, 0.3 g NH_2CONH_2, 1.0 g Proteose peptone No. 2, 0.2 mL Tween-80, 5 mg $FeSO_4 \cdot 7H_2O$, 1.6 mg $MnSO_4 \cdot H_2O$, 1.4 mg $ZnSO_4 \cdot 7H_2O$, 2.0 mg $CoCl_2$ per L of deionized water 1 L [16]. To generate the fungal inocula, spore suspensions containing 1×10^6 to 1×10^7 of spores/mL were transferred to 50-mL centrifuge tubes and incubated in an incubator shaker (Innova® 43, New Brunswick Scientific Co. Inc., Enfield, CT, USA) at 30 °C, 150 rpm for two days.

2.3. Solid-State Pre-Fermentation Stage

In the solid-state pre-fermentation stage, the fungi were grown on milled biomass substrates under static conditions. The moisture content of both substrates was adjusted to 70% (wet basis) using modified Mandel's medium. These were then sterilized using a laboratory autoclave (MLS-3780, SANYO Electric Co., Ltd., Osaka, Japan) at 121 °C for 15 min. Solid-state pre-cultures containing mixed *T. reesei* and *A. niger* in soybean hulls were prepared by adding 10% (*v/w*) of two-day-old inoculum cultures of *T. reesei* in a 250-mL Erlenmeyer flask containing 7.5 g (d.b., dry basis) sterilized soybean hull. *A. niger* was inoculated into the same flask one day after *T. reesei* addition to ensure adequate growth of *T. reesei* before *A. niger* addition and thus balanced cellulase and β-glucosidase activities in this co-culture. Two-day-old *P. chrysosporium* inocula (10% *w/v*) were grown separately in 500-mL flasks containing 22.5 g (dry basis) birch wood chips to induce ligninolytic activity. All flasks were incubated in a humidified incubator (HIS33SD, Powers Scientific Inc., Warminster, PA, USA) at 30 °C, 95% relative humidity for seven days.

2.4. Slurry Fermentation Stage

After the solid-state pre-cultivation period, the static co-cultures of *T. reesei* and *A. niger* in soybean hulls were suspended in 50.0 mL 0.05 M sodium acetate buffer (pH 4.8). The resulting mixture was mixed thoroughly for 20 min before transferring to the 500-mL baffled flask containing the *P. chrysosporium* pre-culture in birch wood chips. The mixture was then agitated to suspend all solids and fungal mycelia. The dry solids content of the combined cultures became 20% (*w/w*) with a birch wood chips to soybean hull mass ratio of 3:1. The flasks containing the combined pre-culture slurry were then incubated in an incubator shaker (Innova® 43, New Brunswick Scientific Co. Inc., Edison, NJ, USA) at 35 °C, 150 rpm for four days.

2.5. Analyses

Samples were collected from the liquid portion of the broth every 12 h and centrifuged at 12,000 rpm for 10 min. The supernatants were diluted with deionized water to 1:5 in 15 mL centrifuge tubes. The diluted supernatant was analyzed for cellulase activities in terms of Filter Paper Units (FPU) according to previously published methods [17]. Briefly, FPU activity was measured by incubating 0.5 mL of diluted supernatant in 1 mL 0.05 M citrate buffer, pH 4.8, and 50 °C for 60 min and measuring the amount of reducing sugar released using the dinitrosalicylic acid (DNS) assay [18]. The absorbance was read using a UV-Vis spectrophotometer (SmartSpec™ 3000, Bio-Rad Laboratories, Inc., Hercules, CA, USA) operated at a wavelength of 540 nm with glucose as standard. Sugars (cellobiose, glucose, and xylose) and organic acids (oxalic acid, citric acid, malic acid, and succinic acid) were analyzed using an Agilent 1100 Series HPLC (Agilent Technologies, Santa Clara, CA, USA) equipped with a BioRad HPX-87H Cation Exchange column and Variable Wave Detector (VWD) for organic acids and Refractive Index (RI) detector for sugars. HPLC analysis was performed at 25 °C and wavelength of 210 nm using 0.005 M H_2SO_4 mobile phase flowing at 0.5 mL/min.

3. Results and Discussion

3.1. Solid-State Pre-Fermentation Stage

The initial solid-state pre-fermentation stage has a two-fold purpose: to induce fungal growth and hydrolytic enzyme activities while biologically pre-treating the biomass in preparation for direct fermentation in the slurry fermentation stage. To induce organic acid overproduction during the succeeding slurry fermentation stage, it is essential that the C-rich (i.e., birch wood chips) and N-rich (soybean hull) biomass substrates be blended to achieve an appropriately high C:N ratio. This would then necessitate inoculating all three fungal species together, simultaneously or sequentially, in the culture. However, preliminary tests showed that fungal inoculation on pre-blended soybean hull and birch wood chips substrate led to overgrowth of *P. chrysosporium* and inhibition of growth and cellulolytic activity of *T. reesei* and *A. niger*. To overcome this negative interaction effect among the fungal species in the mixed culture, it was suggested to separately pre-ferment the soybean hulls and birch wood chips before combining these substrates for the slurry fermentation stage.

Protein-rich soybean hulls are ideal substrates for inducing fungal cellulase enzyme production and were sequentially inoculated with *T. reesei* followed by *A. niger*. *A. niger* was inoculated one day after *T. reesei* to allow the proliferation of slower-growing *T. reesei*, prevent overgrowth and domination by the faster-growing *A. niger*, and establish balanced and sustained cellulase and β-glucosidase activities [8]. Figure 2a shows expansive greenish mycelia present on the surface of the soybean hull biomass with underlying black mycelia, indicating the dominant growth of *T. reesei* over *A. niger* after a one-day delay time of *A. niger* inoculation after *T. reesei*. Preliminary tests also showed that inoculating both species simultaneously resulted in *A. niger* proliferating the soybean hull surface, which reduced the chances of *T. reesei* establishing growth in a co-culture with *A. niger*.

The birch wood chips contain up to 20% (*w*/*w*) lignin [15]. Pretreatment is necessary to remove the lignin to make the cellulose and hemicellulose more accessible to enzymatic hydrolysis [19,20]. Biological pretreatment using fungi was preferred in this study over thermochemical pretreatment as it demonstrates improvement in biomass digestibility with reduced severity [21]. This study employed biological pretreatment of birch wood chip using *P. chrysosporium*, a white-rot fungi. Several studies showed that *P. chrysosporium* could significantly reduce lignin content of various agricultural residues [10,11]. Evidence of white mycelial *P. chrysosporium* growth on birch wood chips is shown in Figure 2b. However, the ligninolytic activity of the *P. chrysosporium* culture in birch wood chips was not quantified.

In addition to pre-saccharifying the substrates prior to direct acidogenic fermentation, separately pre-fermenting the soybean hulls and birch wood chips in the described manner was expected to allow growth and activation of hydrolytic enzyme activities by specialized cellulolytic and ligninolytic fungi. Hostile interactions between these fungi are prevented to allow the fungal niches to establish and sustain cellulolytic and ligninolytic activities well into the slurry fermentation stage.

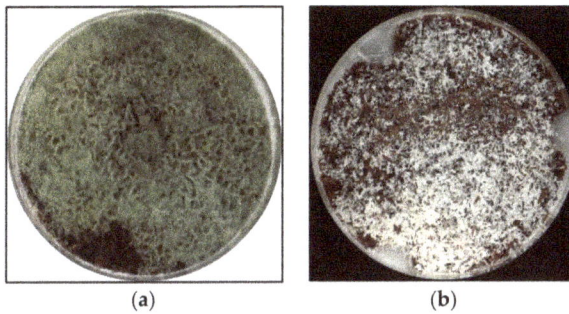

Figure 2. Solid-state fermentation after seven days (**a**) a mixed culture of *Trichoderma reesei* and *Aspergillus niger* on soybean hull; (**b**) *Phanerochaete chrysosporium* on birch wood chip.

3.2. Slurry Fermentation Stage

During the slurry fermentation stage, the separate solid-state pre-cultures of N-rich soybean hulls with *T. reesei* and *A. niger* and C-rich birch wood chips with *P. chrysosporium* were combined, submerged into a buffered medium, and incubated under batch conditions. Kinetic data of cellulase (filter paper-ase or FPase) activities, sugars (cellobiose, glucose, and xylose), succinic acid, and pH in a batch experiment were obtained to understand the underlying bioconversion processes and identify process variables and operating conditions that can be optimized to improve succinic acid production yields for future investigations.

Initial FPase was measured at the end of the solid-state pre-fermentation stage or immediately before transitioning to the slurry fermentation stage. FPase increased within the first 12 h of slurry fermentation and was constant between 12 and 24 h (Figure 3a). Glucose levels in the slurry culture followed this trend very closely (Figure 3b), showing a significant increase between 0 and 12 h and remained steady from 12–24 h. On the other hand, cellobiose increased after 12 h and remained fairly constant at 1 g/L for the next 24 h before it declined after 84 h. Xylose also increased to around 1 g/L after 12 h and remained constant before it declined at 72 h and was depleted after 84 h. FPase increased once more after 24 h and reached its peak at 60 h. Additionally, cellobiose doubled in concentration at 48 h which coincided with high FPase activity. However, residual glucose levels declined after 24 h and were depleted at 60 h. High FPase activity did not translate to high residual sugar present in the biomass as the generated fermentable sugars appear to have been immediately consumed for succinic acid production. At 72 h, FPase began to decline, and no residual glucose was detected. At the end of slurry fermentation, FPase went back to its original level at the start of the slurry fermentation phase, yet no further glucose production was observed. It is possible that at this period, cellulose groups susceptible to enzymatic attack may have been expended.

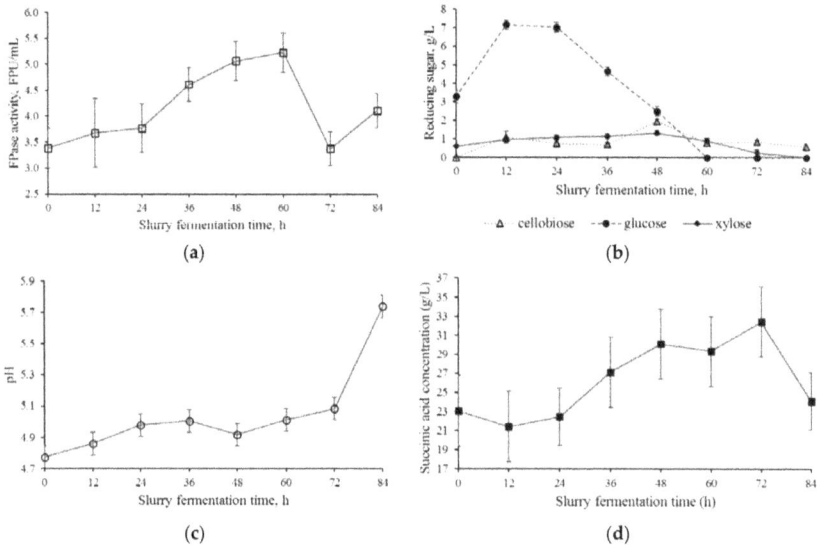

Figure 3. Time profiles of (**a**) filter paper cellulase activity, (**b**) residual sugars, (**c**) pH, and (**d**) succinic acid concentration from a mixed culture of *Trichoderma reesei*, *Aspergillus niger*, and *Phanerochaete chrysosporium* on 3:1 birch wood chip–soybean hull biomass during 84 h of slurry fermentation. Error bars represent pooled standard deviations in the data sets.

The culture pH was relatively stable during the first 72 h of slurry fermentation with values ranging from pH 4.77 to pH 5.08 (Figure 3c). For crude cellulase enzyme broth, enzymatic hydrolysis is usually conducted at pH 4.8 and 4.5 to 50 °C [22]. The pH range in the first 72 h of slurry fermentation is close to the optimum pH; however, the operating temperature in this study is lower than the reported optimum temperatures. Increasing the temperature can adversely affect the growth of the mixed fungal culture, so the culture temperature was kept close to lower recommended incubation temperatures. After 72 h, the pH increased drastically to 5.8 at 84 h, which corresponded to complete depletion of fermentable sugars and minimized filter paper cellulase activity.

As shown in Figure 3d, a substantial amount of succinic acid concentration was already produced during the solid-state pre-fermentation stage before the start of the slurry fermentation stage. At time zero, immediately after suspending the pre-cultures in the buffered media, around 23 g of succinic acid per liter of the media was detected. Unfortunately, it was not experimentally determined from which pre-culture (i.e., soybean hulls with *T. reesei* and *A. niger* vs. birch wood chips with *P. chrysosporium*) it originated. The only other organic acid that was detected was oxalic acid, which was produced at a very low concentration of 0.02 g/L. No significant increase in succinic acid levels was observed in the first 24 h (Figure 3b). Succinic acid concentrations started to increase 24 h after the maximum glucose production was reached (12 h) and continued between 24 to 48 h, after which it remained steady up to 60 h and then increased to its maximum level (32.4 g/L) at 72 h. Succinic acid production occurred concurrently with glucose consumption. The observed increase in FPase up to 60 h could indicate continuous generation of glucose through cellulolysis but this glucose was immediately consumed for acidogenesis. Beyond 72 h, succinate levels declined substantially as there was no glucose available for consumption. Other organic acids produced in minor levels during the slurry fermentation stage include citric acid and oxalic acid, which were generated at 0.89 g/L and 0.52 g/L, respectively. Both citric acid and oxalic acid were not detected after 108 h, but 0.56 g/L of malic acid was observed.

The highest succinic acid concentration obtained in the batch process under this study (32.06 g/L, after 72 h of slurry fermentation) is comparable to succinic acid levels produced by bacteria utilizing pure fermentable sugars derived from pretreatment and hydrolysis of lignocellulosic substrates. One study obtained 23.8 g/L succinic acid using *Anaerobiospirillum succiniciproducens* from wood hydrolysate derived from steam explosion pretreatment of oak wood, and supplemented with corn steep liquor in a 32-h batch fermentation [23]. Another study produced 22.5 g/L succinic acid using *Actinobacillus succinogenes* from hemicellulose hydrolysate produced from acid pretreatment of sugarcane bagasse in a 24-h batch fermentation [24]. Both studies employed additional chemical and physicochemical pretreatment methods to obtain fermentable sugars from lignocellulosic biomass. In this study, however, direct fermentation of the lignocellulosic biomass was performed as the fungal co-culture grew directly on biomass, released enzymes that liberated fermentable sugars from the lignocellulose matrix, and converted the released fermentable sugars to succinic acid in an integrated process without severe pretreatment of the lignocellulosic biomass substrates. Succinic acid production by filamentous fungi has not yet been fully studied. High succinic acid concentration, productivity, low byproduct formation, and tolerance to low pH are some characteristics of promising fungal strains for succinic acid production [25]. Although this study mostly yielded comparable if not higher succinic acid concentrations than those reported in bacterial fermentations, its volumetric productivity of 0.45 g/L/h is still low compared to conventional bacterial succinic acid production. However, this value is higher than the reported succinic acid volumetric productivity of 0.14 g/L/h from mutant *A. niger* GCMC-7 with black strap molasses as a substrate, wherein the succinic acid is a secondary product of citric acid fermentation [13]. Bacterial succinic fermentation operates at a higher pH of 6.0 to 6.5 [23,24] while the fungal co-culture process described in this study operated at a relatively lower pH, which reduces the need for pH control at near-neutral levels.

3.3. Effect of Adding Fresh Substrate

This study also conducted preliminary runs to see the effect of introducing fresh C-rich substrate during the onset of the slurry fermentation stage on succinic acid production. It was hypothesized that adding fresh C-rich substrate could further increase the available carbon in the broth during slurry fermentation. Additional untreated milled birch wood chips (< 500 µm) were added at the start of slurry fermentation. This addition resulted in an increase in the birch wood chips-to-soybean hull ratio of 4:1 from 3:1. These fine particle birch wood chips were obtained from the undersized fraction of the milling process. All other slurry fermentation conditions such as 20% (w/v) solids loading, initial pH, temperature, and agitation speed were held constant. A maximum succinic acid concentration of 61.12 g/L was achieved after 36 h of slurry fermentation as shown in Figure 4. This amounts to a ~26 g/L succinate production relative to initial levels (i.e., transition from solid-state to slurry fermentation stage) compared to ~10 g/L production in the batch process. The addition of fresh substrate increased the succinic acid production from 40 to 77% during slurry fermentation. Additionally, the time to reach the maximum succinic acid was reduced. A maximum reducing sugar of 10.5 g/L was achieved at 24 h. When the residual reducing sugar started to decline at 36 h, the maximum succinic acid concentration was observed. At this point, the accumulated reducing sugar produced during the first 24 h of slurry fermentation was probably directly consumed to produce succinic acid. After 36 h, succinic acid and residual reducing sugar levels in the broth declined. From this preliminary result, the volumetric productivity of succinic acid production becomes 1.70 g/L·h which is 278% higher than our previous result of 0.45 g/L·h. Future studies will focus on optimizing the timing of addition of fresh substrate to further improve succinic acid production.

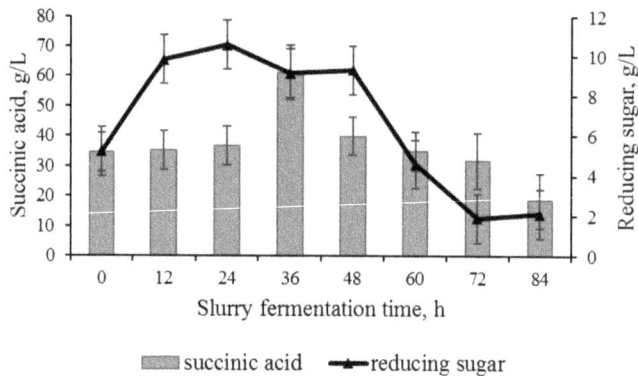

Figure 4. Time profile of succinic acid concentration and residual sugar from a mixed culture of *Trichoderma reesei*, *Aspergillus niger*, and *Phanerochaete chrysosporium* on 4:1 birch wood chip–soybean hull biomass during 84 h of slurry fermentation following the addition of fresh C-rich substrate.

4. Conclusions

This study demonstrated the feasibility of directly producing succinic acid from minimally-pretreated biomass using a new consolidated bioprocessing technique involving sequential solid-state and slurry fermentation with a mixed cellulolytic–acidogenic fungal culture. Under batch conditions, a succinic acid concentration of 32.43 g/L was achieved after 72 h of slurry fermentation, corresponding to ~10 g/L succinic acid production. An overall succinic acid yield of 13.0 g per 100 g dry substrate was also obtained. When the system was converted to fed-batch, wherein fresh substrate was introduced at the onset of the slurry fermentation stage, a succinic acid concentration of 61.12 g/L was reported, corresponding to ~26 g/L succinic acid production. This high succinic acid concentration corresponds to the volumetric productivity of 1.70 g/L·h and a yield of 24.45 g succinic acid per 100 g substrate. The next challenge is to increase the volumetric productivity of succinic acid production using a fungal co-culture for it to be competitive with bacterial succinic acid and to further minimize the cost of production. Factors that may affect fungal growth and succinic acid production such as pH, solids loading, fungal spore loading, fungal species combination, carbon substrate concentration, and others need to be investigated in future studies. Optimizing the consolidated bioprocessing and shifting to fed-batch fermentation could potentially lead to an increase in succinic acid concentration and volumetric productivity.

Acknowledgments: This work was supported by a grant (W2016-006) from the Faculty Research and Creative Activities Award (FRACAA) by the Office of the Vice President for Research, Western Michigan University.

Author Contributions: Andro Mondala, Jerico Alcantara, and Shaun Shields conceived and designed the experiments; Jerico Alcantara and Logan Hughey performed the experiments; Jerico Alcantara and Andro Mondala analyzed the data; Jerico Alcantara and Andro Mondala wrote the paper.

Conflicts of Interest: The authors declare no conflict of interest. The funding sponsors had no role in the design of the study; in the collection, analyses, or interpretation of data; in the writing of the manuscript, and in the decision to publish the results.

References

1. Werpy, P.; Petersen, G. *Top Value Added Chemicals From Biomass: Volume I—Results of sCreening Potential Candidates from Sugars And Synthesis Gas*; US Department of Energy: Washington, DC, USA, 2004.
2. Zeikus, J.G.; Jain, M.K.; Elankovan, P. Biotechnology of succinic acid production and markets for derived industrial products. *Appl. Microbiol. Biotechnol.* **1999**, *51*, 545–552. [CrossRef]

3.   Cornils, B.; Lappe, P. Dicarboxylic acids, aliphatic. In *Ullmann's Encyclopedia of Industrial Chemistry*; Wiley-VCA: Weinheim, Germany, 2010. [CrossRef]
4.   Song, H.; Lee, S.Y. Production of succinic acid by bacterial fermentation. *Enzym. Microb. Technol.* **2006**, *39*, 352–361. [CrossRef]
5.   Thakker, C.; Martinez, I.; San, K.Y.; Bennett, G.N. Succinate production in *Escherichia coli*. *Biotechnol. J.* **2012**, *7*, 213–224. [CrossRef] [PubMed]
6.   Gutierrez-Correa, M.; Portal, L.; Moreno, P.; Tengerdy, R.P. Mixed culture solid substrate fermentation of *Trichoderma reesei* with *Aspergillus niger* on sugar cane bagasse. *Bioresour. Technol.* **1999**, *68*, 173–178. [CrossRef]
7.   Dhillon, G.S.; Oberoi, H.S.; Kaur, S.; Bansal, S.; Brar, S.K. Value-addition of agricultural wastes for augmented cellulase and xylanase production through solid-state tray fermentation employing mixed-culture of fungi. *Ind. Crop Prod.* **2011**, *34*, 1160–1167. [CrossRef]
8.   Fang, H.; Zhao, C.; Song, X.; Chen, M.; Chang, Z.; Chu, J. Enhanced cellulolytic enzyme production by the synergism between *Trichoderma reesei* RUT-C30 and *Aspergillus niger* NL02 and by the addition of surfactants. *Biotechnol. Bioprocess Eng.* **2003**, *18*, 390–398. [CrossRef]
9.   Lu, J.; Weerasiri, R.R.; Liu, Y.; Wang, W.; Ji, S.; Lee, I. Enzyme production by the mixed fungal culture with nano-shear pretreated biomass and lignocellulose hydrolysis. *Biotechnol. Bioeng.* **2013**, *110*, 2123–2130. [CrossRef] [PubMed]
10.   Shi, J.; Chinn, M.S.; Sharma-Shivappa, R.R. Microbial pretreatment of cotton stalks by solid state cultivation of *Phanerochaete chrysosporium*. *Bioresour. Technol.* **2008**, *99*, 6556–6564. [CrossRef] [PubMed]
11.   Zeng, J.; Singh, D.; Chen, S. Biological pretreatment of wheat straw by *Phanerochaete chrysosporium* supplemented with inorganic salts. *Bioresour. Technol.* **2011**, *102*, 3206–3214. [CrossRef] [PubMed]
12.   Bercovitz, A.; Peleg, Y.; Battat, E.; Rokem, J.S.; Goldberg, I. Localization of pyruvate carboxylase in organic acid producing *Aspergillus* strains. *Appl. Environ. Microbiol.* **1990**, *56*, 1594–1597. [PubMed]
13.   Ikram-ul, H.; Ali, S.; Qadeer, M.A.; Iqbal, J. Citric acid production by selected mutants of *Aspergillus niger* from cane molasses. *Bioresour. Technol.* **2004**, *93*, 125–130. [CrossRef] [PubMed]
14.   Mielenz, J.R.; Bardsley, J.S.; Wyman, C.E. Fermentation of soybean hulls to ethanol while preserving protein value. *Bioresour. Technol.* **2009**, *100*, 3532–3539. [CrossRef] [PubMed]
15.   Lai, Y.Z. Chemical degradation. In *Wood and Cellulosic Chemistry*; Hon, D.N.-S., Shiraishi, N., Eds.; Marcel Dekker Inc.: New York, NY, USA, 1991; pp. 455–524.
16.   Tangnu, S.K.; Blanch, H.W.; Wilke, C.R. Enhanced production of cellulase, hemicellulose, and β-glucosidase by *Trichoderma reesei* (Rut C-30). *Biotechnol. Bioeng.* **1981**, *23*, 1837–1849. [CrossRef]
17.   Ghose, T.K. Measurement of cellulase activities. *Pure Appl. Chem.* **1987**, *59*, 257–268. [CrossRef]
18.   Miller, G.L. Use of dinitrosalicylic acid reagent for determination of reducing sugar. *Anal. Chem.* **1959**, *31*, 426–428. [CrossRef]
19.   McMillan, J.D. Pretreatment of lignocellulosic biomass. In *Enzymatic Conversion of Biomass for Fuels Production*; Himmel, M.E., Baker, J.O., Overend, R.P., Eds.; American Chemical Society: Washington, DC, USA, 1994; Volume 566, pp. 292–324.
20.   Mosier, N.; Wyman, C.; Dale, B.; Elander, R.; Lee, Y.Y.; Holtzapple, M.; Ladisch, M. Features of promising technologies for pretreatment of lignocellulosic biomass. *Bioresour. Technol.* **2005**, *96*, 673–686. [CrossRef] [PubMed]
21.   Keller, F.A.; Hamilton, J.E.; Nguyen, Q.A. Microbial pretreatment of biomass: potential for reducing severity of thermochemical biomass pretreatment. *Appl. Biochem. Biotechnol.* **2003**, *105*, 27–41. [CrossRef]
22.   Duff, S.J.B.; Murray, W.D. Bioconversion of forest products industry waste cellulosics to fuel ethanol: A review. *Bioresour. Technol.* **1996**, *55*, 1–33. [CrossRef]
23.   Lee, P.C.; Lee, S.Y.; Hong, S.H.; Chang, H.N.; Park, S.C. Biological conversion of wood hydrolysate to succinic acid by *Anaerobiospirillum succiniciproducens*. *Biotechnol. Lett.* **2003**, *25*, 111–114. [CrossRef] [PubMed]

24. Borges, E.R.; Pereira, N., Jr. Succinic acid production from sugarcane bagasse hemicellulose hydrolysate by *Actinobacillus succinogenes*. *J. Ind. Microbiol. Biotechnol.* **2011**, *38*, 1001–1011. [CrossRef] [PubMed]

25. Magnuson, J.K.; Lasure, L.L. Organic acid production by filamentous Fungi. In *Advances in Fungal Biotechnology for Industry, Agriculture, and Medicine*; Tkacz, J.S., Lange, L., Eds.; Kluwer Academic/Plenum Publishers: New York, NY, USA, 2004; pp. 307–340.

fermentation

MDPI

Article

Wheat and Sugar Beet Coproducts for the Bioproduction of 3-Hydroxypropionic Acid by *Lactobacillus reuteri* DSM17938

Julien Couvreur [1,2], Andreia R. S. Teixeira [1,3], Florent Allais [1,2], Henry-Eric Spinnler [2], Claire Saulou-Bérion [2] and Tiphaine Clément [1,2,*]

[1]   Chaire Agro Biotechnologies Industrielles (ABI)-AgroParisTech, 3 rue des Rouges Terres, F-51110 Pomacle, France; julien.couvreur@agroparistech.fr (J.C.); Andreia.teixeira@agroparistech.fr (A.R.S.T.); florent.allais@agroparistech.fr (F.A.)
[2]   UMR 782 GMPA, AgroParisTech, Institut National de la Recherche Agronomique, Université Paris-Saclay, F-78850 Thiverval-Grignon, France; eric.spinnler@agroparistech.fr (H.-E.S.); claire.saulou-berion@inra.fr (C.S.-B.)
[3]   UMR 1145 GENIAL, AgroParisTech, Institut National de la Recherche Agronomique, Université Paris-Saclay, F-91300 Massy, France
*   Correspondence: tiphaine.clement@agroparistech.fr; Tel.: +33-(0)3-5262-0468

Received: 9 June 2017; Accepted: 28 June 2017; Published: 6 July 2017

Abstract: An experimental design based on Response Surface Methodology (RSM) was used for the formulation of a growth medium based on sugar beet and wheat processing coproducts adapted to the cultivation of *Lactobacillus reuteri* (*L. reuteri*) DSM17938. The strain was cultivated on 30 different media varying by the proportions of sugar beet and wheat processing coproducts, and the concentration of yeast extract, tween 80 and vitamin B12. The media were used in a two-step process consisting of *L. reuteri* cultivation followed by the bioconversion of glycerol into 3-hydroxypropionic acid by resting cells. The efficiency of the formulations was evaluated according to the maximal optical density at the end of the growth phase (ΔOD_{620nm}) and the ability of the resting cells to convert glycerol into 3-hydroxypropionic acid, a platform molecule of interest for the plastic industry. De Man, Rogosa, and Sharpe medium (MRS), commonly used for the cultivation of lactic bacteria, was used as the control medium. The optimized formulation allowed increasing the 3-HP production.

Keywords: *Lactobacillus reuteri*; agroindustrial coproducts; glycerol bioconversion; 3-hydroxypropionic acid

1. Introduction

The global trend to develop a more sustainable economy based on renewable resources increases the demand for new innovative processes needed to obtain bio-based chemicals from biomass. The development of biorefineries, where biomass is converted to energy and various biomaterials is gaining ground, leading to an increased need to valorize the generated coproducts [1,2].

Wheat and beetroot are widespread crops, which are industrially converted to sugar, food additives, or other components for non-food applications. These processes lead to coproducts, which are generally valorized through animal feed or microbial fermentation for the production of biofuels.

Employing fermentation processes, for the bioconversion of biomass into green building blocks present many advantages, such as using of water as solvent and working at mild temperature. Using agro-industrial coproducts in growth media formulations is a good way to lower manufacturing costs of processes involving microbial systems.

Sugar beet molasses is mainly composed of sugar (~500 g·L^{-1}, glucose and fructose) and nitrogen compounds, but also contains minerals (e.g., calcium, magnesium, iron, zinc, copper, manganese) and group B vitamins (i.e., thiamin, niacin, riboflavin, and B6), which are necessary nutrients for microbial growth. On the other hand, many components such as organic salts, nitrites and phenolic compounds may induce inhibitory effects on the growth of microorganisms. However, several studies have shown the suitability of incorporating beet molasses in growth media for the production of chemical building blocks by employing different microorganisms, such as *Aspergillus niger* to synthesize citric acid [3], *Bacillus polymyxa* to produce polysaccharides [4], and *Lactobacillus delbrueckii* to generate lactic acid [5,6].

In this work we investigated the possibility of using combined beet and wheat coproducts as substrates to formulate a growth medium for *L. reuteri* DSM17938. This lactic acid bacteria (LAB) strain is of particular interest for its probiotic properties which are used in infant nutrition [7]. Furthermore, *L. reuteri* is capable of naturally converting glycerol into three molecules of industrial interest, 3-hydroxypropionaldehyde (3-HPA), 1,3-propanediol (1,3-PDO), and 3-hydroxypropionic acid (3-HP), which are used as building blocks for the production of superabsorbent polymers and diverse composite materials. In this work, considering the growing interest of this molecule as a building block of particular industrial interest, we will focus on 3-HP as the target molecule [8].

The bioconversion of glycerol to 3-HP by *L. reuteri* occurs in two steps: glycerol is first dehydrated to 3-HPA by a co-enzyme B12-dependent glycerol dehydratase. The synthesized 3-HPA can then either be excreted, reduced to 1,3-PDO via the NADH$_2$-dependent 1,3-propanediol oxidoreductase, or transformed to 3-HP via an oxidative pathway involving a NAD$^+$-dependent propionaldehyde dehydrogenase, a phosphotransacetylase and a propionate kinase [9]. *L. reuteri* is not able to use glycerol as a carbon source for growth, which limits the coproducts formed during the bioconversion of glycerol by resting cells. The use of resting cells in a restricted medium containing only glycerol is thus interesting as it facilitates the downstream processing [10,11].

In this case the glycerol bioconversion process requires a preliminary biomass production step. The growth of lactic acid bacteria is particularly demanding in terms of nutrient availability and environmental parameters (e.g., temperature, pH) and formulating appropriate growth media for their cultivation is challenging. LAB cultivation is generally conducted in MRS (De Man, Rogosa and Sharpe) broth as the reference growth medium. MRS is composed of ten different ingredients among others polypeptones and meat and yeast extracts, which are quite expensive. Pertinent studies have been conducted to develop less expensive media supporting the growth of particular species of lactobacillus, such as *Lactobacillus casei* [12], *Lactobacillus delbrueckii* [5,6], or *Lactobacillus plantarum* [13] but, to the best of our knowledge, such a study involving *L. reuteri* is lacking.

Processes involving microbial production are complex and the culture medium composition must be adapted to the microbial strain as well as to the expected product. Response Surface Methodology (RSM) is an efficient analytical method used to design experiments and optimize a complex set of factors for a specific response, and adapted to growth medium optimization [14]. The method enables to obtain a large amount of data from a reduced number of experiments and to study the influence of different input variables on a specific response, including the interactions between the studied factors. RSM has been successfully used for the optimization of culture media for different microorganisms including lactobacillus species [15–19].

The aim of this study is to elaborate a medium formulation based on agroindustrial coproducts, for the cultivation of *L. reuteri* and to evaluate the impact of this formulation on the bioconversion of glycerol by resting cells.

2. Materials and Methods

2.1. Microorganism

L. reuteri DSM 17938 was purchased from BioGaia AB (Stockholm, Sweden) and stored on MRS medium with 20% *w/v* glycerol at −80 °C. Before inoculation the strain was defrosted and grown overnight at 37 °C in MRS medium (Biokar, France).

2.2. Media

Low purity sugar beetroot syrup (LPS), was sampled at Cristal-Union (Pomacle, France) and wheat extract (WE) was supplied by Chamtor (Pomacle, France) and filtered before use to eliminate the suspended solid particles. The total fermentable sugar concentrations (glucose and fructose) were analyzed by HPLC. In LPS, the concentrations of glucose and fructose were identical (290 g·L^{-1}) while WE contained 140 g·L^{-1} of glucose and 5.8 g·L^{-1} of fructose. LPS also contained glycerol at a concentration of 17.4 g·L^{-1}. Both LPS and WE were used as sugar sources in the cultivation medium and the total final sugar content was adjusted to 30 g·L^{-1}. The contribution of LPS compared to WE for the sugar concentration was expressed as R. For example, R at 20% means that among the 30 g·L^{-1} of sugar, 20% come from LPS and 80% from WE.

The growth medium was supplemented with yeast extract (YE, Fisher Scientific, Springfield Township, NJ, USA), Tween 80 (T80, Sigma Aldrich, Saint Louis, MO, USA) and cyanocobalamine (vitamin B12, Fisher Scientific, Springfield, NJ, USA). The initial pH was set to 6.8 by adding HCl 2N (Fisher Scientific, Loughborough, UK).

MRS medium (Biokar, Beauvais, France) was used for the standard cultivations and the precultures. The commercial mixture was supplemented with glucose (Fisher, Loughbourough, UK) to a final concentration of 30 g·L^{-1}.

Glycerol bioconversion was conducted in a 10 g·L^{-1} glycerol (Fisher, Loughbourough, UK) solution in distilled water.

All media were sterilized for 20 min at 110 °C and cooled down to room temperature before inoculation.

2.3. Cells Cultivation and Glycerol Bioconversion with Resting Cells

Small-volume cultivations were conducted in 15-mL Falcon tubes, with 12 mL medium, at 37 °C in a static incubator, for 8 h. The kinetics of bacterial growth was followed by the variation of the optical density measured at 620 nm (ΔOD_{620nm}) with an Agilent (Santa Clara, CA, USA) spectrometer, with the cultivation medium before inoculation as reference.

For the design of experiment as well as for the comparison of carbon sources, the cultivation media were inoculated with the same preculture to a starting ΔOD_{620nm} of 0.1. After 8 h, at the end of the growth phase, the bacteria cells were harvested by centrifugation, 10 min at 5000× *g* and 10 °C. The supernatant was discarded and the pellet was washed twice in potassium phosphate buffer (pH 6) and re-suspended in 10 g·L^{-1} glycerol to a final cell concentration of 1×10^{10} cell·mL^{-1}. Then, the tubes were maintained at 37 °C overnight in an orbital shaker under gentle stirring (110 rpm) for the bioconversion step. At the end of the bioconversion, the bacteria were pelleted by centrifugation, 10 min at 5000× *g* and 10 °C, and the supernatants were analyzed for the quantification of the glycerol degradation products (3-HP, 3-HPA and 1,3-PDO).

Cultivations in 2 L were conducted in Biostat B bioreactor (Sartorius, Goettingen, Germany). Cultivations in the optimized medium (WE-LPS) and MRS were done in parallel, under gentle stirring (100 rpm). The temperature was maintained at 37 °C with water circulation. For the bioconversion kinetics experiments, the bacterial cultivation was conducted overnight with a starting ΔOD_{620nm} of 0.001. The bacterial biomass was then harvested after 15 h as described above, resuspended in 10 g·L^{-1} glycerol to a concentration of 1×10^{10} cell·mL^{-1} and the bioconversion was followed for 5 h by sampling the bioconversion medium every 30 min.

2.4. Substrate and Products Quantification

Glycerol, 3-HP, 3-HPA and 1,3-PDO were analyzed by HPLC on an Aminex 87H column (300 mm × 7.8 mm, Bio-Rad, Richmond, VA, USA) equipped with a cation H^+ Micro-Guard column (30 mm × 4.6 mm, Bio-Rad) thermostated to 50 °C. H_2SO_4 (Fisher, Loughbourough, UK) 4 mM was used as mobile phase. The elution flow rate was set at 0.6 mL·min^{-1}. 3-HP was detected with a UV detector (Dionex, Sunnyvale, CA, USA) 210 nm; retention time: 13.1 min), residual glycerol, 3-HPA, and 1,3-PDO were detected with a refractometer (RI-101, Shodex, Japan) (RT: 13.3, 15.0 and 17.5 min respectively). Citric acid (0.5 g·L^{-1}) was added as an internal standard and samples were filtered on 0.22 μm syringe-filter.

Quantification was performed using five-point standard curves obtained under the same conditions of analysis from standard samples prepared from pure products. Glycerol was purchased from Fisher Chemicals (Loughbourough, UK), 3-HP from TCI (Tokyo, Japan) and 1,3-PDO from Sigma-Aldrich (Saint Louis, MO, USA). 3-HPA was synthesized according to Burgé et al. [20].

2.5. Experimental Design and Statistical Analysis

Response surface methodology based on four-factor-multi level D-optimal design was used to optimize the growth medium composition. The different formulations were compared according to two responses, the first one being the growth of *L. reuteri* assessed as the ΔOD_{620nm} of the broth at the end of the cultivation phase, and the second one being the concentration of 3-HP in the supernatant at the end of the bioconversion phase. MODDE 8 software (Umetrics, AB, Umeå, Sweden) was used for statistical analysis of the experimental design [21]. The factors tested were the LPS/WE ratio (R) (from 10% to 90%), and the concentrations of YE (from 0 to 35 g·L^{-1}), B12 (from 0 to 200 μg·L^{-1}), and T80 (from 0 to 5 mL·L^{-1}).

Growth on MRS medium supplemented with glucose (30 g·L^{-1}) was used as a standard to compare the results obtained with the tested media.

The three independent variables levels used for the experimental design are presented in Table 1. The runs set by the D-optimal design and the respective experimental responses obtained for ΔOD_{620nm} (Y1) and 3-HP production (Y2) are also presented in Table 1. Thirty experiments with three replications at the center point were designed for the above mentioned variables. The experiments were made in two different blocks, e.g., two series made by two different operators at two different dates, to eliminate a potential influence of the operator on the results. The data were fitted using multiple linear regressions (MLR). The significant parameters in the model were determined by analysis of variance (ANOVA) for each response. The model validation was based on R^2, Q^2, and lack of fit test. R^2 (0.991 for ΔOD_{620nm} and 0.977 for 3-HP) expresses the percentage of the variation of the response explained by the model, whereas Q^2 (0.982 for ΔOD_{620nm} and 0.940 for 3-HP) expresses the percentage of the variation predicted by the model according to cross validation. The probability for lack of fit were, respectively, 0.079 for ΔOD_{620nm} and 0.102 for 3-HP and the model validity was evaluated to 0.354 for ΔOD_{620nm} and 0.495 for 3-HP, a minimal value of 0.25 being requested to consider the model as valid.

The experimental data were further analyzed using multiple regressions and a second order polynomial model fitted for predicting optimal levels was expressed as follows:

$$Yi = \beta_0 + \sum \beta_i X_i + \sum \beta_{ii} X_i^2 + \sum \beta_{ij} X_i X_j + \varepsilon$$

Yi is the predicted response, β_0 is the intercept coefficient of the *y* axis, βi is a linear coefficient, βij and βii are the quadratic coefficients of the model, and *Xi* and *Xj* are the independent variables and ε the residual error.

Table 1. Experimental design and responses. R: percentage of LPS in the carbon source, YE: yeast extract concentration, T80: Tween 80 concentration, B12: cyanocobalamine concentration, ΔOD_{620nm}: optical density at 620 nm, 3-HP: 3-hydroxypropionic acid.

Exp No.	Exp Name	R (%)	YE (g·L^{-1})	T80 (mL·L^{-1})	B12 (µg·L^{-1})	Block	ΔOD_{620nm}	3-HP (g·L^{-1})
1	N1	90	0	0	0	B1	0.3	0.0
2	N2	10	35	0	0	B1	5.3	1.5
3	N3	10	0	5	0	B1	0.6	0.0
4	N4	90	35	5	0	B1	12.8	0.5
5	N5	90	15	0	100	B1	5.1	1.0
6	N6	10	15	2	100	B1	8.3	1.8
7	N7	90	0	5	100	B1	0.3	0.0
8	N8	10	0	0	200	B1	0.7	0.0
9	N9	90	35	0	200	B1	6.8	1.0
10	N10	90	0	2	200	B1	0.3	0.0
11	N11	50	15	5	200	B1	10.2	1.9
12	N12	10	35	5	200	B1	9.0	2.0
13	N13	50	15	2	100	B1	10.0	2.3
14	N14	50	15	2	100	B1	10.3	2.1
15	N15	50	15	2	100	B1	10.7	1.9
16	N16	10	0	0	0	B2	0.6	0.0
17	N17	90	35	0	0	B2	11.1	0.5
18	N18	50	0	2	0	B2	0.4	0.0
19	N19	90	0	5	0	B2	3.3	0.2
20	N20	90	15	5	0	B2	12.0	0.6
21	N21	10	35	5	0	B2	9.5	1.3
22	N22	50	35	5	100	B2	10.3	2.0
23	N23	90	0	0	200	B2	0.4	0.0
24	N24	10	35	0	200	B2	5.7	1.8
25	N25	10	0	2	200	B2	0.8	0.0
26	N26	10	0	5	200	B2	1.8	0.6
27	N27	90	35	5	200	B2	10.8	1.4
28	N28	50	15	2	100	B2	12.6	1.6
29	N29	50	15	2	100	B2	12.3	1.6
30	N30	50	15	2	100	B2	12.3	1.7

3. Results and Discussion

3.1. Influence of Medium Composition on Bacterial Growth and 3-HP Production

A growth medium based on wheat (WE) and sugar beet (LPS) coproducts was formulated for the cultivation of *L. reuteri* DSM17938 with the perspective of using the resting cells for the bioconversion of glycerol into 3-HP.

A design of experiment was set up as a screening using a four-factor three-level factorial experiment design with three replications at the central point. Twenty-eight different growth medium formulations were tested, varying in the proportion of LPS versus WE to reach the total sugar concentration of 30 g·L^{-1}, the concentration of yeast extract, Tween 80, and vitamin B12. The performance of the tested media was assessed according to two responses (i) the maximal optical density (ΔOD_{620nm}) obtained after the cultivation phase and (ii) the 3-HP production at the end of the bioconversion phase.

The maximum yield of biomass production was obtained from the culture condition of run number 4 (ΔOD_{620nm} 12.8) while the maximal production of 3-HP (2.3 g·L^{-1}) was obtained for run number 13 (Table 1).

In parallel, three control fermentations were conducted in MRS under the same conditions. The average ΔOD obtained was 6.7 ± 0.8 and the 3-HP production at the end of the bioconversion phase was 1.3 ± 0.1 g·L^{-1}.

The contributions of each parameter on both responses were different (Figure 1). The highest contribution to both responses was due to the concentration of yeast extract, which impacts positively both the final ΔOD and the 3-HP production.

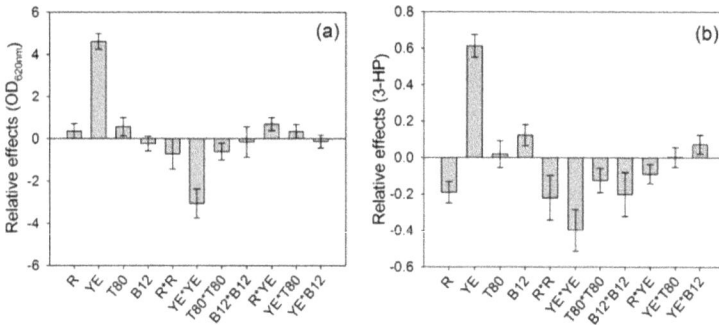

Figure 1. Relative and squared effects of the growth medium parameters on the final optical density at the end of the growth phase (**a**) and the 3-HP concentration (**b**) after the bioconversion of glycerol by *L. reuteri*.

The 3-HP production was also positively impacted by the addition of vitamin B12 to the culture broth during the preliminary cultivation phase (Figure 1b), although no significant impact was noticeable on ΔOD (Figure 1a). This positive influence of vitamin B12 on the 3-HP production is surprising, considering that *L. reuteri* is able to synthesize de novo vitamin B12 from glutamate or glycine [22]. Furthermore, this characteristic has been shown to be directly associated with the ability of the strain to produce 3-HPA [23].

On the other hand, the LPS/WE ratio (R) had a negative impact on the 3-HP production, which means that a lower proportion of LPS compared to WE as carbon source led to a higher 3-HP synthesis, while ΔOD was not significantly influenced by this parameter. Besides, the squared effects had a negative impact on the results, which suggests that the range of tested values included the optimal value for the parameters (Figure 1a,b).

Interestingly, when compared to the control data obtained after bacterial cultivation in MRS medium, thirteen formulations tested led to a higher production of 3-HP, but were not necessarily associated with higher bacterial growth (Figure 2). Conversely, three formulations that allowed higher biomass production led to lower levels of 3-HP when compared to the control cultivation.

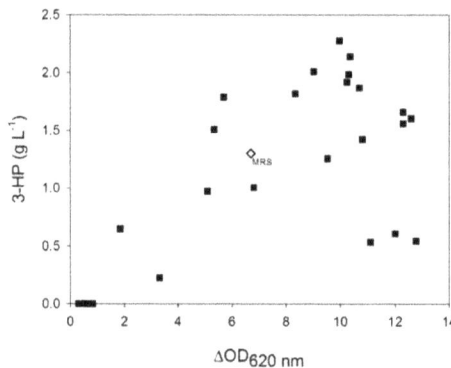

Figure 2. Concentration of 3-HP after the bioconversion phase as a function of the optical density at 620 nm measured at the end of the growth phase. Each point corresponds to a different formulation tested. Blank diamond: values obtained with MRS medium. Black squares: results obtained with the formulations of both design.

These results highlight that the ability of *L. reuteri* to convert glycerol into 3-HP highly depends on the cultivation phase, in particular on the composition of the cultivation medium. In addition, a high bacterial biomass production is not sufficient to ensure a good production of 3-HP.

3.2. Growth Medium Optimization

Response Surface Methodology (RSM) was used to determine the optimized growth medium formulation with the goal to maximise the 3-HP production. Thus, the software prediction tool MODDE8 determined the optimal composition of the medium (WE-LPS medium) as follows (g for 100 g medium): WE: 10.3; LPS: 2.6 yeast extract, 1.5; Tween 80, 0.5; and vitamin B12, 0.01. According to the model, a maximal 3-HP titre of 2.3 g·L^{-1} can be expected when using WE-LPS as cultivation medium for the production of bacterial biomass.

3.3. Interest of Using Both LPS and WE as Carbon Sources in the Medium Formulation

Control fermentations were carried out in order to assess the influence of LPS and WE and the relevance of utilizing both ingredients in the medium formulation employed for the synthesis of 3-HP. Four media were prepared using LPS, WE or commercial glucose as the carbon source. In every formulation the final total sugar concentration was adjusted to 30 g·L^{-1}. The other ingredients were added at the same concentration in the four formulations (Table 2). The performance of the media in terms of final ΔOD were compared to the optimized medium (WE-LPS) and to the MRS medium, in 15 mL Falcon tubes containing 12 mL. All experiments were done in triplicate.

Table 2. Products of glycerol bioconversion obtained after the growth of *L. reuteri* in different culture media containing the same sugar concentration, but differing in their carbon source, compared to the standard medium MRS. The yield of glycerol conversion to 3-HP ($Y_{3-HP/gly}$) and 3-HPA ($Y_{3-HPA/gly}$) are expressed in g of product per g of glycerol consumed. Mean values and standard deviation of three replicates. WE-LPS: carbon source based on a mix of LPS and wheat extract; WE: wheat extract as the only carbon source; LPS: Low purity sugar beet syrup as the only carbon source; Glucose: commercial glucose as the only carbon source; YE: Yeast extract.

Measured or Calculated Variables		MRS	WE-LPS	WE	LPS	Glucose
LPS	(mL·L^{-1})	/	26.0	/	51.7	/
WE	(mL·L^{-1})	/	103.0	195.0	/	/
Glucose	(g·L^{-1})	/	/	/	/	30.0
YE	(g·L^{-1})	/		15		
T80	(mg·L^{-1})	/		2		
B12	(mg·L^{-1})	/		0.1		
ΔOD$_{620nm}$		6.7	11.6	12.4	3.7	4.3
3-HP	(g·L^{-1})	1.3 ± 0.1	2.1 ± 0.2	1.7 ± 0.2	0.8 ± 0.1	0.9 ± 0.1
3-HPA	(g·L^{-1})	6.6 ± 0.9	0.9 ± 0.1	4.0 ± 0.8	0.0 ± 0.1	2.8 ± 0.3
1,3-PDO	(g·L^{-1})	1.1 ± 0.1	2.0 ± 0.2	1.4 ± 0.1	1.1 ± 0.1	0.7 ± 0.1
Total consumed glycerol	(g·L^{-1})	9.7 ± 0.5	5.3 ± 0.4	8.0 ± 0.7	1.9 ± 0.1	5.1 ± 0.4
$Y_{3-HP/gly}$	(g·g^{-1})	0.13 ± 0.02	0.40 ± 0.03	0.21 ± 0.01	0.42 ± 0.01	0.18 ± 0.02
$Y_{3-HPA/gly}$	(g·g^{-1})	0.68 ± 0.08	0.17 ± 0.02	0.50 ± 0.08	0.00 ± 0.01	0.55 ± 0.06

Furthermore, after the bioconversion phase, bioconversion broths were analyzed for residual glycerol and the products of glycerol bioconversion, 3-HP, 3-HPA, and 1,3-PDO (Table 2).

As many aldehydes, 3-HPA is known to display antimicrobial properties and is suspected to cause a rapid loss of glycerol conversion activity during the 3-HP production process [24]. Thus, the limitation of the 3-HPA titre in the conversion broth is of utmost importance for the 3-HP production process. Therefore, the performance of the tested media were evaluated according to three parameters i.e., (i) the total consumed glycerol ($Y_{gly/gly}$); (ii) the glycerol conversion yield to 3-HP ($Y_{3-HP/gly}$); and (iii) the glycerol conversion yield to 3-HPA ($Y_{3-HPA/gly}$). Regarding these criteria, bacteria growth on LPS medium exhibited the best performances. However, 19% of the glycerol was consumed (Table 2) which therefore led to the lowest 3-HP concentration recorded across all tested media. When considering the total products of glycerol bioconversion (3-HP, 3-HPA and 1,3-PDO), the WE-LPS medium distinguished itself from the others, with a lower production of 3-HPA, whereas the synthesis of 3-HP and 1,3-PDO was up to 70% higher compared to standard medium MRS.

Concerning the total glycerol consumption, glycerol was better converted by bacteria cultivated in MRS and WE medium, but 3-HPA was the main final product, with 0.68 and 0.5 g 3-HPA per g of consumed glycerol, respectively (Table 2). Interestingly, after cultivation in WE-LPS medium, the total glycerol consumption was 45% lower compared to the control (MRS), but the bacteria grown in this medium displayed the highest $Y_{3\text{-HP/gly}}$ (0.40 g·g^{-1}) and the lowest $Y_{3\text{-HPA/gly}}$ (0.17 g·g^{-1}). Furthermore, the highest final 3-HP titre was obtained after growth in WE-LPS medium

The comparison of the five cultivation conditions highlighted the influence of the cultivation medium on both glycerol uptake by the cells and the conversion of 3-HPA to 3-HP and 1,3-PDO. While the biomass cultivated on WE medium was the most efficient in terms of glycerol conversion yield, the consumed glycerol was mainly converted into 3-HPA. Finally, the combination of WE and LPS provided the best balance in terms of glycerol uptake and 3-HPA conversion to 3-HP and 1,3-PDO. Thus, the combination of LPS and WE provides a real added value to the 3-HP production process.

This work clearly shows that the metabolism of glycerol in resting cells is highly influenced by the growth medium composition.

The complementation of beet molasses with other industrial crop products to formulate bacterial growth media was reported in a few studies. Beet molasses as a sole fermentation medium generally leads to low biomass growth, due to its high salt concentration and pH, and the presence of betaine as the main nitrogen source [4]. In combination with wheat stillage as nitrogen sources, however, adding beet molasses led to higher bacterial biomass [13]. The authors attributed the increase of LAB biomass to the rise of sugar concentration. In our case, the total sugar concentration was kept constant across the different formulations, but LPS and WE provided fructose to the medium which in addition to glucose has been shown to enhance the growth performance of *L. reuteri* ATCC 55730 by allowing a better redox balance [25]. However, the presence of fructose in the growth medium does not explain the high biomass density (ΔOD) obtained after cultivation in media containing WE (WE-LPS and WE, Table 2), as the biomass density obtained after growth on LPS alone was four times as low as the one obtained in medium containing WE (Table 2).

The influence of the bacteria cultivation medium on the bioconversion step using resting cells is difficult to explain at this stage. The presence of glycerol in LPS may provide an advantage for strain growth and its adaptation to the bioconversion medium [26,27]. Additionally, the process used here for the bioconversion of glycerol involves a succession of steps, which represent environmental changes likely to weaken the cells. The microorganisms cultivated in a complex medium derived from agro-industrial coproducts containing potential inhibitory molecules may develop adaptation mechanisms, providing the cells with a technological advantage for the bioconversion step [28,29].

3.4. Impact of Growth Medium on the Glycerol Bioconversion Pathway

To validate the model and its predictions, *L. reuteri* cultivation in WE-LPSP medium and glycerol bioconversion were successively conducted in 2 L bioreactors using the same conditions as during the previous small-scale setup. The kinetics of glycerol bioconversion was followed by sampling the supernatant every thirty minutes for the first two hours and then every hour. Glycerol and its products were then quantified by HPLC. As for the previous small-scale experiments, bacterial growth in MRS medium, followed by glycerol bioconversion, was performed as a control. Experiments were done in triplicate.

As in small scale experiments, the concentration of 3-HP at the end of the bioconversion in the medium was 70% higher as after a growth phase in MRS (Figure 3b). The 3-HPA and the 1,3-PDO concentrations were also impacted; the production of 1,3-PDO increasing by 60% (Figure 3b) while the maximum 3-HPA concentration was greatly reduced from 4.9 to 1.0 g·L^{-1} (Figure 3a) compared to the results obtained after growth on MRS medium.

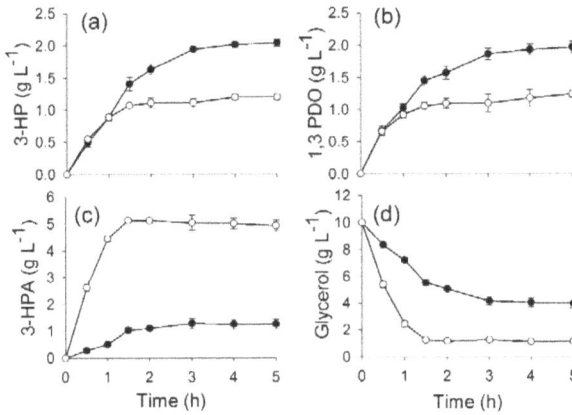

Figure 3. Kinetics of 3-HP (**a**), 1,3-PDO (**b**), and 3-HPA (**c**) accumulation in the extracellular medium and glycerol concentration (**d**) in the bioconversion medium during the bioconversion phase by *L. reuteri*, after growth on optimized medium (WE-LPS) (black circles) or MRS (blank circles). Mean values and standard deviation of three replicates.

Interestingly, the 3-HPA production was greatly impacted by the growth condition (Figure 3a). After growth on MRS, 3-HPA accumulated quickly in the bioconversion medium, with a production rate of 70.7 ± 3.8 mmol·L^{-1}·h^{-1}, while after growth on WE-LPS medium, the production rate at the beginning of the bioconversion phase was ten times lower (7.5 ± 0.9 mmol·L^{-1}·h^{-1}). On the other hand, the 3-HP and 1,3-PDO production rates measured at the onset of the bioconversion process were the same for both growth media (11.0 mmol·L^{-1}·h^{-1} for 3-HP and 17 mmol·L^{-1}·h^{-1} for 1,3-PDO). However, after growth in MRS medium, the production rates decreased quickly and the glycerol bioconversion stopped after only 90 min, while the bioconversion activity was maintained for 180 min after growth on WE-LPS medium.

The metabolic fluxes through the different pathways, presented in Figure 4, were calculated at the onset of the bioconversion phase and are representative of the glycerol dehydratase (GDH) activity of the bacterial population [27].

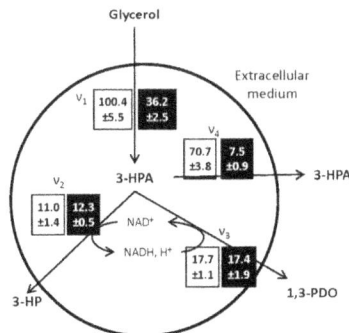

Figure 4. Metabolic fluxes (mmol·L^{-1}·h^{-1}) through the glycerol bioconversion pathway in *L. reuteri* after growth on MRS (white boxes) or WE-LPS medium (black boxes). Mean values of three replicates.

The main impact of the growth medium composition on the glycerol bioconversion was the limitation of the 3-HPA synthesis flux from glycerol (v1). After growth on MRS medium, v1, was, respectively, 9 and 6 times faster than the subsequent oxidative (v2) and reductive (v3) pathways leading to the synthesis of 3-HP and 1,3-PDO. These results are consistent with the ones reported by Dishisha [10,11] using *L. reuteri* DSM 20016. Besides, in the same study, the limitation of the glycerol consumption rate using a fed-batch approach allowed maintaining a low concentration of 3-HPA while increasing the 3-HP and 1,3-PDO production rates. The antimicrobial activity of 3-HPA has been demonstrated for several micro-organisms [30,31]. Indeed, the in situ complexation of 3-HPA with bisulfites increases the productivity and the lifetime of microorganisms [32]. Although less widely studied, 3-HP can also display a toxic activity towards a range of microorganisms [33]. In the case of growth in MRS medium, the drop in bioconversion activity can be attributed to a decrease in cell viability due to the rapid accumulation of 3-HPA as observed by Burgé [24]. Here, the limitation of the 3-HPA production rate, allowed maintaining the cell activity and increased the glycerol conversion yield to 3-HP and 1,3-PDO from 0.1 $g·g^{-1}$ in the control condition to 0.4 $g·g^{-1}$ after growth in WE-LPS medium. The reason for this limitation after growth on WE-LPS medium is still to clarify. Considering the glycerol consumption flux, the restriction of glycerol intake through the cell membrane could be a factor. On the other hand, the addition of vitamin B12 to the growth medium has a positive influence on the 3-HP production, as observed in the first part of the study. According to the link between the synthesis of both vitamin B12 and 3-HPA by *L. reuteri*, the presence of an exogenous source of vitamin B12 might moderate the expression of the glycerol dehydratase, reducing the 3-HPA production flux [23].

4. Conclusions

A growth medium based on two agro-industrial coproducts, derived from wheat and sugar beet processing and suitable for the cultivation of *L. reuteri*, was successfully formulated. Response Surface Methodology allowed optimizing the formulation for the bioconversion of glycerol by resting cells. Unexpectedly, cultivating *L. reuteri* on the optimized medium led to a 70% increase of the 3-HP production yield while lowering the concentration of the microbial inhibitor 3-HPA obtained during the bioconversion step. These results may be attributed to the limitation of the glycerol dehydratase activity. These observations highlight the importance of the cultivation conditions on the bioconversion performance of resting cells.

Acknowledgments: The authors thank Chamtor (Pomacle, France) for providing us with wheat extract sample and Cristal Union (Pomacle, France) for providing LPS from sugar beet. This work was funded by Region Champagne-Ardennes, Conseil général de la Marne and Reims Métropole.

Author Contributions: Julien Couvreur and Tiphaine Clément designed and performed the experiments; Andreia R. S. Teixeira contributed to the design of experiment; Tiphaine Clément, Florent Allais, Claire Saulou-Bérion, and Henry-Eric Spinnler analyzed the data and prepared the paper.

Conflicts of Interest: The authors declare no conflict of interest. The founding sponsors had no role in the design of the study; in the collection, analyzes, or interpretation of data; in the writing of the manuscript, and in the decision to publish the results.

References

1. Budzianowski, W.M.; Postawa, K. Total Chain Integration of sustainable biorefinery systems. *Appl. Energy* **2016**. [CrossRef]
2. Ragauskas, A.J.; Williams, C.K.; Davison, B.H.; Britovsek, G.; Cairney, J.; Eckert, C.A.; Frederick, W.J.; Hallett, J.P.; Leak, D.J.; Liotta, C.L.; et al. The path forward for biofuels and biomaterials. *Science* **2006**, *311*, 484–489. [CrossRef] [PubMed]
3. Wang, J. Improvement of citric acid production by *Aspergillus niger* with addition of phytate to beet molasses. *Bioresour. Technol.* **1998**, *65*, 243–245. [CrossRef]
4. Han, Y.W.; Watson, M.A. Production of microbial levan from sucrose, sugarcane juice and beet molasses. *J. Ind. Microbiol.* **1992**, *9*, 257–260. [CrossRef]

5. Kotzamanidis, C.; Roukas, T.; Skaracis, G. Optimization of lactic acid production from beet molasses by *Lactobacillus delbrueckii* NCIMB 8130. *World J. Microbiol. Biotechnol.* **2002**, *18*, 441–448. [CrossRef]

6. Dumbrepatil, A.; Adsul, M.; Chaudhari, S.; Khire, J.; Gokhale, D. Utilization of molasses sugar for lactic acid production by *Lactobacillus delbrueckii subsp.* delbrueckii Mutant Uc-3 in batch fermentation. *Appl. Environ. Microbiol.* **2008**, *74*, 333–335. [CrossRef] [PubMed]

7. Savino, F.; Fornasero, S.; Ceratto, S.; De Marco, A.; Mandras, N.; Roana, J.; Tullio, V.; Amisano, G. Probiotics and gut health in infants: A preliminary case–control observational study about early treatment with *Lactobacillus reuteri* DSM 17938. *Clin. Chim. Acta* **2015**. [CrossRef] [PubMed]

8. Sauer, M.; Porro, D.; Mattanovich, D.; Branduardi, P. Microbial production of organic acids: Expanding the markets. *Trends Biotechnol.* **2008**, *26*, 100–108. [CrossRef] [PubMed]

9. Luo, L.H.; Seo, J.-W.; Baek, J.-O.; Oh, B.-R.; Heo, S.-Y.; Hong, W.-K.; Kim, D.-H.; Kim, C.H. Identification and characterization of the propanediol utilization protein PduP of *Lactobacillus reuteri* for 3-hydroxypropionic acid production from glycerol. *Appl. Microbiol. Biotechnol.* **2011**, *89*, 697–703. [CrossRef] [PubMed]

10. Dishisha, T.; Pyo, S.-H.; Hatti-Kaul, R. Bio-based 3-hydroxypropionic- and acrylic acid production from biodiesel glycerol via integrated microbial and chemical catalysis. *Microb. Cell Fact.* **2015**, *14*, 200. [CrossRef] [PubMed]

11. Dishisha, T.; Pereyra, L.P.; Pyo, S.-H.; Britton, R.A.; Hatti-Kaul, R. Flux analysis of the *Lactobacillus reuteri* propanediol-utilization pathway for production of 3-hydroxypropionaldehyde, 3-hydroxypropionic acid and 1,3-propanediol from glycerol. *Microb. Cell Fact.* **2014**, *13*, 76. [CrossRef] [PubMed]

12. Aguirre-Ezkauriatza, E.J.; Aguilar-Yáñez, J.M.; Ramírez-Medrano, A.; Alvarez, M.M. Production of probiotic biomass (*Lactobacillus casei*) in goat milk whey: Comparison of batch, continuous and fed-batch cultures. *Bioresour. Technol.* **2010**, *101*, 2837–2844. [CrossRef] [PubMed]

13. Krzywonos, M.; Eberhard, T. High density process to cultivate *Lactobacillus plantarum* biomass using wheat stillage and sugar beet molasses. *Electron. J. Biotechnol.* **2011**, *14*. [CrossRef]

14. Maddox, I.S.; Richert, S.H. Use of response surface methodology for the rapid optimization of microbiological media. *J. Appl. Bacteriol.* **1977**, *43*, 197–204. [CrossRef] [PubMed]

15. Chang, C.P.; Liew, S.L. Growth medium optimization for biomass production of a probiotic bacterium, *Lactobacillus rhamnosus* ATCC 7469. *J. Food Biochem.* **2012**. [CrossRef]

16. Kumari, A.; Mahapatra, P.; Banerjee, R. Statistical optimization of culture conditions by response surface methodology for synthesis of lipase with *Enterobacter aerogenes*. *Braz. Arch. Biol. Technol.* **2009**, *52*, 1349–1356. [CrossRef]

17. Polak-Berecka, M.; Waśko, A.; Kordowska-Wiater, M.; Podleśny, M.; Targoński, Z.; Kubik-Komar, A. Optimization of medium composition for enhancing growth of *Lactobacillus rhamnosus* PEN using response surface methodology. *Pol. J. Microbiol.* **2010**, *59*, 113–118. [PubMed]

18. Rafigh, S.M.; Yazdi, A.V.; Vossoughi, M.; Safekordi, A.A.; Ardjmand, M. Optimization of culture medium and modeling of curdlan production from *Paenibacillus polymyxa* by RSM and ANN. *Int. J. Biol. Macromol.* **2014**, *70*, 463–473. [CrossRef] [PubMed]

19. Zárate-Chaves, C.A.; Romero-Rodríguez, M.C.; Niño-Arias, F.C.; Robles-Camargo, J.; Linares-Linares, M.; Rodríguez-Bocanegra, M.X.; Gutiérrez-Rojas, I. Optimizing a culture medium for biomass and phenolic compounds production using *Ganoderma lucidum*. *Braz. J. Microbiol.* **2013**, *44*, 215–223. [CrossRef] [PubMed]

20. Burgé, G.; Flourat, A.L.; Pollet, B.; Spinnler, H.E.; Allais, F. 3-Hydroxypropionaldehyde (3-HPA) quantification by HPLC using a synthetic acrolein-free 3-hydroxypropionaldehyde system as analytical standard. *RSC Adv.* **2015**, *5*, 92619–92627. [CrossRef]

21. Eriksson, L. *Design of Experiments: Principles and Applications; Umetrics Academy-Training in Multivariate Technology*; Umetrics AB: Umea, Sweden, 2000.

22. Taranto, M.P.; Vera, J.L.; Hugenholtz, J.; De Valdez, G.F.; Sesma, F. *Lactobacillus reuteri* CRL1098 produces cobalamin. *J. Bacteriol.* **2003**, *185*, 5643–5647. [CrossRef] [PubMed]

23. Santos, F.; Vera, J.L.; van der Heijden, R.; Valdez, G.; de Vos, W.M.; Sesma, F.; Hugenholtz, J. The complete coenzyme B12 biosynthesis gene cluster of *Lactobacillus reuteri* CRL1098. *Microbiology* **2008**, *154*, 81–93. [CrossRef] [PubMed]

24. Burgé, G.; Saulou-Bérion, C.; Moussa, M.; Pollet, B.; Flourat, A.; Allais, F.; Athès, V.; Spinnler, H.E. Diversity of *Lactobacillus reuteri* strains in converting glycerol into 3-Hydroxypropionic ccid. *Appl. Biochem. Biotechnol.* **2015**, *177*, 923–939. [CrossRef] [PubMed]

25. Arsköld, E.; Lohmeier-Vogel, E.; Cao, R.; Roos, S.; Rådström, P.; van Niel, E.W.J. Phosphoketolase pathway dominates in *Lactobacillus reuteri* ATCC 55730 containing dual pathways for glycolysis. *J. Bacteriol.* **2008**, *190*, 206–212. [CrossRef] [PubMed]

26. Santos, F.; Spinler, J.K.; Saulnier, D.M.A.; Molenaar, D.; Teusink, B.; de Vos, W.M.; Versalovic, J.; Hugenholtz, J. Functional identification in *Lactobacillus reuteri* of a PocR-like transcription factor regulating glycerol utilization and vitamin B12 synthesis. *Microb. Cell Fact.* **2011**, *10*, 55–66. [CrossRef] [PubMed]

27. Krauter, H.; Willke, T.; Vorlop, K.-D. Production of high amounts of 3-hydroxypropionaldehyde from glycerol by *Lactobacillus reuteri* with strongly increased biocatalyst lifetime and productivity. *New Biotechnol.* **2012**, *29*, 211–217. [CrossRef] [PubMed]

28. Corcoran, B.M.; Stanton, C.; Fitzgerald, G.; Ross, R.P. Life under stress: The probiotic stress response and how it may be manipulated. *Curr. Pharm. Des.* **2008**, *14*, 1382–1399. [CrossRef] [PubMed]

29. Senz, M.; van Lengerich, B.; Bader, J.; Stahl, U. Control of cell morphology of probiotic *Lactobacillus acidophilus* for enhanced cell stability during industrial processing. *Int. J. Food Microbiol.* **2015**, *192*, 34–42. [CrossRef] [PubMed]

30. Barbirato, F.; Grivet, J.P.; Soucaille, P.; Bories, A. 3-Hydroxypropionaldehyde, an inhibitory metabolite of glycerol fermentation to 1,3-propanediol by enterobacterial species. *Appl. Environ. Microbiol.* **1996**, *62*, 1448–1451. [PubMed]

31. Cleusix, V.; Lacroix, C.; Vollenweider, S.; Duboux, M.; Blay, G. Le Inhibitory activity spectrum of reuterin produced by *Lactobacillus reuteri* against intestinal bacteria. *BMC Microbiol.* **2007**, *7*, 101. [CrossRef] [PubMed]

32. Sardari, R.R.R.; Dishisha, T.; Pyo, S.-H.; Hatti-Kaul, R. Biotransformation of glycerol to 3-hydroxypropionaldehyde: Improved production by in situ complexation with bisulfite in a fed-batch mode and separation on anion exchanger. *J. Biotechnol.* **2013**, *168*, 534–542. [CrossRef] [PubMed]

33. Sebastianes, F.L.S.; Cabedo, N.; El Aouad, N.; Valente, A.M.M.P.; Lacava, P.T.; Azevedo, J.L.; Pizzirani-Kleiner, A.A.; Cortes, D. 3-hydroxypropionic acid as an antibacterial agent from endophytic fungi *Diaporthe phaseolorum. Curr. Microbiol.* **2012**, *65*, 622–632. [CrossRef] [PubMed]

fermentation

MDPI

Article

Biotechnological Production of Fumaric Acid: The Effect of Morphology of *Rhizopus arrhizus* NRRL 2582

Aikaterini Papadaki [1], Nikolaos Androutsopoulos [1], Maria Patsalou [2], Michalis Koutinas [2], Nikolaos Kopsahelis [1,3], Aline Machado de Castro [4], Seraphim Papanikolaou [1] and Apostolis A. Koutinas [1,*]

[1] Department of Food Science and Human Nutrition, Agricultural University of Athens, Iera Odos 75, 11855 Athens, Greece; kpapadaki@aua.gr (A.P.); nik.androutsopoulos@gmail.com (N.A.); kopsahelis@upatras.gr (N.K.); spapanik@aua.gr (S.P.)
[2] Department of Environmental Science & Technology, Cyprus University of Technology, 30 Archbishop Kyprianou Str., 3036 Limassol, Cyprus; mx.patsalou@edu.cut.ac.cy (M.P.); michail.koutinas@cut.ac.cy (M.K.)
[3] Department of Food Technology, Technological Educational Institute (TEI) of Ionian Islands, Argostoli 28100, Kefalonia, Greece
[4] Biotechnology Division, Research and Development Centre, PETROBRAS, Rio de Janeiro 20000-000, Brazil; alinebio@petrobras.com.br
* Correspondence: akoutinas@aua.gr; Tel.: +30-210-529-4629

Received: 18 April 2017; Accepted: 14 June 2017; Published: 8 July 2017

Abstract: Fumaric acid is a platform chemical with many applications in bio-based chemical and polymer production. Fungal cell morphology is an important factor that affects fumaric acid production via fermentation. In the present study, pellet and dispersed mycelia morphology of *Rhizopus arrhizus* NRRL 2582 was analysed using image analysis software and the impact on fumaric acid production was evaluated. Batch experiments were carried out in shake flasks using glucose as carbon source. The highest fumaric acid yield of 0.84 g/g total sugars was achieved in the case of dispersed mycelia with a final fumaric acid concentration of 19.7 g/L. The fumaric acid production was also evaluated using a nutrient rich feedstock obtained from soybean cake, as substitute of the commercial nitrogen sources. Solid state fermentation was performed in order to produce proteolytic enzymes, which were utilised for soybean cake hydrolysis. Batch fermentations were conducted using 50 g/L glucose and soybean cake hydrolysate achieving up to 33 g/L fumaric acid concentration. To the best of our knowledge the influence of *R. arrhizus* morphology on fumaric acid production has not been reported previously. The results indicated that dispersed clumps were more effective in fumaric acid production than pellets and renewable resources could be alternatively valorised for the biotechnological production of platform chemicals.

Keywords: fumaric acid; *Rhizopus*; fungal morphology; soybean cake; enzymatic hydrolysis; *Aspergillus oryzae*; biorefinery; image analysis

---

1. Introduction

Fumaric acid is considered a platform chemical with several applications in bio-based chemical and polymer production. It is used mainly as a food acidulant and as chemical feedstock for the production of paper resins, unsaturated polyester resins, alkyd resins, plasticizers, and miscellaneous industrial products [1]. The global fumaric acid market demand was estimated as 225.2 kt in 2012 and it is expected to be over 300 kt in 2020 [2]. The production of fumaric acid is based on the isomerization of maleic acid derived originally from n-butene [3]. The petrochemical method has the advantage of high production yields, compared with the biotechnological production through fermentation. The fermentation process yields around 85% w/w, using glucose as a carbon source, which is considered cheaper than the raw material of the chemical process [4]. Since there is an increasing concern over sustainability, the utilization of microbial strains for fumaric acid production has been appraised with the aim of its biotechnological production instead of the chemical route [4].

Fermentative production of fumaric acid has been investigated using various fungal strains of *Rhizopus arrhizus* and *Rhizopus oryzae*. The oxidative TCA cycle is required for fungal growth, while fumaric acid production involves the reductive TCA cycle [1]. Many *Rhizopus* species have been reported to produce fumaric acid in greater or lesser extents. Some of them have shown a capability of producing significant amounts of fumaric acid (>90 g/L) [5]. The species *R. oryzae* and *R. arrhizus* are the two most studied species for fumaric acid production, however not all strains within these species can be used to produce fumaric acid. In particular, *R. arrhizus* NRRL 2582 and *R. oryzae* NRRL 1526 have presented the highest final concentrations and conversion yields [6–9].

Several studies have shown that fungal morphology is a significant factor affecting fumaric acid concentration and yield [10–13]. The fungal morphology of *Rhizopus* species is influenced by various factors such as inoculum size, agitation, working volume, pH, temperature, nutrients, and metal ions [10–16]. Liao et al. reported that each factor affecting fungal morphology has different importance to the growth morphology of individual strains [10]. The formation of different fungal morphologies, such as clumps, pellets, and filaments, in submerged fermentations is affected by the growth conditions. The main disadvantage of filamentous and clump morphologies, during submerged fermentations, is the tendency to grow on bioreactor baffles, shaft and wall, which causes low oxygen transfer rates. The formation of fungal pellets promotes mass transfer due to lower medium viscosity [11,12]. However, the centre of large pellets could undergo autolysis due to nutrient limitation, which affects negatively fumaric acid production [13]. Other studies have reported that small pellets enhanced fumaric acid production in comparison to clump morphology [11,16].

The fungus *R. arrhizus* is considered as an important strain for the production of fumaric acid [17]. Hence, many studies have been carried out focusing on the morphology of several *Rhizopus* species [10,12,14–16]. Still, studies dealing with the *R. arrhizus* NRRL 2582 strain are scarcely found in the literature. In this work, two different fungal morphologies of *R. arrhizus*, pellets and dispersed mycelia, were accessed in submerged fermentations using commercial nitrogen sources and glucose. The effect of ammonium sulphate concentrations on the production of fumaric acid was also evaluated. Subsequently, submerged fermentations were carried out using a nutrient-rich feedstock derived from soybean cake in order to replace the commercial nitrogen sources. Soybean cake feedstock was obtained through enzymatic hydrolysis. The soybean cake was initially utilised as a substrate for the production of a crude enzyme consortia by *Aspergillus oryzae* through solid state fermentation. The crude enzymes were then used for the hydrolysis process of soybean cake. Soybean cake is a renewable resource that could be efficiently utilised for the sustainable biotechnological production of platform chemicals. A previous study reported the utilization of a chemically hydrolysed soybean meal for fumaric acid production by *R. oryzae* ATCC 20344 [12].

2. Materials and Methods

2.1. Microorganisms and Growth Media

Fumaric acid production was studied using the fungal strain *Rhizopus arrhizus* NRRL 2582, which was purchased from the ARS Culture Collection (NRRL) (Peoria, IL, USA). The fungus was cultivated in slopes containing 50 g/L soybean cake and 20 g/L agar, for six days at 30 °C and further propagated in Erlenmeyer flasks of 250 mL, which contained the same growth medium as in slopes, in order to form adequate quantities of spores. In particular, spores from slants were washed with 5 mL of sterilised water supplemented with 0.01% (v/v) of Tween 80 (Sigma-Aldrich, St. Louis, MO, USA) and then 2 mL were aseptically transferred to sporulation medium following incubation at 30 °C for three days. Fungal spore suspension was obtained by adding 50 mL sterilised water with 0.01% (v/v) Tween

80 and glass beads and the spores were liberated by vigorous shaking of the flasks [18]. The spore suspensions were collected and stored in cryovials, filled with pure glycerol, at −80 °C. The vials containing spore suspensions were used to inoculate pre-culture medium. All media were autoclaved at 121°C for 20 min.

The fungal strain *Aspergillus oryzae* was kindly provided by Prof. C. Webb (University of Manchester, UK) and was utilised for the production of crude proteolytic enzymes. The strain was originally isolated and purified by Wang et al. [19], from a soy sauce industry (Amoy Food Ltd., Hong Kong, China). Storage, maintenance, sporulation, and inoculum preparations of the fungal strain *A. oryzae* have been described in previous studies [18,20].

2.2. Raw Materials Used as Fermentation Media

Soybean cake, a by-product of the biodiesel production process, was kindly provided by Petrobras (Rio de Janeiro, Brazil).

2.3. Crude Enzyme Production by Solid State Fermentation

Solid-state fermentations (SSF) were carried out in 250 mL Erlenmeyer flasks, containing 5 g (on dry basis, db) of soybean cake and sterilized at 121 °C for 20 min. Fungal spore suspension of *Aspergillus oryzae* was used to inoculate the SSF so as the inoculum spore concentration was approximately 2×10^6 spores/mL and the moisture content of the substrate was adjusted at 65% (w/w, db) [18,20]. All SSFs were incubated at 30 °C for up to 48 h. Then, the fermented solids were suspended in sterilised water, macerated using a kitchen blender and filtered in order to obtain the crude enzyme consortia suspension, which was subsequently used in soybean cake hydrolysis.

2.4. Enzymatic Hydrolysis of Soybean Cake

Enzymatic hydrolysis was performed in 1 L Duran bottles, containing a total concentration of 45 g/L of soybean cake using an initial proteolytic activity of 3.9 U/mL. Duran bottles containing soybean cake were sterilized and then the crude enzyme consortia suspension from SSF was added. The hydrolysis was carried out at different temperatures under agitation and uncontrolled pH. Samples were collected at regular intervals the first 10 h of hydrolysis and then every 24 h. The solids were separated via centrifugation ($3000 \times g$, 10 min) and the supernatant was used for the analysis of free amino nitrogen (FAN) and inorganic phosphorus (IP). Hydrolysis yield was expressed as the percentage of total Kjeldahl nitrogen (TKN) to FAN conversion. The experiments were carried out in duplicates and the results represent the mean values obtained.

After the hydrolysis, remaining solids were removed by centrifugation ($3000 \times g$, 5 °C, 15 min) following vacuum filtration. The pH of the hydrolysate was adjusted to 6 with 5 M KOH and then was filter-sterilized using a 0.2 μm filter unit (Polycap TM AS, Whatman Ltd., Maidstone, UK). The hydrolysate was further utilised as substitute of the commercial nitrogen sources for fumaric acid production.

2.5. Control of Fungal Morphology

The impact of two different cell morphologies (pellets and dispersed mycelium) of *R. arrhizus* on fumaric acid production was evaluated. The different media compositions and conditions were as follows.

2.5.1. Production of Pelletized Biomass

Pellets were formed in baffled flasks of 1 L filled with 50 mL of the pre-culture medium as depicted in Table 1. The pre-culture was incubated at 30 °C for 24 h in a rotary shaker with an agitation speed of 180 rpm. The inoculation of the fermentation medium was performed using a 10% (v/v) of the pre-culture.

Table 1. Composition of different pre-culture media controlling the fungal morphology.

Constituents	Pellets [1]	Dispersed Clumps [2]
Glucose (g/L)	10	25
Peptone (g/L)	2	1.6
Urea (g/L)	–	1
$(NH_4)_2SO_4$ (g/L)	–	–
KH_2PO_4 (g/L)	–	0.6
$MgSO_4 \cdot 7H_2O$ (g/L)	–	0.4
$ZnSO_4 \cdot 7H_2O$ (g/L)	–	0.044
$FeCl_3 \cdot 6H_2O$ (g/L)	–	0.016
Tartaric acid (g/L)	–	–
Methanol (mL/L)	–	–
Corn steep liquor (mL/L)	–	3
Corn starch (g/L)	–	30
Agar (g/L)	–	1

[1] Byrne and Ward [14]; [2] modification of Rhodes et al. [17].

2.5.2. Production of Dispersed Mycelia

Uniformly dispersed mycelia were produced in Erlenmeyer flasks of 250 mL filled with 50 mL of the pre-culture medium, based on a modification of Rhodes et al. [17] pre-culture medium synthesis (Table 1). The pre-culture was incubated at 30 °C for 24 h in a rotary shaker with an agitation speed of 180 rpm and dispersed mycelia was inoculated (10%, v/v) in the fermentation medium.

2.6. Fumaric Acid Production

Fumaric acid production by *R. arrhizus* was studied in 250 mL Erlenmeyer flasks through submerged fermentations. Shake flasks contained 50 mL of a medium, with the following composition (g/L): glucose at initial concentration of 25 or 50; $CaCO_3$, 80% of the initial carbon source concentration used in each fermentation; KH_2PO_4, 0.6; $MgSO_4 \cdot 7H_2O$, 0.4; $ZnSO_4 \cdot 7H_2O$, 0.044; $FeCl_3 \cdot 6H_2O$, 0.016; tartaric acid, 0.0075; corn steep liquor, 0.5 mL and methanol 15 mL [11,12]. The medium was autoclaved at 121 °C for 20 min. Corn steep liquor and ammonium sulphate were autoclaved separately. Methanol was filter sterilized and added to the sterilized medium under aseptic conditions. The pH was adjusted to 6 using 5 M KOH or 5 M HCl prior to sterilization. The effect of the initial ammonium sulphate concentration at 0.2 and 0.04 g/L on fumaric acid production was also evaluated. The fermentations were incubated at 30 °C with an agitation rate of 180 rpm. The experiments were carried out in duplicates and the results represent the mean values.

2.7. Fumaric Acid Production Using Soybean Cake Hydrolysate

The effect of soybean cake hydrolysate, as substitute of commercial nitrogen sources, was evaluated by adding different FAN concentrations in the pre-culture medium. The pre-culture medium without the nitrogen sources was autoclaved at 121 °C for 20 min and then the filter-sterilized hydrolysate was aseptically added. Incubation conditions were described in Section 2.2. Subsequently, batch fermentations were conducted in shake flasks as mentioned in Section 2.3.

2.8. Analytical Methods

The TKN (total Kjeldahl nitrogen) concentration was measured using a Kjeltek™ 8100 distillation Unit (Foss, Hillerød, Denmark). Ash, lipid, acid detergent fibre (ADF), acid detergent lignin (ADL), and neutral detergent fiber (NDF) content of soybean cake were also determined [21,22]. The results of ADL, ADF, and NDF were expressed as cellulose (ADF-ADL), hemicellulose (NDF-ADF), and lignin (ADL) content.

The enzyme activity of proteases during the SSF was determined according to the method reported by Kachrimanidou et al. [18]. Briefly, fermented solids obtained from SSF were mixed with citric acid-Na_2HPO_4 (0.2 M, pH 6) buffer solution for the extraction of proteases. After centrifugation ($3000 \times g$, 4 °C, 15 min) extracts were used for the determination of enzymatic activities. Protease activity was quantified by the FAN production during hydrolysis of 7.5 g/L of casein at 55 °C. One unit of protease activity (U) was defined as the protease required for the production of 1 μg FAN in one minute at 55 °C and pH 6.0. The results were expressed as U/g of fermented solids.

FAN and IP concentrations were determined according to the ninhydrin colorimetric method and the ammonium molybdate spectrophotometric method, respectively [23,24].

An inductively-coupled plasma mass spectrometer (ICP-MS) (Thermo X-Series II, Thermo Fischer Scientific, Schwerte, Germany) was used for elemental determination of the soybean cake hydrolysate. Calibration curves with at least eight points in the range of 0.5–100 μg/L were prepared for 33 trace elements: Ar Cl, As, Ba, Be, Br, Ca, Cd, Co, Cu, Cr, Fe, Ga, Hg, In, K, Kr, Li, Lu, Mo, Mn, Mg, Na, Ni, P, Pb, S, Sb, Se, Sr, Ti, Tl, V, and Zn. The calibration curve with the highest correlation coefficient was used for each element.

Fumaric acid determination in culture broths was carried out via dilution with deionized water and 3 M H_2SO_4 solution to dissolve the residual $CaCO_3$ and reduce the pH to less than 1.0, following heating at 80 °C until the broth become clear [25]. Suspensions were filtered and the biomass was washed with deionized water. Samples from the clear filtrate were obtained and sugar consumption, fumaric acid and other by-products were monitored by HPLC equipped with a Bio-rad Aminex HPX-87H column with 300 mm length and 7.8 mm internal diameter. A 10 mM H_2SO_4 solution was used as mobile phase with 0.6 mL/min flow rate and 45 °C column temperature. The fumaric acid production was expressed as g/L. The yield was defined as grams of fumaric acid produced per gram of total glucose added to the fermentation medium.

The morphology of fungal biomass was monitored by image analysis software (Image-Pro Plus 7.0, Media Cybernetics, Warrendale, PA, USA) and the number of pellets or clumps as well as the mean diameter were determined. The mean diameter is defined as the average length of diameters measured at two degree intervals and passing through the centroid of the object. The data presented are the mean values of two replicates and error bars represent the standard deviation of those values (as calculated in Microsoft Excel).

3. Results and Discussion

The fungus *R. arrhizus* was studied in shake flask fermentations using two different pre-culture media in order to obtain two morphologies: pellets and uniformly-dispersed mycelia. The formation of pelletized biomass was initially carried out in flasks of 250 mL, but since the biomass was aggregated, baffled flasks of 1 L were employed in order to achieve pelletized biomass. Uniformly dispersed clumps were produced in a high-starch concentration medium, which led to a viscous liquid that inhibits the aggregation of the biomass of *R. arrhizus*.

The effect of the two types of morphologies on fumaric acid production was evaluated through submerged experiments using glucose as a carbon source. Moreover, the influence of ammonium sulphate was also assessed at 0.04 and 0.2 g/L initial concentrations. The formation of each morphology was studied by scanning the biomass at 48 h of each fermentation and the scanned images was processed by image analysis software.

3.1. Effect of Pelletized Mycelia on Fumaric Acid Production

The fumaric acid production using pelletized biomass of *R. arrhizus* is depicted in Figure 1. The highest fumaric acid production of 7 g/L with a yield of 0.29 g/g and a productivity of 0.10 g/L $\times$ h^{-1} was produced at 0.04 g/L initial ammonium sulphate concentration (Figure 1a). Fumaric acid production, yield, and productivity were increased two-fold when the ammonium sulphate concentration increased to 0.2 g/L (Figure 1b). Particularly, a final concentration of

14.7 g/L fumaric acid was observed with a yield of 0.60 g/g and a productivity of 0.20 g/L × h^{-1}. The by-products succinic acid and ethanol were also produced up to 1.28 g/L and 0.2 g/L, respectively.

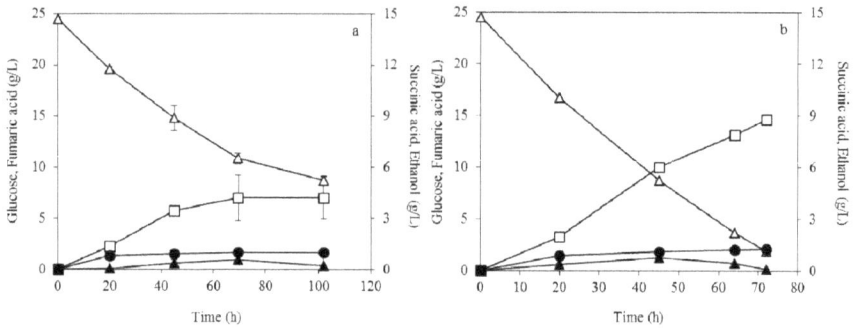

Figure 1. Effect of pellet morphology on fumaric acid (□), ethanol (▲) and succinic acid (●) production during submerged fermentation of *R. arrhizus* NRRL 2582 at 25 g/L initial glucose concentration (△). The initial ammonium sulphate concentration was (**a**) 0.04 g/L and (**b**) 0.2 g/L.

The distribution of the pellets formed during fermentation containing 0.04 and 0.2 g/L of ammonium sulphate is shown in Figure 2. The pellet diameter was mainly ranged from 0.25–1 mm and 0.25–1.5 mm, while the average pellet diameter was 0.37 mm and 0.46 mm in the case of 0.04 and 0.2 g/L ammonium sulphate concentration, respectively. The results from image analysis showed that a total number of 168 ± 22 cells/mL and 268 ± 2 cells/mL were formed at 0.04 g/L and 0.2 g/L initial concentration of ammonium sulphate, respectively.

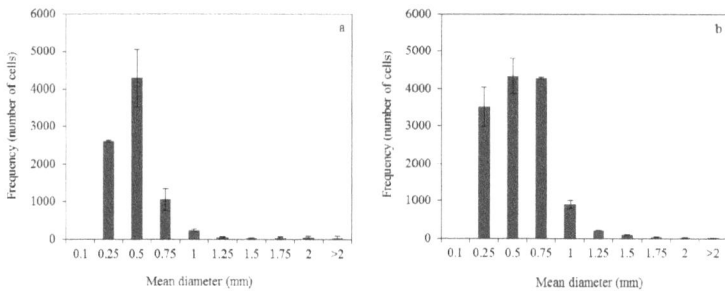

Figure 2. Pellet diameter distribution at 48 h of submerged fermentation of *R. arrhizus* NRRL 2582 using 25 g/L initial glucose concentration. The initial ammonium sulphate concentration was (**a**) 0.04 g/L and (**b**) 0.2 g/L.

The initial concentration of ammonium sulphate seems to affect the metabolism of *R. arrhizus*, as the higher concentration enhanced fumaric acid production as well as the yield and the productivity. This could be related to the higher number of cells formed in the case of 0.2 g/L ammonium sulphate.

Pellet morphology has been utilised in several studies for fumaric acid production. Liao et al. [10] and Zhou et al. [11] have mentioned that fungal growth in pellet form enhanced fumaric acid production. Roa Engel et al. [13] optimised the conditions in order to achieve small uniform pellets of *R. oryzae* ATCC 20344 with diameter less than 1 mm and achieved 30.2 g/L fumaric acid with a yield of 0.28 g/g. Zhou et al. [11] found that pellet morphology of *R. delemar* NRRL 1526 produced higher concentration of fumaric acid (38.9 g/L) instead of suspended mycelia (31.8 g/L) using 100 g/L of glucose.

3.2. Effect of Dispersed Mycelia on Fumaric Acid Production

Figure 3 illustrates the kinetic profile of glucose consumption, fumaric acid and by-products formation during the fermentation using dispersed mycelia of *R. arrhizus*. A final concentration of 19.7 g/L fumaric acid was produced in the case of 0.04 g/L ammonium sulphate (Figure 3a). The corresponding yield and productivity were 0.84 g/g and 0.18 g/L × h^{-1}, respectively. Similar results were observed during the fermentation at 0.2 g/L ammonium sulphate (Figure 3b). Particularly, a high fumaric acid production of 19.4 g/L was determined with a yield of 0.78 g/g and a productivity of 0.27 g/L × h^{-1}. In the case of fermentation using dispersed mycelia, only the productivity of fumaric acid was affected by the different initial concentration of ammonium sulphate.

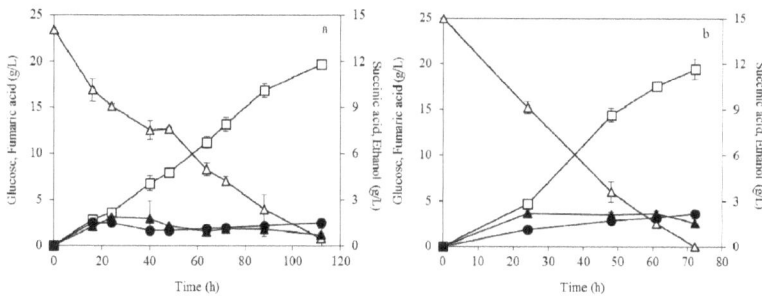

Figure 3. Effect of dispersed mycelia morphology on fumaric acid (□), ethanol (▲) and succinic acid (●) production during submerged fermentation of *R. arrhizus* NRRL 2582 at 25 g/L initial glucose concentration (△). The initial ammonium sulphate concentration was (**a**) 0.04 g/L and (**b**) 0.2 g/L.

These results suggest that the utilisation of dispersed mycelia of *R. arrhizus* was more effective in fumaric acid production as the final concentration was significantly increased when compared with the results presented in Section 3.1. The fungus was able to consume glucose with higher efficiency and the yield was increased by 40%. The opposite effect has been reported in the case of *R. delemar* NRRL 1526 [11].

The distribution of dispersed mycelia diameter (Figure 4) had wider range than the corresponding one of pellets. Specifically, in the fermentation with 0.04 g/L ammonium sulphate concentration, the cells diameter range was 0.25–2 mm with an average diameter of 0.86 mm (Figure 4a). When ammonium sulphate concentration of 0.2 g/L was utilised, the diameter of the dispersed mycelia was in the range of 0.25–1.5 mm and the average diameter was calculated at 0.43 mm (Figure 4b). The number of cells was determined as 58 ± 1 cells/mL at 0.04 g/L ammonium sulphate concentration and 332 ± 49 cells/mL at 0.2 g/L ammonium sulphate concentration.

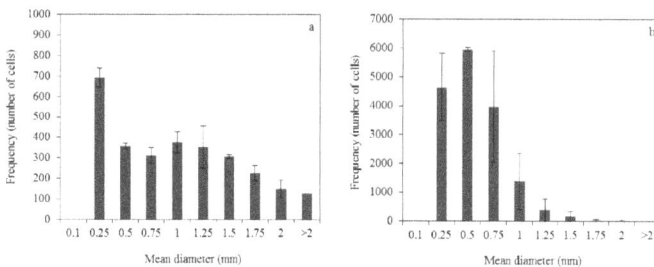

Figure 4. Dispersed clumps diameter distribution at 48 h of submerged fermentation of *R. arrhizus* NRRL 2582 using 25 g/L initial glucose concentration. The initial ammonium sulphate concentration was (**a**) 0.04 g/L and (**b**) 0.2 g/L.

The dispersed mycelia formation of *R. arrhizus* has been efficiently utilised in fermentations for fumaric acid production by Rhodes et al. [6,17]. The dispersed mycelia of the fungus *R. arrhizus* has been previously cultivated in bioreactors, producing high concentrations (73–107 g/L) of fumaric acid with yields ranging from 0.72 g/g to 0.82 g/g using glucose as carbon source [6,8,26]. The small and uniform mycelial clumps led to higher fumaric acid production than compact pellets maybe due to better oxygen transfer [12].

The potential of the dispersed mycelia fungal morphology to produce high concentrations of fumaric acid was evaluated at fermentations containing higher glucose concentration. Figure 5 presents the results during fermentation of *R. arrhizus* at 45 g/L glucose. The highest fumaric acid production of 30.1 g/L with a yield of 0.66 g/g were observed at initial concentration of 0.04 g/L ammonium sulphate. The corresponding results were lower in the case of 0.2 g/L of ammonium sulphate (25.4 g/L fumaric acid with a yield of 0.53 g/g). The image analysis showed that the mean diameter and the number of the formed cells was almost the same as in fermentations with 25 g/L initial glucose concentration.

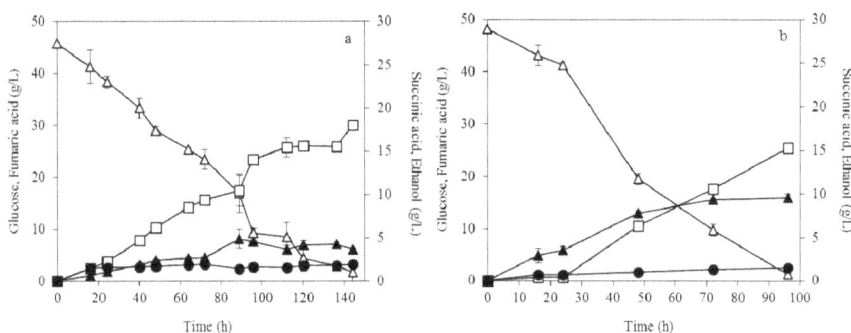

Figure 5. Effect of dispersed mycelia morphology on fumaric acid (□), ethanol (▲) and succinic acid (●) production during submerged fermentation of *R. arrhizus* NRRL 2582 at 45 g/L initial glucose concentration (△). The initial ammonium sulphate concentration was (**a**) 0.04 g/L and (**b**) 0.2 g/L.

3.3. Enzymatic Hydrolysis of Soybean Cake

As depicted in Table 2, soybean cake is rich in protein thus it could be utilised as an alternative raw material for producing a nutrient-rich feedstock for microbial fermentations. The fungal strain *A. oryzae* was selected in the present study due to the efficient production of proteolytic enzymes on various substrates [18,20,27]. The initial moisture content and pH value have been optimised in previous studies [18,20], thus the SSF on soybean cake was employed at an initial moisture content of 65% and uncontrolled pH. The kinetic profile of proteolytic activity during the SSF on soybean cake was evaluated and it is presented in Figure 6. The proteolytic activity increased during the fermentation reaching up to 205 U/g at 70 h. The SSF was utilised for hydrolysis process, when the proteolytic activity had reached the maximum value.

Table 2. Composition of soybean cake.

Composition	% (Dry Basis)
Moisture (%, wet basis)	13 ± 0.4
Protein (TKN × 6.25)	47 ± 1.2
Ash	6.5 ± 0.09
Lipids	2.2 ± 0.2
Cellulose	24.1 ± 0.9
Hemicellulose	18.1 ± 1.6
Lignin	2.1 ± 0.08

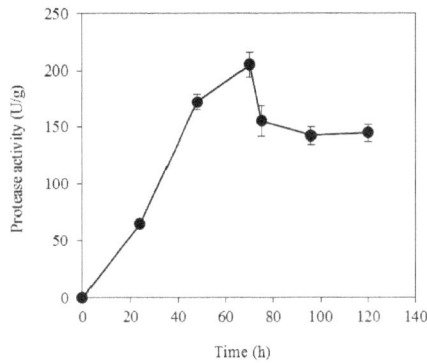

Figure 6. Proteolytic activity (U/g of fermented solids) of *A. oryzae* during solid-state fermentation of soybean cake.

The hydrolysis of soybean cake was carried out at different temperatures in the range of 40 °C to 55 °C (Figure 7). The activity of proteases was enhanced at 45 °C during the hydrolysis of 50 g/L soybean cake, producing about 1.4 g/L FAN (free amino nitrogen) at 28 h of hydrolysis (Figure 7a). The IP production was similar at the different temperatures of hydrolysis reaching up to 113 mg/L at 45 °C (Figure 7b). The results showed that soybean cake was effectively hydrolysed by the proteases produced during SSF by *A. oryzae*. In particular, a TKN to FAN conversion yield of 43% was achieved.

Kachrimanidou et al. reported FAN production of around 0.6 g/L and 1.5 g/L at 45 °C when crude proteolytic enzymes of *A. oryzae* were employed for the hydrolysis of 45 g/L and 90 g/L of sunflower meal, respectively, using an initial proteolytic activity of 6.4 U/mL [18]. Higher FAN production of 2.3 g/L was achieved when the initial proteolytic activity increased to 16 U/mL [18]. Tsouko et al. achieved 451.6 mg/L FAN concentration and 161.3 mg/L IP concentration during hydrolysis of 98.7 g/L of palm kernel cake at 50 °C, using 6 U/mL initial proteolytic activity produced by *A. oryzae* [27]. The high FAN production obtained in this study was due to the high protein content of the soybean cake.

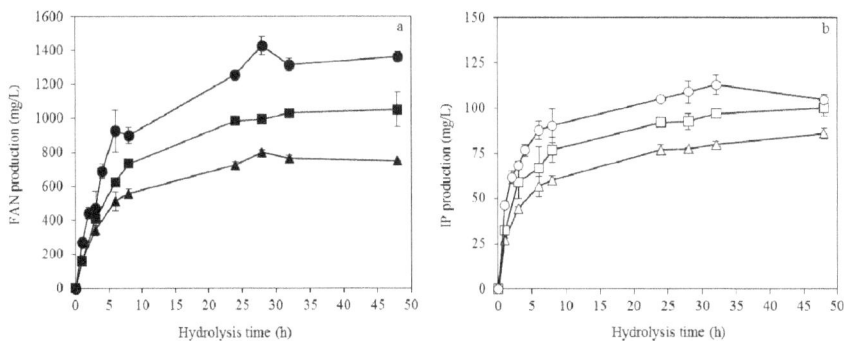

Figure 7. FAN (**a**) and IP (**b**) production during hydrolysis of 45 g/L initial solids (soybean cake and fermented solids from SSF) at 45 °C (circles), 50 °C (squares) and 55 °C (triangles) using initial proteolytic activity of 3.9 U/mL.

3.4. Fumaric Acid Production Using Soybean Cake Hydrolysate

The potential utilization of soybean cake hydrolysate as nutrient and nitrogen source for fumaric acid production by *R. arrhizus* strain was evaluated using dispersed mycelia morphology. The study focused on the replacement of the commercial nitrogen sources of the preculture by the soybean cake hydrolysate. Figure 8 presents the results during fermentation of *R. arrhizus* using 50 g/L initial glucose concentration and 0.2 g/L initial ammonium sulphate concentration. The preculture was conducted by substituting urea and peptone with soybean cake hydrolysate at initial FAN concentrations of 200 mg/L (Figure 8a) and 400 mg/L (Figure 8b). The results showed that the substitution of peptone and urea by the soybean cake hydrolysate influenced positively fumaric acid production. Particularly, the highest fumaric acid concentration of 33 g/L was achieved, when the initial FAN concentration of soybean cake was 200 mg/L. The corresponding yield was 0.69 g/g and the final productivity was 0.19 g/L × h^{-1}.

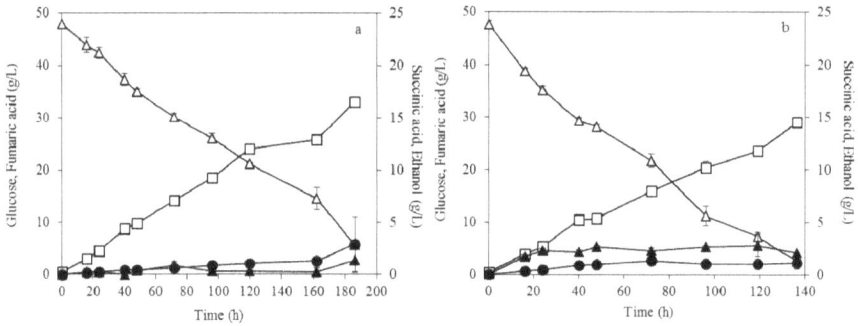

Figure 8. Fumaric acid (□), ethanol (▲) and succinic acid (●) production during submerged fermentation of R. arrhizus NRRL 2582 at ~50 g/L of initial glucose (△) concentration using (a) 200 mg/L and (b) 400 mg/L FAN of soybean cake hydrolysate as nitrogen source in the preculture medium. The initial ammonium sulphate concentration was 0.2 g/L.

The fumaric acid production was improved when soybean cake hydrolysate was utilised as nitrogen source. The beneficial effect of commercial soybean peptone on the morphology of *R. oryzae* ATCC 20344 and *R. oryzae* NRRL 1526 has been reported in other studies [10,11,14]. Zhang et al. [12] showed that chemical hydrolysate derived from soybean meal can be utilised in the pre-culture medium of *R. oryzae* ATCC 20344 for fumaric acid production. Particularly, the utilisation of soybean meal hydrolysate led to 50 g/L fumaric acid production using 80 g/L glucose as carbon source, with a yield of 0.72 g/g and 0.36 g/L × h^{-1} productivity [12].

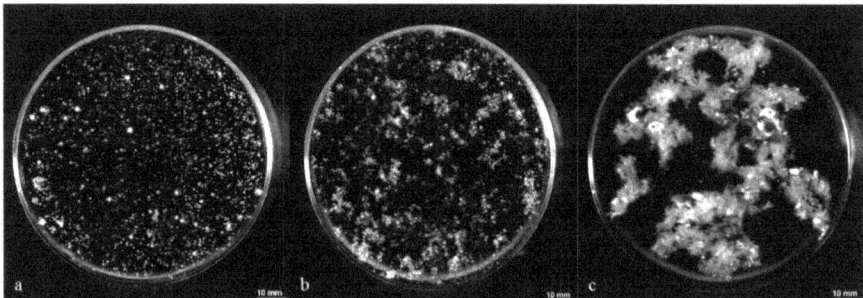

Figure 9. Images of (**a**) pelletized biomass, (**b**) dispersed mycelia using commercial nitrogen sources, and (**c**) dispersed mycelia using 200 mg/L FAN of soybean cake hydrolysate as nitrogen sources. Images were taken at 48 h of submerged fermentation of *R. arrhizus* NRRL 2582 using 25 g/L of initial glucose and 0.2 g/L initial ammonium sulphate concentration.

The formed pelletized biomass and dispersed mycelia used in this study are presented in Figure 9. Although the significant differences found in morphology during image analysis, between the fermentations with commercial nitrogen sources (Figure 9a,b) and soybean cake hydrolysate (Figure 9c), the results indicated that fumaric acid production was also significantly affected by the nutritional parameters. Specifically, the dispersed mycelia was more uniform in fermentation with commercial nitrogen sources (Figure 9b) than soybean cake hydrolysate (Figure 9c), but the addition of soybean cake led to the highest fumaric acid production. ICP-MS analysis was performed in order to study the composition of soybean cake hydrolysate. The results showed that the hydrolysate was rich in microelements such as Mg (41.9 mg/L), Ca (11.5 mg/L), Fe (7.4 mg/L), and Zn (1.8 mg/L). Additionally, lower quantities (less than 0.2 mg/L) of Mn was detected.

4. Conclusions

The present study focused on the impact of the different morphologies on the fumaric acid production by the fungal strain *R. arrhizus*. The effect of fungus morphology on the fumaric acid production using different *Rhizopus* strains and on other microbial products such as citric acid [28] has been studied in the past, but there is not any published research regarding the impact of *R. arrhizus* NRRL 2582 morphology on fumaric acid production. The biomass in the form of pellets and dispersed mycelia was produced and utilised in a synthetic medium for fumaric acid production. The results showed that uniformly dispersed mycelia led to higher fumaric acid concentrations and yields, compared with the results obtained using the pelletized biomass. Additionally, ammonium sulphate concentration in the fermentation medium had a key role in the metabolism of *R. arrhizus*, since the productivity was significantly improved at higher concentrations of ammonium sulphate. The potential substitution of commercial nitrogen sources by an enzymatic soybean cake hydrolysate was also evaluated. The soybean cake hydrolysate was effectively utilised by dispersed mycelia for high production of fumaric acid. This study demonstrated that agro-industrial biomass could be valorised for the sustainable biotechnological production of platform chemicals such as fumaric acid.

Acknowledgments: The authors would like to thank I. Mandala for providing access to Image Pro Plus Analysis Software. The work presented in this study has been funded by National Agency of Petroleum (ANP), Petrobras (Brazil) (project 00320-2/2012) and the National Council for Scientific and Technological Development of the Ministry of Science, Technology and Innovation (CNPq/MCTI) through the Special Visiting Researcher fellowship (process number: 313772/2013-4).

Author Contributions: Aikaterini Papadaki, Seraphim Papanikolaou, and Apostolis A. Koutinas conceived and designed the experiments; Nikolaos Androutsopoulos and Aikaterini Papadaki performed the experiments; Nikolaos Androutsopoulos carried out image analysis; Maria Patsalou and Michalis Koutinas contributed and performed the ICP-MS analysis; Aikaterini Papadaki, Nikolaos Kopsahelis, and Apostolis A. Koutinas were responsible for writing the paper, Aline Machado de Castro and Apostolis A. Koutinas revised the manuscript.

Conflicts of Interest: The authors declare no conflict of interest.

References

1. Koutinas, A.A.; Vlysidis, A.; Pleissner, D.; Kopsahelis, N.; Garcia, I.L.; Kookos, I.K.; Papanikolaou, S.; Kwan, T.H.; Lin, C.S. Valorization of industrial waste and by-product streams via fermentation for the production of chemicals and biopolymers. *Chem. Soc. Rev.* **2014**, *43*, 2587–2627. [CrossRef] [PubMed]
2. Fumaric Acid Market Analysis by Application (Food & Beverages, Rosin Paper Sizes, UPR, Alkyd Resins) and Segment Forecasts to 2020. 2015. Available online: http://www.grandviewresearch.com/industry-analysis/fumaric-acid-market (accessed on 26 March 2017).
3. Zhang, K.; Zhang, B.; Yang, S.-T. Production of citric, itaconic, fumaric and malic acids in filamentous fungal fermentation. In *Bioprocessing Technologies in Biorefinery for Sustainable Production of Fuels, Chemicals, and Polymers*, 1st ed.; Yang, S.-T., El-Enshasy, H.A., Thongchul, N., Eds.; John Wiley & Sons, Inc.: Hoboken, NJ, USA, 2013.
4. Roa Engel, C.A.; Straathof, A.J.J.; Zijlmans, T.W.; van Gulk, W.M.; van der Wielen, L.A.M. Fumaric acid production by fermentation. *Appl. Microbiol. Biotechnol.* **2008**, *78*, 379–389. [CrossRef] [PubMed]

5. Magnuson, J.K.; Lasure, L.L. Organic acid production by filamentous fungi. In *Advances in Fungal Biotechnology for Industry, Agriculture and Medicine*; Tracz, J.S., Lange, L., Eds.; Kluwer/Plenum: New York, NY, USA, 2004; pp. 307–340.

6. Rhodes, R.A.; Lagoda, A.A.; Jackson, R.W.; Misenhei, T.J.; Smith, M.L.; Anderson, R.F. Production of fumaric acid in 20 liter fermentors. *Appl. Microbiol.* **1962**, *10*, 9–15. [PubMed]

7. Kenealy, W.; Zaady, E.; Dupreez, J.C.; Stieglitz, B.; Goldberg, I. Biochemical aspects of fumaric acid accumulation by *Rhizopus arrhizus*. *Appl. Environ. Microbiol.* **1986**, *52*, 128–133. [PubMed]

8. Gangl, I.C.; Weigand, W.A.; Keller, F.A. Economic comparison of calcium fumarate and sodium fumarate production by *Rhizopus arrhizus*. *Appl. Biochem. Biotechnol.* **1990**, *24–25*, 663–677. [CrossRef]

9. Petruccioli, M.; Angiani, E.; Federici, F. Semi continuous fumaric acid production by *Rhizopus arrhizus* immobilized in polyurethane sponge. *Process Biochem.* **1996**, *31*, 463–469. [CrossRef]

10. Liao, W.; Liu, Y.; Frear, C.; Chen, S. A new approach of pellet formation of a filamentous fungus–*Rhizopus oryzae*. *Bioresour. Technol.* **2007**, *98*, 3415–3423. [CrossRef] [PubMed]

11. Zhou, Z.; Du, G.; Hua, Z.; Zhou, J.; Chen, J. Optimisation of fumaric acid production by *Rhizopus delemar* based on the morphology formation. *Bioresour. Technol.* **2011**, *102*, 9345–9349. [CrossRef] [PubMed]

12. Zhang, K.; Yu, C.; Yang, S.-T. Effects of soybean meal hydrolysate as the nitrogen source on seed culture morphology and fumaric acid production by *Rhizopus oryzae*. *Process Biochem.* **2015**, *50*, 173–179. [CrossRef]

13. Roa Engel, C.A.; van Gulk, W.M.; Marang, L.; van der Wielen, L.A.M.; Straathof, A.J.J. Development of low pH fermentation strategy for fumaric acid production by *Rhizopus oryzae*. *Enzyme Microb. Technol.* **2011**, *48*, 39–47. [CrossRef] [PubMed]

14. Byrne, G.S.; Ward, O.P. Effect of nutrition on pellet formation by *Rhizopus arrhizus*. *Biotechnol. Bioeng.* **1989**, *33*, 912–914. [CrossRef] [PubMed]

15. Zhou, Y.; Du, J.X.; Tsao, G.T. Mycelial pellet formation by *Rhizopus oryzae* ATCC 20344. *Appl. Biochem. Biotechnol.* **2000**, *86*, 779–789. [CrossRef]

16. Das, R.K.; Brar, S.K. Enhanced fumaric acid production from brewery wastewater and insight into the morphology of *Rhizopus oryzae* 1526. *Appl. Biochem. Biotechnol.* **2014**, *172*, 2974–2988. [CrossRef] [PubMed]

17. Rhodes, R.A.; Moyer, A.J.; Smith, M.L.; Kelley, S.E. Production of fumaric acid by *Rhizopus arrhizus*. *Appl. Microbiol.* **1959**, *7*, 74–80. [PubMed]

18. Kachrimanidou, V.; Kopsahelis, N.; Chatzifragkou, A.; Papanikolaou, S.; Yanniotis, S.; Kookos, I.; Koutinas, A.A. Utilisation of by-products from sunflower-based biodiesel production processes for the production of fermentation feedstock. *Waste Biomass Valoriz.* **2013**, *4*, 529–537. [CrossRef]

19. Wang, R.; Law, R.C.S.; Webb, C. Protease production and conidiation by *Aspergillus oryzae* in flour fermentation. *Proc. Biochem.* **2005**, *40*, 217–227. [CrossRef]

20. Dimou, C.; Kopsahelis, N.; Papadaki, A.; Papanikolaou, S.; Kookos, I.K.; Mandala, I.; Koutinas, A.A. Wine lees valorization: Biorefinery development including production of a generic fermentation feedstock employed for poly(3-hydroxybutyrate) synthesis. *Food Res. Int.* **2015**, *73*, 81–87. [CrossRef]

21. American Association of Cereal Chemists Inc. (AACC). *Approved Methods of the American Association of Cereal Chemists*, 8th ed.; American Association of Cereal Chemists Inc.: St. Paul, MN, USA, 1983.

22. Association of Official Analytical Chemists (AOAC). *Official Method 973.18, Fiber (Acid Detergent) and Lignin in Animal Feed*, 15th ed.; Association of Official Analytical Chemists, Inc.: Arlington, VA, USA, 1990.

23. Lie, S. The EBC-ninhydrin method for determination of free alpha amino nitrogen. *J. Inst. Brew.* **1973**, *79*, 37–41. [CrossRef]

24. Harland, B.F.; Harland, J. Fermentative reduction of phytate in rye, white and whole wheat breads. *Cereal Chem.* **1980**, *57*, 226–229.

25. Goldberg, I.; Lonberg-Holm, K.; Bagley, E.A.; Stieglitz, B. Improved conversion of fumarate to succinate *Escherichia coli* strains amplified for fumarate reductase. *Appl. Environ. Microbiol.* **1983**, 1838–1847.

26. Ng, T.K.; Hesser, R.J.; Stieglitz, B.; Griffiths, B.S.; Ling, L.B. Production of tetrahydrofuran/1,4-butanediol by a combined biological and chemical process. *Biotechnol. Bioeng. Symp.* **1986**, *17*, 344–363.

27. Tsouko, E.; Kachrimanidou, V.; Dos Santos, A.F.; do Nascimento Vitorino Lima, M.E.; Papanikolaou, S.; de Castro, A.M.; Freire, D.M.; Koutinas, A.A. Valorization of By-Products from Palm Oil Mills for the Production of Generic Fermentation Media for Microbial Oil Synthesis. *Appl. Biochem. Biotechnol.* **2016**, *181*, 1241–1256. [CrossRef] [PubMed]

28. Ashraf, S.; Sikander, A.; Haq, I. Acidic pre-treatment of sugarcane molasses for molasses for citric acid production by *Aspergillus niger* NG-4. *Int. J. Curr. Microbiol. Appl. Sci.* **2015**, *4*, 584–595.

MDPI AG

St. Alban-Anlage 66

4052 Basel, Switzerland

Tel. +41 61 683 77 34

Fax +41 61 302 89 18

http://www.mdpi.com

Fermentation Editorial Office

E-mail: fermentation@mdpi.com

http://www.mdpi.com/journal/fermentation

www.ingramcontent.com/pod-product-compliance
Lightning Source LLC
Chambersburg PA
CBHW051900210326
41597CB00033B/5964